THEORETICAL
HIGH ENERGY
PHYSICS

Previous Proceedings in the Series of MRST Conferences

Year		Held in	Publisher	ISBN
2000	22nd	Rochester, New York, USA	AIP Conf. Proceedings Vol. 541	1-56396-966-1
1999	21st	Ottawa, Ont. Canada	AIP Conf. Proceedings Vol. 488	1-56396-902-5
1998	20th	Montreal, Que. Canada	AIP Conf. Proceedings Vol. 452	1-56396-845-2

Other Related Titles from AIP Conference Proceedings

594 Hadrons and Nuclei: First International Symposium
Edited by Il-Tong Cheon, Taekeun Choi, Seung-Woo Hong, and Su Houng Lee, November 2001, 0-7354-0037-7

589 New Developments in Fundamental Interaction Theories: 37th Karpacz Winterschool of Theoretical Physics
Edited by Jerzy Lukierski and Jakub Rembieliński, October 2001, 0-7354-0029-6

570 SPIN 2000; 14th International Spin Physics Symposium
Edited by Kichiji Hatanaka, Takashi Nakano, Kenichi Imai, and Hiroyasu Ejiri, June 2001, 0-7354-0008-3

564 Quantum Electrodynamics and Physics of the Vacuum: QED 2000, Second Workshop
Edited by Giovanni Cantatore, May 2001, 0-7354-0000-8

562 Particles and Fields: Ninth Mexican School
Edited by Gerardo Herrera Corral and Lukas Nellen, April 2001, 1-56396-998-X

540 Particle Physics and Cosmology: Second Tropical Workshop
Edited by José F. Nieves, October 2000, 1-56396-965-3

539 Symmetries in Subatomic Physics: 3rd International Symposium
Edited by X.-H. Guo, A. W. Thomas, and A. G. Williams, October 2000, 1-56396-964-5

531 Particles and Fields: Seventh Mexican Workshop
Edited by Alejandro Ayala, Guillermo Contreras, and Gerardo Herrera, July 2000, 1-56396-954-8

490 Particles and Fields: Eighth Mexican School
Edited by Juan Carlos D'Olivo, Gabriel López Castro, and Myriam Mondragón, November 1999, 1-56396-895-9

To learn more about these titles, or the AIP Conference Proceedings Series, please visit the webpage http://proceedings.aip.org

THEORETICAL HIGH ENERGY PHYSICS

MRST 2001: A Tribute to Roger Migneron

London, Ontario, Canada   15–18 May 2001

EDITORS
V. Elias
D. G. C. McKeon
V. A. Miransky
The University of Western Ontario
London, Ontario, Canada

Melville, New York, 2001
AIP CONFERENCE PROCEEDINGS ■ VOLUME 601

Editors:

V. Elias
D. G. C. McKeon
V. A. Miransky

Department of Applied Mathematics
The University of Western Ontario
London, Ontario N6A 5B7
CANADA

E-mail: velias@uwo.ca
dgmckeo2@uwo.ca
vmiransk@uwo.ca

The article on pp. 197-205 was originally published in the European Physics Journal **C19**, 677, © 2001 Springer-Verlag, and is reprinted here with permission.

Authorization to photocopy items for internal or personal use, beyond the free copying permitted under the 1978 U.S. Copyright Law (see statement below), is granted by the American Institute of Physics for users registered with the Copyright Clearance Center (CCC) Transactional Reporting Service, provided that the base fee of $18.00 per copy is paid directly to CCC, 222 Rosewood Drive, Danvers, MA 01923. For those organizations that have been granted a photocopy license by CCC, a separate system of payment has been arranged. The fee code for users of the Transactional Reporting Service is: 0-7354-0045-8/01/$18.00.

© 2001 American Institute of Physics

Individual readers of this volume and nonprofit libraries, acting for them, are permitted to make fair use of the material in it, such as copying an article for use in teaching or research. Permission is granted to quote from this volume in scientific work with the customary acknowledgment of the source. To reprint a figure, table, or other excerpt requires the consent of one of the original authors and notification to AIP. Republication or systematic or multiple reproduction of any material in this volume is permitted only under license from AIP. Address inquiries to Office of Rights and Permissions, Suite 1NO1, 2 Huntington Quadrangle, Melville, N.Y. 11747-4502; phone: 516-576-2268; fax: 516-576-2450; e-mail: rights@aip.org.

L.C. Catalog Card No. 2001096791
ISBN 0-7354-0045-8
ISSN 0094-243X
Printed in the United States of America

CONTENTS

Preface..ix

MIGNERON TRIBUTE SESSION I

Progress in M-Theory 3
    M. J. Duff
Perturbative SUSYM vs. AdS/CFT: A Brief Review 19
    G. W. Semenoff
**Production of Heavy Quarks in Deep-Inelastic Lepton-Hadron
Scattering**..40
    W. L. van Neerven
**Spinor and Supersymmetry in Spaces of Various Dimensions and
Signatures**..60
    D. G. C. McKeon and T. N. Sherry
Dynamically Generating the Quark-Level SU(2) Linear Sigma Model 66
    M. D. Scadron

GRAVITY/GEOMETRY

Oscillating Metrics and the Cosmological Constant 75
    B. Holdom
Hiding a Cosmological Constant in a Warped Extra Dimension...............82
    H. Collins
**5-Dimensional Warped Cosmological Solutions with
Radius Stabilization by a Bulk Scalar**..88
    J. M. Cline and H. Firouzjahi

B-PHYSICS

**Determining $|V_{ub}|$ from the $\bar{B} \to X_u l \bar{\nu}$ Dilepton Invariant
Mass Spectrum** 97
    C. W. Bauer, Z. Ligeti, and M. Luke
$B \to X_s l^+ l^-$ in the Vectorlike Quark Model 105
    M. R. Ahmady, M. Nagashima, and A. Sugamoto
Renormalon Analysis of Heavy-Light Exclusive B Decays 115
    A. R. Williamson

MIGNERON TRIBUTE SESSION II

**Composites in Color Superconducting Phase of Dense QCD with Two
Quark Flavors**...129
    V. A. Miransky

Infrared Dynamics in Vector-Like Gauge Theories: QCD and Beyond 140
    V. Elias
Topics on Neutrino Physics 148
    G. Karl and V. Novikov
QCD Sum-Rule Bounds on the Light Quark Masses 151
    T. G. Steele

QUARKS, GLUONS, AND MESONS

String Model Building at Low String Scale: Towards the
Standard Model 161
    R. G. Leigh
Lepton Pair Production in a Charged Quark Gluon Plasma 168
    A. Majumder and C. Gale
$O(\alpha_s^3)$ Estimate for the Longitudinal Cross Section in e^+e^-
Annihilation to Hadrons 175
    F. A. Chishtie
Exploring Pseudoscalar Meson Scattering in Linear Sigma Models 182
    D. Black, A. H. Fariborz, S. Moussa, S. Nasri, and J. Schechter
Probing Scalar Mesons Using a Toy Model in the Linear
Sigma Model 188
    D. Black, A. H. Fariborz, S. Moussa, S. Nasri, and J. Schechter

FIELD THEORY I

Cutoff Dependence of the Casimir Effect 197
    C. R. Hagen
Pair Production of Arbitrary Spin Particles with EDM and AMM and
Vacuum Polarization 206
    S. I. Kruglov
Analyzing the 't Hooft Model on a Light-Cone Lattice 212
    J. S. Rozowsky
Fuzzy Non-trivial Gauge Configurations 219
    B. Ydri
Generalized Coherent State Approach to Star Products and
Applications to the Fuzzy Sphere 226
    G. Alexanian, A. Pinzul, and A. Stern

FIELD THEORY II

Spontaneous CPT Violation in Confined QED 235
    E. J. Ferrer, V. de la Incera, and A. Romeo
Issues on Radiatively Induced Lorentz and CPT Violation in
Quantum Electrodynamics 242
    W. F. Chen

Thermal Conductivity of the 2+1-Dimensional NJL Model in an External Magnetic Field 253
    E. J. Ferrer, V. A. Gusynin, and V. de la Incera

Example of an Asymptotically Free Matrix Model 259
    A. Agarwal and S. G. Rajeev

Variational Principle for Large N Matrix Models 267
    L. Akant, G. S. Krishnaswami, and S. G. Rajeev

BRANES, STRINGS, AND THINGS

Analytic Semi-classical Quantization of a QCD String with Light Quarks 277
    T. J. Allen, C. Goebel, M. G. Olsson, and S. Veseli

String Webs from Field Theory 282
    P. Argyres and K. Narayan

Geometry of Large Extra Dimensions versus Graviton Emission 287
    F. Leblond

Quantum Myers Effect and its Supergravity Dual for D0/D4 Systems 295
    P. J. Silva

AFTERWORD

Roger Migneron (1937–1999): Reminiscences on Sharing Life with a Physicist 305
    I. P. Migneron

List of Participants 311

Author Index 313

PREFACE

MRST 2001 was the 23rd of a series of meetings in theoretical high energy physics that normally rotate between McGill University, The University of Toronto, The University of Rochester, and Syracuse University. MRST meetings attract participants from other universities generally (though not exclusively) from the Northeastern United States and Eastern Canada. Past meetings in this series have always provided a congenial setting for junior and senior researchers to interact fruitfully, as well as a respected forum for speakers to disseminate their research findings to a knowledgeable, engaged and supportive audience. Such was certainly the case for MRST 2001.

In holding MRST 2001 at The University of Western Ontario (UWO), a "non-founding" institution, we broke with tradition in order to honour the memory of Roger Migneron, a frequent participant in past MRST meetings and a strong contributor to elementary particle physics in Canada. MRST 2001 had two special Migneron Tribute sessions in which exciting advances in theoretical high energy physics were presented by several outstanding physicists ordinarily out of the (in-itself quite distinguished) orbit of usual MRST participants. At the meeting banquet, tributes to Roger's memory were presented by Willy van Neerven, Roger's principal research collaborator, and Ina Pakkert Migneron, his widow. Ina's moving account of twenty five years shared with a passionate research scientist is the final contribution to this Proceedings – I urge you to read it even if you did not know Roger personally. Most of us will recognise something of ourselves in her words.

MRST 2001 received very generous sponsorship support from the new Perimeter Institute for Theoretical Physics (Waterloo, Ontario) and The Fields Institute for Research in the Mathematical Sciences (Toronto, Ontario). We gratefully acknowledge additional support from the Institute for Particle Physics (Ottawa, Ontario) and from the Faculties of Science and Graduate Studies of The University of Western Ontario. Logistical support from The Department of Applied Mathematics, including substantial secretarial contributions from Gayle McKenzie-Foster and Pat Malone, was vital for the success of our meeting. We are grateful as well to Huron College of The University of Western Ontario for providing a lovely and well-equipped conference venue at very modest cost.

Finally, special thanks are owed to Audrey Kager, our conference secretary, computer expert, and financial officer, who did everything from designing our web-page and formatting proceedings articles to manning the Registration Table after-hours, mastering all the registration and accommodation fee book-keeping, and helping to make the conference organisers appear much more competent than they really were.

Victor Elias
The University of Western Ontario

MIGNERON TRIBUTE SESSION I

PROGRESS IN M-THEORY*

M. J. DUFF[†]
Michigan Center for Theoretical Physics
Randall Laboratory, Department of Physics, University of Michigan
Ann Arbor, MI 48109-1120, USA

> After reviewing how M-theory subsumes string theory, we report on some new and interesting developments, focusing on the "brane-world": circumventing no-go theorems for supersymmetric brane-worlds and complementarity of the Maldacena and Randall-Sundrum pictures. We also discuss the quantum $M \to 0$ discontinuity of massive gravity with a Λ term.

My talk is in three parts. In section 1 we briefly review M-theory, in section 2 we discuss some recent developments on the brane-world and finally in section 3 we address the issue of whether the graviton can have a mass.

1. The story so far

1.1. *M-theory and dualities*

Not so long ago it was widely believed that there were five different superstring theories each competing for the title of "Theory of everything," that all-embracing theory that describes all physical phenomena. See Table 1.

Moreover, on the (d, D) "branescan" of supersymmetric extended objects with d worldvolume dimensions moving in a spacetime of D dimensions, all these theories occupied the same $(d = 2, D = 10)$ slot. See table 2. The orthodox wisdom was that while $(d = 2, D = 10)$ was the Theory of Everything, the other branes on the scan were Theories of Nothing. All that has now changed. We now know that there are not five different theories at all but, together with $D = 11$ supergravity, they form merely six different corners of a deeper, unique and more profound theory called "M-theory" where M stand for Magic, Mystery or Membrane. M-theory involves all of the other branes on the branescan, in particular the eleven-dimensional membrane $(d = 3, D = 11)$ and eleven-dimensional fivebrane $(d = 6, D = 11)$, thus resolving the mystery of why strings stop at ten dimensions while supersymmetry allows eleven[2].

Although we can glimpse various corners of M-theory, the big picture still eludes us. Uncompactified M-theory has no dimensionless parameters, which is good from the uniqueness point of view but makes ordinary perturbation theory impossible since there are no small coupling constants to provide the expansion parameters. A

---
*Research supported in part by DOE Grant DE-FG02-95ER40899.
[†]mduff@umich.edu

	Gauge Group	Chiral?	Supersymmetry charges
Type I	$SO(32)$	yes	16
Type IIA	$U(1)$	no	32
Type IIB	–	yes	32
Heterotic	$E_8 \times E_8$	yes	16
Heterotic	$SO(32)$	yes	16

Table 1: The Five Superstring Theories

low energy, E, expansion is possible in powers of E/M_P, with M_P the Planck mass, and leads to the familiar $D = 11$ supergravity plus corrections of higher powers in the curvature. Figuring out what governs these corrections would go a long way in pinning down what M-theory really is.

Why, therefore, do we place so much trust in a theory we cannot even define? First we know that its equations (though not in general its vacua) have the maximal number of 32 supersymmetry charges. This is a powerful constraint and provides many "What else can it be?" arguments in guessing what the theory looks like when compactified to $D < 11$ dimensions. For example, when M-theory is compactified on a circle S^1 of radius R_{11}, it leads to the Type IIA string, with string coupling constant g_s given by

$$g_s = R_{11}{}^{3/2} \qquad (1)$$

We recover the weak coupling regime only when $R_{11} \to 0$, which explains the earlier illusion that the theory is defined in $D = 10$. Similarly, if we compactify on a line segment (known technically as S^1/Z_2) we recover the $E_8 \times E_8$ heterotic string. Moreover, although the corners of M-theory we understand best correspond to the weakly coupled, perturbative, regimes where the theory can be approximated by a string theory, they are related to one another by a web of dualities, some of which are rigorously established and some of which are still conjectural but eminently plausible. For example, if we further compactify Type IIA string on a circle of radius R, we can show rigorously that it is equivalent to the Type IIB string compactified on a circle or radius $1/R$. If we do the same thing for the $E_8 \times E_8$ heterotic string we recover the $SO(32)$ heterotic string. These well-established relationships which remain within the weak coupling regimes are called *T-dualities*. The name *S-dualities* refers to the less well-established strong/weak coupling relationships. For example, the $SO(32)$ heterotic string is believed to be S-dual to the $SO(32)$ Type I string, and the Type IIB string to be self-S-dual. If we compactify more dimensions, other dualities can appear. For example, the heterotic string compactified on a six-dimensional torus T^6 is also believed to be self-S-dual. There is also the phenomenon of *duality of dualities* by which the T-duality of one theory is the S-duality of another. When M-theory is compactified on T^n, these S and T dualities are combined into what are termed U-dualities. All the consistency checks we have been able to think of (and after 5 years there dozens of them) have worked and

D ↑	0	1	2	3	4	5	6	7	8	9	10	11
11	.			S			T					
10	.	V	S/V	V	V	V	S/V	V	V	V	V	
9	.	V	S			V	S					
8	.				S							
7	.			S			T					
6	.	V	S/V	V	S/V	V	V					
5	.	V	S	S								
4	.	V	S/V	S/V	V							
3	.	S/V	S/V	V								
2	.	S										
1	.											
0	.											
	0	1	2	3	4	5	6	7	8	9	10	11 $d \to$

Table 2: The branescan, where S, V and T denote scalar, vector and antisymmetric tensor multiplets.

convinced us that all these dualities are in fact valid. Of course we can compactify M-theory on more complicated manifolds such as the four-dimensional $K3$ or the six-dimensional Calabi-Yau spaces and these lead to a bewildering array of other dualities. For example: the heterotic string on T^4 is dual to the Type II string on $K3$; the heterotic string on T^6 is dual to the the Type II string on Calabi-Yau; the Type IIA string on a Calabi-Yau manifold is dual to the Type IIB string on the mirror Calabi-Yau manifold. These more complicated compactifications lead to many more parameters in the theory, known to the mathematicians as *moduli*, but in physical uncompactified spacetime have the interpretation as expectation values of scalar fields. Within string perturbation theory, these scalar fields have flat potentials and their expectation values are arbitrary. So deciding which topology Nature actually chooses and the values of the moduli within that topology is known as the *vacuum degeneracy problem*.

1.2. Branes

In the previous section we outlined how M-theory makes contact with and relates the previously known superstring theories, but as its name suggest, M-theory also relies heavily on membranes or more generally p-branes, extended objects with $p = d - 1$ spatial dimensions (so a particle is a 0-brane, a string is a 1-brane, a membrane is a 2-brane and so on). In $D = 4$, a charged 0-brane couples naturally to an Maxwell vector potential A_μ, with field strength $F_{\mu\nu}$ and carries an electric charge

$$Q \sim \int_{S^2} * F_2 \tag{2}$$

and magnetic charge

$$P \sim \int_{S^2} F_2 \tag{3}$$

where F_2 is the Maxwell 2-form field strength, $*F_2$ is its 2-form dual and S^2 is a 2-sphere surrounding the charge. This idea may be generalized to p-branes in D dimensions. A p-brane couples to $(p+1)$-form potential $A_{\mu_1\mu_2...\mu_{p+1}}$ with $(p+2)$-form field strength $F_{\mu_1\mu_2...\mu_{p+2}}$ and carries an "electric" charge per unit p-volume

$$Q \sim \int_{S^{D-p-2}} *F_{D-p-2} \tag{4}$$

and "magnetic" charge per unit p-volume

$$P \sim \int_{S^{p+2}} F_{p+2} \tag{5}$$

where F_{p+2} is the $(p+2)$-form field strength, $*F_{D-p-2}$ its $(D-p-2)$-form dual and S^n is an n-sphere surrounding the brane. A special role is played in M-theory by the BPS (Bogomolnyi-Prasad-Sommerfield) branes whose mass per unit p-volume, or tension T, is equal to the charge per unit p-volume

$$T \sim Q \tag{6}$$

This formula may be generalized to the cases where the branes carry several electric and magnetic charges. The supersymmetric branes shown on the branescan are always BPS, although the converse is not true. M-theory also makes use of non-BPS and non-supersymmetric branes not shown on the branescan, but the supersymmetric ones do play a special role because they are guaranteed to be stable.

The letters S, V, and T on the branescan refer to scalar, vector and antisymmetric tensor supermultiplets of fields that propagate on the worldvolume of the brane. Historically, these points on the branescan were discovered in three different ways. The S branes were classified by writing down spacetime supersymmetric worldvolume actions that generalize the Green-Schwarz actions on the superstring worldsheet[5]. By contrast, the V and T branes were shown to arise as soliton solutions[a] of the underlying supergravity theories[6]. However, the solitonic V branes found this way were bound by $p \leq 7$. The 8-brane and 9-brane slots were included on the scan only because they were allowed by supersymmetry[6]. Subsequently, all the V-branes were given a new interpretation as Dirichlet p-branes, called D-branes, surfaces of dimension p on which open strings can end and which carry R-R (Ramond-Ramond) charge[9]. The IIA theory has D-branes with $p = 0, 2, 4, 6, 8$ and the IIB theory has D-branes with $p = 1, 3, 5, 7, 9$. They are related to one another by T-duality. In terms of how their tensions depend on the string coupling g_s, the

---

[a]The 3-brane soliton of Type IIB supergravity was an early candidate for a 'brane-world', firstly because of its dimensionality[7,8] and secondly because gauge fields propagate on its worldvolume[8]. See section 2.2.

D-branes are mid-way between the fundamental (F) strings and the solitonic (S) fivebranes:

$$T_{F1} \sim m_s^2, \quad T_{Dp} \sim \frac{m_s^{p+1}}{g_s}, \quad T_{S5} \sim \frac{m_s^6}{g_s^2} \qquad (7)$$

Since they are BPS, there is a no-force condition between the branes that allows us to have many branes of the same charge parallel to one another. The gauge group on a single D-brane is an abelian $U(1)$. If we stack N such branes on top of one another, the gauge group is the non-abelian $U(N)$. As we separate them this decomposes into its subgroups, so in fact there is a Higgs mechanism at work whereby the vacuum expectation values of the Higgs fields are related to the separation of the branes. For example the theory that lives on a stack of N Type IIB $D3$ branes is a four-dimensional $U(N)$ $n = 4$ super Yang-Mills theory. In the limit of large N the geometry of this configuration tends to the product of five dimensional anti-de Sitter space and a five dimensional sphere, $AdS_5 \times S^5$.

In $D = 11$, M-theory has two BPS branes, an electric 2-brane and its magnetic dual which is a 5-brane. Their tensions are related to each other and the Planck mass by

$$T_2^3 \sim T_5 \sim M_P^6 \qquad (8)$$

if we stack N such branes on top of one another, the M2-brane geometry tends in the large N limit to $AdS_4 \times S^7$ and the M5-brane geometry to $AdS_7 \times S^4$. In addition there are two other objects in $D = 11$, the plane wave and the Kaluza-Klein monopole, which though not branes are still BPS. When spacetime is compactified a p-brane may remain a p-brane or else become a $(p-k)$-brane if it wraps around k of the compactified directions. For example, the Type IIA fundamental string emerges by wrapping the M2-brane around S^1 and shrinking its radius to zero, and the Type IIA 4-brane emerges in a similar way from the $M5$-brane.

1.3. *Spin-offs of M-theory*

What do we now know with M-theory that we did not know with old-fashioned string theory? Here are a few examples, references to which may be found in Ref. 2.

1) Electric-magnetic (strong/weak coupling) duality in $D = 4$ is a consequence of string/string duality in $D = 6$ which in turn is a consequence of membrane/fivebrane duality in $D = 11$.

2) *Exact* electric-magnetic duality, first proposed for the maximally supersymmetric conformally invariant $n = 4$ super Yang-Mills theory, has been extended to *effective* duality by Seiberg and Witten to non-conformal $n = 2$ theories: the so-called Seiberg-Witten theory. This has been very successful in providing the first proofs of quark confinement (albeit in the as-yet-unphysical super QCD) and in generating new pure mathematics on the topology of four-manifolds. Seiberg-Witten theory and other $n = 1$ dualities of Seiberg may, in their turn, be derived from M-theory.

3) Indeed, it seems likely that all supersymmetric quantum field theories with any gauge group, and their spontaneous symmetry breaking, admit a geometrical

interpretation within M-theory as the worldvolume fields that propagate on the common intersection of stacks of p-branes wrapped around various cycles of the compactified dimensions, with the Higgs expectation values given by the brane separations.

4) In perturbative string theory, the vacuum degeneracy problems arises because there are billions of Calabi-Yau vacua which are distinct according to classical topology. Like higher-dimensional Swiss cheeses, each can have different number of p-dimensional holes. This results in many different kinds of four-dimensional gauge theories with different gauge groups, numbers of families and different choices of quark and lepton representations. Moreover, M-theory introduces new non-perturbative effects which allow many more possibilities, making the degeneracy problem apparently even worse. However, most (if not all) of these manifolds are in fact smoothly connected in M-theory by shrinking the p-branes that can wrap around the p-dimensional holes in the manifold and which appear as black holes in spacetime. As the wrapped-brane volume shrinks to zero, the black holes become massless and effect a smooth transition from one Calabi-Yau manifold to another. Although this does not yet cure the vacuum degeneracy problem, it puts it in a different light. The question is no longer why we live in one topology rather than another but why we live in one particular corner of the unique topology. This may well have a dynamical explanation.

5) Ever since the 1970's, when Hawking used macroscopic arguments to predict that black holes have an entropy equal to one quarter the area of their event horizon, a microscopic explanation has been lacking. But treating black holes as wrapped p-branes, together with the realization that Type II branes have a dual interpretation as Dirichlet branes, allows the first microscopic prediction in complete agreement with Hawking. The fact that M-theory is clearing up some long standing problems in quantum gravity gives us confidence that we are on the right track.

6) It is known that the strengths of the four forces change with energy. In supersymmetric extensions of the standard model, one finds that the fine structure constants $\alpha_3, \alpha_2, \alpha_1$ associated with the $SU(3) \times SU(2) \times U(1)$ all meet at about 10^{16} GeV, entirely consistent with the idea of grand unification. The strength of the dimensionless number $\alpha_G = GE^2$, where G is Newton's constant and E is the energy, also almost meets the other three, but not quite. This near miss has been a source of great interest, but also frustration. However, in a universe of the kind envisioned by Witten, spacetime is approximately a narrow five dimensional layer bounded by four-dimensional walls. The particles of the standard model live on the walls but gravity lives in the five-dimensional bulk. As a result, it is possible to choose the size of this fifth dimension so that all four forces meet at this common scale. Note that this is much less than the Planck scale of 10^{19} GeV, so gravitational effects may be much closer in energy than we previously thought; a result that would have all kinds of cosmological consequences.

So what is *M*-theory?

There is still no definitive answer to this question, although several different proposals have been made. By far the most popular is M(atrix) theory[10]. The matrix models of M-theory are $U(N)$ supersymmetric gauge quantum mechanical models with 16 supersymmetries. Such models are also interpretable as the effective action of N coincident Dirichlet 0-branes.

The theory begins by compactifying the eleventh dimension on a circle of radius R, so that the longitudinal momentum is quantized in units of $1/R$ with total P_L N/R with $N \to \infty$. The theory is *holographic* in that it contains only degrees of freedom which carry the smallest unit of longitudinal momentum, other states being composites of these fundamental states. This is, of course entirely consistent with their identification with the Kaluza-Klein modes. It is convenient to describe these N degrees of freedom as $N \times N$ matrices. When these matrices commute, their simultaneous eigenvalues are the positions of the 0-branes in the conventional sense. That they will in general be non-commuting, however, suggests that to properly understand M-theory, we must entertain the idea of a fuzzy spacetime in which spacetime coordinates are described by non-commuting matrices. In any event, this matrix approach has had success in reproducing many of the expected properties of M-theory such as $D = 11$ Lorentz covariance, $D = 11$ supergravity as the low-energy limit, and the existence of membranes and fivebranes.

It was further proposed that when compactified on T^{d-1}, the quantum mechanical model should be replaced by an d-dimensional $U(N)$ Yang-Mills field theory defined on the dual torus $\tilde{T}^{d-1}$. Another test of this M(atrix) approach, then, is that it should explain the U-dualities. For $d = 4$, for example, this group is $SL(3, Z) \times SL(2, Z)$. The $SL(3, Z)$ just comes from the modular group of T^3 whereas the $SL(2, Z)$ is the electric/magnetic duality group of four-dimensional $n = 4$ Yang-Mills. For $d > 4$, however, this picture looks suspicious because the corresponding gauge theory becomes non-renormalizable and the full U-duality group has still escaped explanation. There have been speculations on what compactified M-theory might be, including a revival of the old proposal that it is really M(embrane)theory. In other words, perhaps $D = 11$ supergravity together with its BPS configurations: plane wave, membrane, fivebrane, KK monopole and the $D = 11$ embedding of the Type IIA eightbrane, are all there is to M-theory and that we need look no further for new degrees of freedom, but only for a new non-perturbative quantization scheme. At the time of writing this is still being hotly debated.

What seems certain, however, is that M-theory is not a string theory. It can be approximated by a string theory only in certain peculiar corners of parameter space. So "string phenomenology" will become an oxymoron unless, for some as yet unknown reason, our universe happens to occupy one of these corners.

1.4. *AdS/CFT and the brane-world*

The year 1998 marked a renaissance in anti de-Sitter space (AdS) brought about by Maldacena's conjectured duality between physics in the bulk of AdS and a conformal field theory on its boundary[11]. For example, M-theory on $AdS_4 \times S^7$ is dual to a non-abelian ($n = 8, d = 3$) superconformal theory, Type IIB string theory on $AdS_5 \times S^5$ is dual to a ($n = 4, d = 4$) $U(N)$ super Yang-Mills theory and M-theory on $AdS_7 \times S^4$ is dual to a non-abelian ((n_+, n_-) = (2, 0), $d = 6$) conformal theory. In particular, as has been spelled out most clearly in the $d = 4$ $U(N)$ Yang-Mills case, there is seen to be a correspondence between the Kaluza-Klein mass spectrum in the bulk and the conformal dimension of operators on the boundary[12,13]. We note that, by choosing Poincaré coordinates on AdS_5, the metric may be written as

$$ds^2 = e^{-2y/L}(dx^\mu)^2 + dy^2, \tag{9}$$

where x^μ, ($\mu = 0, 1, 2, 3$), are the four-dimensional brane coordinates. In this case the superconformal Yang-Mills theory is taken to reside at the boundary $y \to -\infty$. The AdS length scale L is given by

$$L^4 = 4\pi\alpha'^2(g_{YM}^2 N) \tag{10}$$

The string coupling g_s and the Yang-Mills coupling g_{YM} are related by

$$g_s = g_{YM}^2 \tag{11}$$

The full quantum string theory on this spacetime is difficult to deal with, but we can approximate it by classical Type IIB supergravity provided

$$L^2 >> \alpha' \tag{12}$$

so that stringy correction to supergravity are small, and that $g_s << 1$ or

$$N \to \infty \tag{13}$$

so that loop corrections can be neglected. There is now overwhelming evidence in favor of this correspondence and it allows us to calculate previously uncalculable strong coupling effects in the gauge theory starting from classical supergravity. Models of this kind, where a bulk theory with gravity is equivalent to a boundary theory without gravity, have also been advocated by 't Hooft[14] and by Susskind[15] who call them *holographic* theories. Many theorists are understandably excited about the AdS/CFT correspondence because of what it can teach us about non-perturbative QCD. In my opinion, however, this is, in a sense, a diversion from the really fundamental question: What is M-theory? So my hope is that this will be a two-way process and that superconformal field theories will also teach us more about M-theory.

The Randall-Sundrum mechanism[16] also involves AdS but was originally motivated, not via the decoupling of gravity from D3-branes, but rather as a possible mechanism for evading Kaluza-Klein compactification by localizing gravity in the presence of an uncompactified extra dimension. This was accomplished by inserting a positive tension 3-brane (representing our spacetime) into AdS_5. In terms of the Poincaré patch of AdS_5 given above, this corresponds to removing the region $y < 0$,

and either joining on a second partial copy of AdS$_5$, or leaving the brane at the end of a single patch of AdS$_5$. In either case the resulting Randall-Sundrum metric is given by
$$ds^2 = e^{-2|y|/L}(dx^\mu)^2 + dy^2, \quad (14)$$
where $y \in (-\infty, \infty)$ or $y \in [0, \infty)$ for a 'two-sided' or 'one-sided' Randall-Sundrum brane respectively.

The similarity of these two scenarios led to the notion that they are in fact closely tied together. To make this connection clear, consider the one-sided Randall-Sundrum brane. By introducing a boundary in AdS$_5$ at $y = 0$, this model is conjectured to be dual to a cutoff CFT coupled to gravity, with $y = 0$, the location of the Randall-Sundrum brane, providing the UV cutoff. This extended version of the Maldacena conjecture[17] then reduces to the standard AdS/CFT duality as the boundary is pushed off to $y \to -\infty$, whereupon the cutoff is removed and gravity becomes completely decoupled. Note in particular that this connection involves a single CFT at the boundary of a single patch of AdS$_5$. For the case of a brane sitting between two patches of AdS$_5$, one would instead require two copies of the CFT, one for each of the patches. A crucial test of this assumed complementarity of the Maldacena and Randall-Sundrum pictures is that both should yield the same corrections to Newton's law. See section 2.3.

A third development in the brane-world has been the idea that the extra dimensions are compact but much larger than the conventional Planck sized dimensions in traditional Kaluza-Klein theories[18]. This is possible if the standard model fields are confined to the $d = 4$ brane with only gravity propagating in the $d > 4$ bulk[18]. We shall not pursue this possibility here.

2. Developments on the Brane-World

2.1. *No-go theorems for supersymmetry*

If we are to give a "top-down" justification of the Randall-Sundrum brane-world by embedding it in string theory or M-theory, it is desirable that the R-S picture be consistent with supersymmetry. Indeed, such a supersymmetric brane-world is necessary if the Maldacena and Randall-Sundrum (R-S) pictures are to stand any chance of being complementary. At first, however, this seemed to be problematical and several papers appeared in the literature suggesting that R-S could not be supersymmetric. Some of these no-go theorems listed below are exactly as they appeared; with others I have taken the liberty of setting up the straw man so as more effectively to knock him down.

1) R-S branes cannot be SUSY because massless supergravity scalars give kink-up and not kink-down potentials which do not bind gravity to the brane.

2) R-S branes cannot be SUSY because their tension is not that of a BPS brane.

3) R-S branes cannot be SUSY because δ-functions are incompatible with susy transformation rules.

4) R-S branes cannot be SUSY because the photon superpartners of the graviton cannot be bound to the brane.

2.2. *Yes-go theorems for Supersymmetry of the brane-world*

In fact, the domain-wall solution of Bremer et al[19] provides a supersymmetric Type IIB Randall-Sundrum realization[20]. See also Refs. 21 and 22. So it is instructive to see how the no-go theorems are circumvented:

1) **The required supergravity scalar is massive, being the breathing mode of the S^5 compactification**[20]. So the negative conclusions about massless scalars in Refs. 23,24, while correct, are not relevant.

2) **The tension comes from two sources: the BPS D3-branes and the kink**[25]. So the observation of Ref. 26 that the D3 brane tension is only 2/3 of the R-S tension, while correct, is not relevant.

3) **The sign flip of the coupling constant across the brane removes the δ-functions in the supersymmetry transformation rules**[27,28,29,20,30]. So the problems raised by Ref. 31, while correct, are not relevant.

4) **Photons can be bound to the brane but their bulk origin is not Maxwell's equations but rather odd-dimensional self-duality equations**[32,33]. So the result of Ref. 37, that photons obeying Maxwell's equations in the bulk cannot be bound to the brane, while correct, is not relevant[b]

An entirely different question is whether a *smooth* domain wall can provide a supersymmetric Randall-Sundrum realization, and here ordinary supergravity seems to fail requiring some kind of higher derivative and presumably stringy corrections[23,38].

2.3. *Complementarity of the Maldacena and Randall-Sundrum pictures*

In his 1972 PhD thesis under Abdus Salam, the author showed that, when one-loop quantum corrections to the graviton propagator are taken into account, the inverse square r^{-2} behavior of Newton's gravity force law receives an r^{-4} correction whose coefficient depends on the number and spins of the elementary particles[39,40]. Specifically, the potential looks like

$$V(r) = \frac{GM}{r}\left(1 + \frac{\alpha G}{r^2}\right), \qquad (15)$$

where G is the four-dimensional Newton's constant, $\hbar = c = 1$ and α is a purely numerical coefficient given, in the case of spins $s \leq 1$, by $45\pi\alpha = 12N_1 + 3N_{1/2} + N_0$, where N_s are the numbers of particle species of spin s going around the loop.

---

[b]The authors of Refs. 34,35,36 showed that, treated as test particles, Maxwell photons could be bound to the brane but their charge would be screened. However, the combined bulk Einstein-Maxwell equations rule out photons on the brane altogether[37].

Now fast-forward to 1999 when Randall and Sundrum proposed that our four-dimensional world is a 3-brane embedded in an infinite five-dimensional universe. Gravity reaches out into the five-dimensional bulk but the other forces are confined to the four-dimensional brane. Contrary to expectation, they showed that an inverse square r^{-2} law for gravity is still possible but with an r^{-4} correction coming from the massive Kaluza-Klein modes whose coefficient depends on the bulk cosmological constant. Their potential looks like

$$V(r) = \frac{GM}{r}\left(1 + \frac{2L^2}{3r^2}\right). \tag{16}$$

where L is the radius of AdS_5. Since (15) was the result of a four-dimensional quantum calculation and (16) the result of a five-dimensional classical calculation, they seem superficially completely unrelated. However, Ref. 41 invokes the AdS/CFT correspondence of Maldacena to demonstrate that the two are in fact completely equivalent ways of describing the same physics. From (15), we see that the contribution of a single $n = 4\, U(N)$ Yang-Mills CFT, with $(N_1, N_{1/2}, N_0) = (N^2, 4N^2, 6N^2)$, is

$$V(r) = \frac{GM}{r}\left(1 + \frac{2N^2 G}{3\pi r^2}\right). \tag{17}$$

Using the AdS/CFT relation $N^2 = \pi L^3/2G_5$ and the one-sided brane-world relation $G = 2G_5/L$, where G_5 is the five-dimensional Newton's constant, we obtain exactly (16).

As discussed in the August 2000 edition of Scientific American[18], experimental tests of deviations from Newton's inverse square law are currently under way.

2.4. *Five versus eleven*

As we have seen M-theory requires eleven dimensions, whereas if the brane-world picture is correct, we really need only five with the other six going along for the ride. Why should Nature behave like this? The only good answer to this question I could find is in Mother Goose's Nursery Rhymes:

Nature requires five,
Custom allows seven,
Idleness takes Nine,
And Wickedness Eleven.

3. A massive graviton?

An old question is whether the graviton has exactly zero mass or perhaps a small but non-zero mass. This issue seemed to have been resolved by van Dam and Veltman[42] and, independently, Zakharov[43] when they noted that there is a discrete difference between the propagator of a strictly massless graviton and that of a graviton with mass M in the $M \to 0$ limit. This difference gives rise to a discontinuity between the corresponding amplitudes involving graviton exchange.

In particular, the bending of light by the sun in the massive case is only 3/4 of the experimentally confirmed massless case, thus ruling out a massive graviton.

Recently, however, the masslessness of the graviton has been called into question by two papers[44,45] pointing out that the van Dam-Veltman-Zakharov discontinuity disappears if, instead of being Minkowski, the background spacetime is anti-de Sitter (AdS). The same result in de Sitter space had earlier been obtained in Ref.46,47. In fact, as shown in Ref.48, this can be extended to any Einstein space satisfying

$$R_{\mu\nu} = \Lambda g_{\mu\nu} \qquad (18)$$

with a non-zero cosmological constant $\Lambda \neq 0$ provided $M^2/\Lambda \to 0$.

Let us define the second-order spin operators acting on (A, B) representations of the Lorentz group [49]: the scalar Laplacian $\Delta(0,0) \equiv -\Box$, the Lichnerowicz operator for symmetric rank-2 tensors $\Delta(1,1)\phi_{\mu\nu} = -\Box\phi + R_{\mu\tau}\phi_\nu^\tau + R_{\nu\tau}\phi_\mu^\tau - 2R_{\mu\rho\nu\tau}\phi^{\rho\tau}$. and the second-order vector operator by $\Delta(1/2, 1/2)\xi_\mu \equiv -\Box\xi_\mu + R_{\mu\nu}\xi^\nu$. We have exploited the Einstein condition (18) for the background metric. Then one finds for $M \neq 0$, that the one-graviton exchange amplitude is given by

$$A[T] = \frac{1}{4}\Big[2\,T^{\mu\nu}\left(\Delta(1,1) - 2\Lambda + M^2\right)^{-1} T_{\mu\nu} \qquad (19)$$
$$-\left(\frac{-2\Lambda + 2M^2}{-2\Lambda + 3M^2}\right) T_\mu^\mu \left(\Delta(0,0) - 2\Lambda + M^2\right)^{-1} T_\mu^\mu\Big],$$

in agreement with the result of Ref. 45, while for $M = 0$

$$A[T] = \frac{1}{4}\Big[2\,T^{\mu\nu}\left(\Delta(1,1) - 2\Lambda\right)^{-1} T_{\mu\nu} \qquad (20)$$
$$-T_\mu^\mu \left(\Delta(0,0) - 2\Lambda\right)^{-1} T_\mu^\mu\Big],$$

So for $\Lambda = 0$ there is a discontinuity as $M \to 0$, but for $\Lambda \neq 0$ the limit is continuous. These results remain surprising, however, since the massive graviton retains five degrees of freedom, while the massless one only has two. Although these extra states decouple from a covariantly conserved stress tensor for $M^2/\Lambda \to 0$, yielding a smooth classical limit, they are nevertheless still present in the theory, suggesting that a discontinuity may remain at the quantum level. In Ref.48, we demonstrate that this is indeed the case by calculating the one loop graviton vacuum amplitude for a massive graviton and showing that it does not reproduce the result for the massless case in the limit. Thus the apparent absence of the discontinuity is only an artifact of the tree approximation and the discontinuity reappears at one loop.

For $M \neq 0$, the one-loop effective action Γ is given by

$$exp(-\Gamma) = \frac{[det(\Delta(1/2,1/2) - 2\Lambda + M^2)]^{1/2}}{[det(\Delta(1,1) - 2\Lambda + M^2]^{1/2}} \qquad (21)$$

while for the $M = 0$ case

$$exp(-\Gamma) = \frac{det(\Delta(1/2,1/2) - 2\Lambda)}{[det(\Delta(1,1) - 2\Lambda)det(\Delta(0,0) - 2\Lambda)]^{1/2}} \qquad (22)$$

The difference in these two expressions reflects the fact that 5 degrees of freedom are being propagated around the loop in the massive case and only 2 in the massless case. Denoting the dimension of the spin (A, B) representation by $D(A, B) = (2A + 1)(2B + 1)$, we count $D(1, 1) - D(1/2, 1/2) = 5$ for the massive case, while $D(1, 1) - 2D(1/2, 1/2) + D(0, 0) = 2$ for the massless one.

It remains to check that there is no conspiracy among the eigenvalues of these operators that would make these two expressions coincide. To show this, it suffices to calculate the coefficients in the heat-kernel expansion for the massive graviton propagator, and compare it with the massless case given in Ref. 49. The coefficient functions $b_k^{(\Lambda)}$ in the expansion

$$\mathrm{Tr} e^{-\Delta^{(\Lambda)} t} = \sum_{k=0}^{\infty} t^{(k-4)/2} \int \delta^4 x \sqrt{g}\, b_k^{(\Lambda)} \tag{23}$$

were calculated in Ref. 49 for general "spin operators" $\Delta^{(\Lambda)}(A, B) \equiv \Delta(A, B) - 2\Lambda$ with the result

$$\begin{aligned}
180(4\pi)^2 b_4^{(\Lambda)}(1, 1) &= 189 R_{\mu\nu\rho\sigma} R^{\mu\nu\rho\sigma} - 756\Lambda^2, \\
180(4\pi)^2 b_4^{(\Lambda)}(1/2, 1/2) &= -11 R_{\mu\nu\rho\sigma} R^{\mu\nu\rho\sigma} + 984\Lambda^2, \\
180(4\pi)^2 b_4^{(\Lambda)}(0, 0) &= R_{\mu\nu\rho\sigma} R^{\mu\nu\rho\sigma} + 636\Lambda^2.
\end{aligned} \tag{24}$$

It is straightforward to extend those results to relevant massive operators $\Delta^{(\Lambda,M)}(A, B) \equiv \Delta(A, B) - 2\Lambda + M^2$. The coefficients $b_k^{(\Lambda,M)}(A, B)$ for these operators are perfectly smooth functions of M^2. Thus, as $M^2 \to 0$, we obtain

$$\begin{aligned}
180(4\pi)^2 & b_4^{(\Lambda,M)}(\text{total}) \\
&= 180(4\pi)^2 \left[b_4^{(\Lambda,M)}(1, 1) - b_4^{(\Lambda,M)}(1/2, 1/2) \right] \\
&\to 200 R_{\mu\nu\rho\sigma} R^{\mu\nu\rho\sigma} - 1740\Lambda^2,
\end{aligned} \tag{25}$$

which clearly differs from the $M^2 = 0$ result

$$\begin{aligned}
180(4\pi)^2 & b_4^{(\Lambda)}(\text{total}) \\
&= 180(4\pi)^2 \left[b_4^{(\Lambda)}(1, 1) - 2 b_4^{(\Lambda)}(1/2, 1/2) + b_4^{(\Lambda)}(0, 0) \right] \\
&= 212 R_{\mu\nu\rho\sigma} R^{\mu\nu\rho\sigma} - 2088\Lambda^2.
\end{aligned} \tag{26}$$

(These one-loop differences between massive and massless spin 2 in the $\Lambda = 0$ case are well-known[51]). Even in the case of backgrounds with constant curvature

$$\begin{aligned}
R_{\mu\nu\rho\sigma} &= \frac{1}{3}\Lambda(g_{\mu\nu} g_{\rho\sigma} - g_{\mu\rho} g_{\nu\sigma}), \\
R_{\mu\nu\rho\sigma} R^{\mu\nu\rho\sigma} &= \frac{8}{3}\Lambda^2,
\end{aligned} \tag{27}$$

there is no cancellation. Thus we conclude that the absence of the discontinuity between the $M^2 \to 0$ and $M^2 = 0$ results for massive spin 2, demonstrated in

Ref. 45,44, is an artifact of the tree approximation and that the discontinuity itself persists at one loop.

That the full quantum theory is discontinuous is not surprising considering the different degrees of freedom for the two cases. However, the three additional longitudinal degrees of freedom of the massive graviton do not couple to a conserved stress tensor. Thus, in the absence of any self-couplings (or at tree-level), the additional longitudinal modes would decouple from matter, yielding a smooth $M^2 \to 0$ limit. Nevertheless, due to these self-couplings (seen here as couplings to the background metric in the linearized approach), these additional modes do not decouple, thus yielding the resulting discontinuity in the massless limit. (This result also suggests that the situation would be similar for the spin-$\frac{3}{2}$ case considered in Ref. 52,53.) Of course, these one loop effects are very small and so experiments such as the bending of light would still not be able to distinguish a massless graviton from a very light graviton in the presence of a non-zero cosmologocal constant.

Similarly, we note that a classical continuity but quantum discontinuity arises in the "partially massless" limit $M^2 \to 2\Lambda/3$ [54] as a result of going from five degrees of freedom to four [50].

We finish with the important caveat that the $M \to 0$ discontinuity for fixed Λ of the massless limit of massive spin-2 we have demonstrated applies to fields described by the Pauli-Fierz action discussed in Ref. 45,44. One may question whether this is a suitable action to describe the interaction of massive gravitons. We are not necessarily ruling out a smooth limit for other actions that might appear in Kaluza-Klein or brane-world models, for example. Indeed one would expect a smooth limit if the mass is acquired spontaneously[55] rather than through an explicit Pauli-Fierz term. In conventional Kaluza-Klein models, however, this limit, though smooth, would also be the decompactification limit and would result in massless gravitons in the higher dimension rather than four dimensions. A closer examination would be necessary to discern the form of the effective action describing the trapped graviton of the brane-world scenario of Refs. 56,57.

Acknowledgments

I am grateful to my collaborators Mirjam Cvetic, Fred Dilkes, James T. Liu, Hong Lu, Chris Pope, Wafic Sabra, Hisham Sati and Kelly Stelle.

References

1. J. H. Schwarz, *Superstrings. The first fifteen years of superstring theory*, (World Scientific, 1985).
2. M. J. Duff, *The world in eleven dimensions: supergravity, supermembranes and M-theory*, (IOP Publishing, 1999).
3. J. H. Schwarz, *Recent progress in superstring theory*, hep-th/0007130.
4. J. H. Schwarz, *Does superstring theory have a conformally invariant limit?*, hep-th/0008009.
5. A. Achucarro, J. Evans, P. Townsend and D. Wiltshire, *Super p-branes*, Phys. Lett.

B198, 441 (1987).
6. M. J. Duff, R. R. Khuri and J. X. Lu, *String solitons*, Phys. Rep. **259**, 213 (1995), [hep-th/9412184].
7. G. T. Horowitz and A. Strominger, *Black strings and p-branes*, Nucl. Phys. **B360** 197 (1991).
8. M. J. Duff and J. X. Lu, *The self-dual Type IIB superthreebrane*, Phys. Lett. **B273** 409 (1991).
9. J. Polchinski, *Dirichlet-branes and Ramond-Ramond charges*, Phys. Rev. Lett. **75**, 4724 (1995).
10. T. Banks, W. Fischler, S. H. Shenker and L. Susskind, *M-theory as a matrix model: a conjecture*, Phys. Rev. **D55**, 5112 (1997).
11. J. Maldacena, *The large N limit of superconformal field theories and supergravity*, Adv. Theor. Math. Phys. **2**, 231 (1998) [hep-th/9711200].
12. S. S. Gubser, I. R. Klebanov and A. M. Polyakov, *Gauge theory correlators from non-critical string theory*, Phys. Lett. **B428**, 105 (1998) [hep-th/9802109].
13. E. Witten, *Anti-de Sitter space and holography*, Adv. Theor. Math. Phys. **2**, 253 (1998) [hep-th/9802150].
14. G. 't Hooft, *Dimensional reduction in quantum gravity*, gr-qc/9310026.
15. L. Susskind, *The world as a hologram*, J. Math. Phys. **36**, 6377 (1995) [hep-th/9409089].
16. L. Randall and R. Sundrum, *An alternative to compactification*, Phys. Rev. Lett. **83**, 4690 (1999) [hep-th/9906064].
17. L. Susskind and E. Witten, *The Holographic Bound in Anti-de Sitter Space*, hep-th/9805114.
18. N. Arkani-Hamed, S. Dimopoulos and G. Dvali, *The universe's unseen dimensions*, Scientific American, August 2000, 62.
19. M. S. Bremer, M. J. Duff, H. Lü, C. N. Pope and K. S. Stelle, *Instanton cosmology and domain walls from M-theory and string theory*, Nucl. Phys. **B543**, 321 (1999) [hep-th/9807051].
20. M. J. Duff, J. T. Liu and K. S. Stelle, *A supersymmetric Type IIB Randall-Sundrum realization*, hep-th/0007120.
21. M. Cvetic, H. Lü and C.N. Pope, *Localized gravity in the singular domain wall background?*, hep-th/0002054.
22. S.P. de Alwis, *Brane world scenarios and the cosmological constant*, hep-th/0002174.
23. R. Kallosh and A. Linde, *Supersymmetry and the brane world*, JHEP **0002**, 005 (2000) [hep-th/0001071].
24. K. Behrndt and M. Cvetic, *Anti-de Sitter vacua of gauged supergravities with 8 supercharges*, Phys. Rev. **D61**, 101901 (2000) [hep-th/0001159].
25. M. Cvetic, M. J. Duff, J. T. Liu, H. Lu, C. N. Pope and K. S. Stelle, *Randall-Sundrum brane tensions*, hep-th/0011167.
26. P. Krauss, *Dynamics of anti-de Sitter domain walls*, JHEP **9912**, 011 (1999) [hep-th/9910149].
27. T. Ghergetta and A. Pomarol, *Bulk fields and supersymmetry in a slice of AdS*, hep-th/0003129.
28. N. Alonso-Alberca, P. Meessen and T. Ortin, *Supersymmetric brane-worlds*, Phys. Lett. **B482**, 400 (2000) [hep-th/0003248].
29. E. Bergshoeff, R. Kallosh and A. Van Proeyen, *Supersymmetry in singular spaces*, JHEP **0010**, 033 (2000) [hep-th/0007044].
30. J. T. Liu and H. Sati, *Breathing mode compactifications and supersymmetry of the brane world*, hep-th/0009184
31. R. Altendorfer, J. Bagger and D. Nemechansky, *Supersymmetric Randall-Sundrum scenario*, hep-th/0003117.

32. M.J. Duff, James T. Liu and W.A. Sabra, *Localization of supergravity on the brane*, hep-th/0009212.
33. H. Lu and C. Pope, *Branes on the Brane*, hep-th/0008050.
34. B. Bajc and G. Gabadadze, *Localization of matter and cosmological constant on a brane in anti de Sitter space*, Phys. Lett. **B474**, 282 (2000) [hep-th/9912232].
35. A. Pomarol, *Grand unified theories without the desert*, hep-th/0052931
36. N. Kaloper, E. Silverstein and L. Susskind, *Gauge Symmetry and Localized Gravity in M Theory*, hep-th/0006192.
37. M.J. Duff and James T. Liu *Hodge duality on the brane*, hep-th/0010171.
38. J. Maldacena, *Supergravity description of field theories on curved manifolds and a no-go theorem*, hep-th/0006085.
39. M. J. Duff, *Problems in the classical and quantum theories of gravitation*, Ph. D. thesis, Imperial College, London (1972).
40. M. J. Duff, *Quantum corrections to the Schwarzschild solution*, Phys. Rev. **D9**, 1837 (1974).
41. M. J. Duff and James T. Liu, *Complementarity of the Maldacena and Randall-Sundrum pictures*, Phys. Rev. Lett. **85**, 2052 (2000) [hep-th/0003237].
42. H. van Dam and M. Veltman, *Massive and massless Yang-Mills and gravitational fields* Nucl. Phys. **B22**, 397 (1970).
43. V.I. Zakharov, JETP Lett. **12**, 312 (1970).
44. I. I. Kogan, S. Mouslopoulos and A. Papazoglou, *The $m \to 0$ limit for massive graviton in dS_4 and AdS_4: How to circumvent the van Dam-Veltman-Zakharov discontinuity* hep-th/0011138.
45. M. Porrati, *No van Dam-Veltman-Zakharov discontinuity in AdS space* hep-th/0011152.
46. A. Higuchi, *Forbidden Mass Range For Spin-2 Field Theory In De Sitter Space-Time* Nucl. Phys. **B282**, 397 (1987).
47. A. Higuchi, *Massive symmetric tensor field in space-times with a positive cosmological constant* Nucl. Phys. **B325**, 745 (1989).
48. F. Dilkes, M. J. Duff, James T. Liu and H. Sati, *Quantum discontinuity between massless and infinitessimal graviton mass with a Λ term* Phys. Rev. Lett. (to appear), hep-th/0102093.
49. S. M. Christensen and M. J. Duff, *Quantizing gravity with a cosmological constant* Nucl. Phys. **B170**, 480 (1980).
50. S. Deser and R. I. Nepomechie, *Gauge Invariance Versus Masslessness In De Sitter Space* Annals Phys. **154**, 396 (1984).
51. S. M. Christensen and M. J. Duff, *Axial and conformal anomalies for arbitrary spin in gravity and supergravity* Phys. Lett. **76B**, 571 (1978).
52. P. A. Grassi and P. van Nieuwenhuizen, *No van Dam-Veltman-Zakharov discontinuity for supergravity in AdS space* hep-th/0011278.
53. S. Deser and A. Waldron, *Discontinuities of massless limits in spin 3/2 mediated interactions and cosmological supergravity* hep-th/0012014.
54. M.J. Duff, James T. Liu and H. Sati, *Quantum $M^2 \to 2\Lambda/3$ discontinuity for massive gravity with a Lambda term*, hep-th/0105008.
55. M. J. Duff, *Dynamical breaking of general covariance and massive spin-two mesons* Phys. Rev. **D12**, 3969 (1975).
56. I. I. Kogan, S. Mouslopoulos and A. Papazoglou, *A new bigravity model with exclusively positive branes* Phys. Lett. **B501**, 140 (2001) [hep-th/0011141].
57. A. Karch and L. Randall, *Locally localized gravity* hep-th/0011156.

Perturbative SUSYM vs. AdS/CFT: a brief review

Gordon W. Semenoff

Department of Physics and Astronomy,
University of British Columbia,
Vancouver, British Columbia, Canada V6T 1Z1

1 Gauge fields, strings and gravity

One of the most interesting products of string theory research during the past five years is the emergence of a deep and unexpected relationship between gauge fields and gravity. It has become apparent that many gauge field theories have a dual description as a gravitational theory and vice-versa. These are typically strong to weak coupling dualities and thus are potentially very useful for understanding both gauge and gravitational phenomena in strong coupling regimes. The duality is best understood for gauge and gravitational theories which are directly related to string theory - supergravity and gauge theory with a high degree of supersymmetry.

A fascinating fact is that this duality is not understood without string theory as an intermediate stage. The gravitational theory and the gauge theory are both obtained from the appropriate limits of string theory. These limits can have an overlapping domain of validity. In that case, within the overlap, they are two different descriptions of the same dynamical system.

Another aspect of this circle of ideas is the duality between gauge field theory and string theory. The idea that a gauge field theory could have a string theory dual has a long history dating back to the dual resonance models of the strong interactions which were formulated in the late 1960's. The best theoretical evidence for this duality comes from 'tHooft's large N expansion of Yang-Mills theory [1] based on the gauge group SU(N). In that expansion, the rank of the gauge group N is taken to infinity holding the 'tHooft coupling $\lambda \equiv g^2 N$ fixed. The orders of the expansion in $1/N^2$ are classified according to the genus of the Riemann surfaces on which one could draw Feynman diagrams without crossing any lines. The leading order is called planar - it is given by all diagrams which can be drawn on a plane. The idea is that, if there is a dual string theory, it is weakly coupled in the limit where N is large and computations could well be more tractable there. In spite of this optimism, over more than a quarter of a century there has been little progress in either solving the large N limit of four dimensional Yang-Mills theory or finding a concrete model for its string theoretical dual.

This is an important problem. Motivation comes from the fact that asymptotically free gauge theories are strongly coupled in the infrared kinematical regime where much important physics occurs. For example, in quantum chromodynamics, the gauge theory of the strong interactions, quark confinement and the formation of the hadron spectrum are non-perturbative phenomena, In a sense, the dimensional transmutation that makes a coupling constant run replaces it with a mass scale and any processes with energies smaller or bigger than that mass scale (depending on whether the theory is asymptotically or infrared free) will be non-perturbative in their coupling constant expansion. The other parameter that a gauge theory can be expanded in is $1/N$. It is not expected that this expansion has ultraviolet divergences beyond those which are removed by the usual counterterms. Thus 1/N is not a running coupling and the orders of its expansion should be perturbative at all scales. It has been argued that the 1/N expansion reproduces many qualitative features of the hadron spectrum, including Regge phenomenology which is also a hint at string-like behavior [2].

2 AdS/CFT

Recently one concrete realization of a gauge theory - string theory duality has emerged. It has been *conjectured* [3] that there is an exact mapping between $\mathcal{N} = 4$ supersymmetric Yang-Mills theory (SYM) with gauge group SU(N) on four dimensional spacetime and IIB superstring theory on background $AdS_5 \times S_5$ with N units of Ramond-Ramond 4-form flux. This is part of a series of conjectures generally referred to as the AdS/CFT correspondence. There are three levels at which the most direct form of this conjecture, due to Maldacena [3], could be correct:

- In its strongest version the correspondence asserts that there is an exact equivalence between four dimensional $\mathcal{N} = 4$ supersymmetric Yang-Mills theory and type IIB superstring theory on the $AdS_5 \times S_5$ background. This also contains the conjecture that the $AdS_5 \times S_5$ background is an exact solution of type IIB superstring theory.

- A weaker version asserts a duality of the 't Hooft limit of the gauge theory, where $N \to \infty$ with the 't Hooft coupling $\lambda = g^2 N$ held fixed, and the classical $g_s \to 0$ limit of type IIA superstring theory on $AdS_5 \times S_5$. In this correspondence, corrections to classical supergravity theory from stringy effects which are of order α' would agree with corrections to the large 't Hooft coupling limit, of order $1/\sqrt{\lambda}$, but higher orders in g_s on the supergravity side and non-planar diagrams on the gauge theory side could disagree.

- An even weaker version is a duality between the 't Hooft limit where one also takes the strong coupling limit $\lambda \to \infty$ and the low energy, supergravity limit of type IIB superstring theory on $AdS_5 \times S_5$. In this case, there would be order α' and g_s corrections to supergravity which might not agree with order $1/N^2$ and $1/\sqrt{\lambda}$ corrections to $\mathcal{N} = 4$ supersymmetric Yang-Mills theory.

Even the last, weakest version of the correspondence has profound consequences. Previous to it, the only successful quantitative tool which could be used to attack supersymmetric Yang-Mills theory was perturbation theory in g^2, the Yang-Mills coupling constant. This is limited to the regime where g^2 and λ are small. Furthermore, although some qualitative features of the large N limit are known, it is not possible to sum planar diagrams explicitly. The AdS/CFT correspondence enables one to compute correlation functions in the large N, large λ limit. This limit contains the highly nontrivial sum of all planar Feynman diagrams, and emphasizes those diagrams which have infinitely many vertices.

It should be emphasized that, in its simplest form, this gauge theory - string theory mapping is not the hoped for one between an asymptotically free and infrared strongly coupled gauge theory to a string theory, but instead a mapping between two conformally invariant theories. (For attempts at more realistic theories, see [4] [5]). $\mathcal{N}=4$ supersymmetric Yang-Mills theory has vanishing beta function and its coupling constant doesn't run. If the coupling is small, it has an asymptotic expansion at weak coupling. One could also consider a double limit of the theory - weak coupling and large N. To a given order in $1/N$, the expansion in the coupling constant is not an asymptotic series in that it should have a non-zero radius of convergence. In this Paper, we are going to review some results about summing perturbative series of planar Feynman diagrams. We will see explicitly that these sums converge.

The best evidence in support of the AdS/CFT correspondence comes from symmetry arguments. The global symmetries of both $\mathcal{N} = 4$ supersymmetric Yang-Mills theory and type IIB string theory on $AdS_5 \times S_5$ are identical. They have the same global super-conformal group $SU(2,2\,|\,4)$ (whose bosonic subgroup is $SO(4,2) \times SU(4)$). Not only are the global symmetries the same, but some of those objects which carry the representations of the symmetry group—the spectrum of chiral operators in the field theory and the fields in supergravity theory—can, to some degree, be matched [6]. Furthermore, both theories are conjectured to have an Montonen-Olive $SL(2,Z)$ duality acting on their coupling constants. Once a correspondence between the maximally symmetric theories is established, one could hope for a similar mapping between a broader class of theories which are obtained by perturbing the models on either side of the correspondence with relevant operators.

3 Perturbative checks of AdS/CFT

Though it has been used for many computations of the strong coupling limits of gauge theory quantities (see refs. [7]-[11] for reviews), it is difficult to obtain an explicit check of the AdS/CFT correspondence. The reason for this is the fact that the correspondence with supergravity computes gauge theory in the large λ limit, with corrections from tree level string effects being suppressed by powers of $1/\sqrt{\lambda}$ and sometimes computable to the next order. On the other hand, the only other analytical tool which can be used systematically in the gauge theory is perturbation theory which is an asymptotic expansion in small λ. Generally, the only quantities for which these expansions have an overlapping range of validity is for quantities which are so protected by supersymmetry that they do not depend on the coupling constant.

There are now a few known examples of a quantities which are non-trivial functions of the coupling constant and whose large N limit is computable and is thought to be known to all orders in perturbation theory in planar diagrams. The first example is the circular Wilson loop. Its expectation value was computed in ref. [12]. The contribution of a subset of all Feynman graphs, the planar rainbow diagrams, were found at each order in λ and the sum of all orders was taken to obtain the result

$$\langle W[\text{circle}] \rangle = \frac{2}{\sqrt{\lambda}} I_1(\sqrt{\lambda}) \approx \sqrt{\frac{2}{\pi}} \frac{e^{\sqrt{\lambda}}}{\lambda^{3/4}} \text{ as } \lambda \to \infty \quad (1)$$

where $I_1(x)$ is a modified Bessel function. It was also shown explicitly that the leading corrections to the sum of rainbow diagrams cancels identically. The computations leading to this conclusion are reviewed in Section 7 below. It was conjectured that this cancellation would also occur at higher orders and the result (1) was thus the exact sum of all planar diagrams. Some support for this conjecture was developed in ref. [13]. They also observed that the sum over Feynman diagrams could be obtained for all orders in the $1/N$ expansion and had a beautiful argument that, in the large λ limit, these higher orders produced the expected higher genus string corrections.

The other quantities that are also probably exactly known have to do with the correlators of the circular Wilson loop operator with so-called chiral primary operators. The chiral primary operator is

$$\mathcal{O}^I_k = \frac{(8\pi^2)^{k/2}}{\sqrt{k}\lambda^{k/2}} C^I_{i_1\ldots i_k} \text{Tr} \Phi^{i_1} \ldots \Phi^{i_k}, \quad (2)$$

In [14] the sum of all ladder and rainbow-like planar Feynman diagrams was found. The result is

$$\frac{\langle W[\text{circle}]\, \mathcal{O}^I_k\rangle}{\langle W[\text{circle}]\rangle} = 2^{k/2-1}\sqrt{k\lambda}\, \frac{I_k(\sqrt{\lambda})}{I_1(\sqrt{\lambda})} \frac{R^k}{L^{2k}} Y^I(\theta) \quad (3)$$

where L is the distance of the operator insertion from the center of the loop and R is the radius of the circular loop. It was also argued in [14] that radiative corrections to this expression cancel exactly. These computations are reviewed in Sections 8-10.

21

In the large λ limit, this correlator agrees with a computation of it using the AdS/CFT correspondence which was originally performed in [15].

4 $\mathcal{N} = 4$ supersymmetric Yang-Mills theory

The Euclidean space action of four dimensional $\mathcal{N} = 4$ supersymmetric Yang-Mills theory in four dimensions is

$$S = \int d^4x \frac{1}{2g^2} \left(\frac{1}{2}(F^a_{\mu\nu})^2 + \left(\partial_\mu \phi^a_i + f^{abc} A^b_\mu \phi^c_i\right)^2 + \bar{\psi}^a i\gamma^\mu \left(\partial_\mu \psi^a + f^{abc} A^b_\mu \psi^c\right) + \right.$$

$$\left. + i f^{abc} \bar{\psi}^a \Gamma^i \phi^b_i \psi^c - \sum_{i<j} f^{abc} f^{ade} \phi^b_i \phi^c_j \phi^d_i \phi^e_j + \partial_\mu \bar{c}^a \left(\partial_\mu c^a + f^{abc} A^b_\mu c^c\right) + \xi(\partial_\mu A^a_\mu)^2 \right) \quad (4)$$

where we use the standard convention for the curvature

$$F^a_{\mu\nu} = \partial_\mu A^a_\nu - \partial_\nu A^a_\mu + f^{abc} A^b_\mu A^c_\nu$$

and f^{abc} are the structure constants of the $SU(N)$ Lie algebra,

$$[T^a, T^b] = if^{abc}T^c.$$

The generators are normalized so that

$$\text{Tr} T^a T^b = \frac{1}{2}\delta^{ab},$$

and obey the identity

$$\sum_{c,d} f^{acd} f^{bcd} = N\delta^{ab}.$$

ψ^a is a sixteen component spinor obeying the Majorana condition

$$\psi(x) = C\psi^*(x), \quad (5)$$

where C is the charge conjugation matrix. $\Gamma^A = (\gamma^\mu, \Gamma^i)$, for $\mu = 1, ..., 4$ and $i = 5, ..., 10$ are ten real 16×16 Dirac matrices (in the 10-dimensional Majorana representation with the Weyl constraint) obeying

$$\text{Tr}\left(\Gamma^A \Gamma^B\right) = 16\delta^{AB}.$$

All fields are $N \times N$ matrices and transform in the adjoint representation of the gauge group. We have chosen the covariant gauge fixing condition,

$$\partial_\mu A^a_\mu = 0, \quad (6)$$

and added the appropriate Faddeev-Popov ghosts to the action. We will work in Feynman gauge, where the gauge parameter is chosen as $\xi = 1$.

In Feynman gauge, the vector field propagator is

$$\Delta^{ab}_{\mu\nu}(p) = g^2 \delta^{ab} \frac{\delta_{\mu\nu}}{p^2},$$

the scalar propagator is

$$D^{ab}_{ij}(p) = g^2 \delta^{ab} \frac{\delta_{ij}}{p^2},$$

the fermion propagator is

$$S^{ab}(p) = g^2 \delta^{ab} \frac{-\gamma \cdot p}{p^2},$$

and the ghost propagator is

$$C^{ab}(p) = g^2 \delta^{ab} \frac{1}{p^2}.$$

The vertices can be easily deduced from the non-quadratic terms in the action (4). Each vertex carries a factor of $1/g^2$ and each propagator carries a factor of g^2.

We will use a particular form of dimensional regularization in explicit computations. For this, we will use position-space propagators in 2ω-dimensions. These can be deduced from the Fourier transform

$$\int \frac{d^{2\omega}p}{(2\pi)^{2\omega}} \frac{e^{ip\cdot x}}{[p^2]^s} = \frac{\Gamma(\omega - s)}{4^s \pi^\omega \Gamma(s)} \frac{1}{[x^2]^{\omega-s}}. \tag{7}$$

By setting $s = 1$ we find the Green function in 2ω dimensions:

$$\Delta(x) = \frac{\Gamma(\omega - 1)}{4\pi^\omega} \frac{1}{[x^2]^{\omega-1}} \quad \text{which satisfies} \quad -\partial^2 \Delta(x) = \delta^{2\omega}(x). \tag{8}$$

4.1 some useful formulae

Some formulae which are useful in reproducing the computations are the following. Euler's B function is defined as

$$B(\mu, \nu) = \int_0^1 dx\, x^{\mu-1}(1-x)^{\nu-1} = \frac{\Gamma(\mu)\Gamma(\nu)}{\Gamma(\mu+\nu)}$$

and the Γ function is

$$\Gamma(n+1) = \int_0^\infty dt\, t^n e^{-t},$$

which satisfies $\Gamma(n+1) = n!$, $\Gamma(1/2) = \sqrt{\pi}$, and the combinatorial formulae

$$\Gamma((2n+1)/2) = (n-1/2)(n-3/2)\ldots(1/2)\sqrt{\pi}, \tag{9}$$

$$\Gamma(n)\Gamma(1/2) = 2^{n-1}\Gamma(n/2)\Gamma((n+1)/2). \tag{10}$$

One loop integrals in 2ω dimensions can be computed using the dimensional regularization formulae:

$$\int d^{2\omega}k (k^2 + 2p\cdot k + m^2)^{-s} = \pi^\omega \frac{\Gamma(s-\omega)}{\gamma(s)}(m^2 - p^2)^{\omega-s} \tag{11}$$

$$\int d^{2\omega}k k_\mu (k^2 + 2p\cdot k + m^2)^{-s} = -p_\mu \pi^\omega \frac{\Gamma(s-\omega)}{\gamma(s)}(m^2 - p^2)^{\omega-s} \tag{12}$$

$$\int d^{2\omega}k k_\mu k_\nu (k^2 + 2p\cdot k + m^2)^{-s} = \pi^\omega \frac{1}{\Gamma(s)}(m^2 - p^2)^{\omega-s} \tag{13}$$

$$\times \left[p_\mu p_\nu \Gamma(s-\omega) - \frac{1}{2} g_{\mu\nu} \Gamma(s-\omega-1)(p^2 + m^2) \right] \tag{14}$$

and the Feynman parameter formula,

$$\prod_i A_i^{-n_i} = \frac{\Gamma(\sum n_i)}{\prod_i \Gamma(n_i)} \int_0^1 dx_1 \cdots dx_k\, x_1^{n_1-1}\cdots x_k^{n_k-1} \frac{\delta(1-\sum x_i)}{[\sum_i A_i x_i]^{\sum n_i}}.$$

4.2 The Wilson Loop

The Wilson loop operator is a phase factor associated with the trajectory of a heavy quark in the fundamental representation of the gauge group, which in our case is $SU(N)$. The loop operator of most interest in the Ads/CFT correspondence is the one which couples to classical quantities in the superstring theory in the simplest way provides a source for a classical string. It is

$$W(C) = \frac{1}{N} \operatorname{Tr} \exp \oint_C d\tau \, (iA_\mu(x)\dot{x}_\mu + \Phi_i(x)y_i), \tag{15}$$

where $x_\mu(\tau)$ is a parameterization of the loop, $y_i = \sqrt{\dot{x}^2}\theta_i$ and θ_i is a point on the five dimensional unit sphere ($\theta^2 = 1$). This operator measures the holonomy of a heavy W-boson whose mass results from spontaneous breaking of $SU(N+1)$ gauge symmetry to $SU(N) \times U(1)$. In the AdS/CFT correspondence, it is computed by finding the area of the world-sheet of the classical string in $AdS_5 \times S_5$ whose boundary is the loop C, which in turn lies on the boundary of AdS_5 [16],[17].

The expectation value of the Wilson loop operator is easy to compute in perturbation theory to the leading order,

$$\langle W(C) \rangle = 1 + \frac{g^2 N}{4\pi^2} \oint_C d\tau_1 \, d\tau_2 \frac{|\dot{x}(\tau_1)||\dot{x}(\tau_2)| - \dot{x}(\tau_1) \cdot \dot{x}(\tau_2)}{|x(\tau_1) - x(\tau_2)|^2} + \cdots. \tag{16}$$

For a loop without cusps or self-intersections, this result is finite. This is because a cancellation occurs between the contributions of the scalar and vector fields which we can see occurring explicitly in the integrand in this formula. The case where the contour has cusps and self-intersections has been discussed in [18].

5 Perturbative computations

Ordinarily it is difficult to find regularizations of a field theory which are both gauge invariant and supersymmetric. Non-Abelian gauge theories are usually regulated using dimensional regularization. This works well because gauge invariance is independent of dimension. However, in supersymetric theories, generally, the number of fermionic degrees of freedom varies with dimension differently than the number of bosonic degrees of freedom and dimensional regularization is not supersymmetric. Here, we will use an alternative to naive dimensional regularization which was first used in [12]. This procedure considers supersymmetric Yang-Mills theory in 2ω dimensions as a dimensional reduction of $\mathcal{N} = 1$ supersymmetric Yang-Mills theory in ten dimensions. In this scheme, the gauge field $A_\mu^a(x)$ is a 2ω component vector field. The index of the scalar field runs over $10 - 2\omega$ values, $i = 1, ..., 10 - 2\omega$. In every dimension, the fermion field has sixteen real components. Regularization by dimensional reduction preserves the sixteen supersymmetries of the ten dimensional Yang-Mills theory. These lead to four conserved four component Majorana spinor supercharges in four dimensions. This regularization scheme provides a supersymmetric regularization of the four dimensional theory. Since the gauge coupling becomes dimensional in any spacetime dimension other than four, the regularization breaks conformal symmetry explicitly. It also modifies the R-symmetry.

5.1 One loop self energy of the vector and scalar fields

Consider the one loop self-energies of the vector and scalar field. The dimension of space-time is $D = 2\omega$. All of the loop integrals are elementary and can be found in reference books such as [19]. We parameterize the Wilson loop by $x(\tau)$, and abbreviate $x^{(i)} = x(\tau^i)$.

The vector field obtains self-energy corrections from:

- N^2 colors of 2ω-component vector fields and ghost fields:

$$= \delta^{ab} g^4 N \frac{\Gamma(2-\omega)\Gamma(\omega)\Gamma(\omega-1)}{(4\pi)^\omega \Gamma(2\omega)} \cdot 2(3\omega - 1) \frac{\delta_{\mu\nu} - p_\mu p_\nu/p^2}{p^{2-2\omega}}$$

- $10 - 2\omega$ real scalar fields in the adjoint representation:

$$= -\delta^{ab}g^4 N \frac{\Gamma(2-\omega)\Gamma(\omega)\Gamma(\omega-1)}{(4\pi)^\omega \Gamma(2\omega)} \cdot (10-2\omega)\frac{\delta_{\mu\nu} - p_\mu p_\nu/p^2}{p^{2-2\omega}}$$

- four flavors of four-component Majorana fermions in the adjoint representation:

$$= -\delta^{ab}g^4 N \frac{\Gamma(2-\omega)\Gamma(\omega)\Gamma(\omega-1)}{(4\pi)^\omega \Gamma(2\omega)} \cdot 16(\omega-1)\frac{\delta_{\mu\nu} - p_\mu p_\nu/p^2}{p^{2-2\omega}}$$

Note that these are the negative of the conventionally defined self-energies. Thus, to one loop order, the propagator for the unrenormalized gluon, in Feynman gauge is

$$\Delta^{ab}_{\mu\nu} = g^2\delta^{ab}\frac{\delta_{\mu\nu}}{p^2} - g^4 N \frac{\Gamma(2-\omega)\Gamma(\omega)\Gamma(\omega-1)}{(4\pi)^\omega \Gamma(2\omega)} \cdot 4(2\omega-1)\delta^{ab}\frac{\delta_{\mu\nu} - p_\mu p_\nu/p^2}{p^{6-2\omega}}. \quad (17)$$

Similarly, we can compute the one loop correction to the scalar propagator. It obtains corrections from:

- the scalar-vector intermediate state:

$$= \delta^{ab}g^4 N \frac{\Gamma(2-\omega)\Gamma(\omega)\Gamma(\omega-1)}{(4\pi)^\omega \Gamma(2\omega)} \cdot 4(2\omega-1)\frac{\delta_{ij}}{p^{6-2\omega}}$$

- and the fermion loop:

$$= -\delta^{ab}g^4 N \frac{\Gamma(2-\omega)\Gamma(\omega)\Gamma(\omega-1)}{(4\pi)^\omega \Gamma(2\omega)} \cdot 8(2\omega-1)\frac{\delta_{ij}}{p^{6-2\omega}}$$

Thus, to one loop order, the (unrenormalized) scalar propagator is

$$D^{ab}_{ij} = g^2\delta^{ab}\frac{\delta_{ij}}{p^2} - g^4 N \frac{\Gamma(2-\omega)\Gamma(\omega)\Gamma(\omega-1)}{(4\pi)^\omega \Gamma(2\omega)} \cdot 4(2\omega-1)\frac{\delta_{ij}\delta^{ab}}{p^{6-2\omega}}. \quad (18)$$

Note that, aside from vector indices, the scalar and vector propagators are identical. This is a result of the gauge and supersymmetric invariance of the regularization.

Also, note that the self-energy corrections have poles at $\omega = 2$ which arise from an ultraviolet divergence. If we were to compute correlators of local renormalized fields, it would be necessary to add a counterterm to the action in order to cancel these ultraviolet singularities. Here, for purpose of computing the Wilson loop, we leave them unrenormalized.

5.2 Ladder diagrams

The ladder-like diagrams to order $g^4 N^2$ contribute

$$\Sigma_1 = \frac{g^4 N^2}{6} \oint_{\tau_1 > \tau_2 > \tau_3 > \tau_4} d\tau_1\, d\tau_2\, d\tau_3\, d\tau_4 \frac{(|\dot{x}^{(1)}||\dot{x}^{(2)}| - \dot{x}^{(1)}\cdot\dot{x}^{(2)})(|\dot{x}^{(3)}||\dot{x}^{(4)}| - \dot{x}^{(3)}\cdot\dot{x}^{(4)})}{(|x^{(1)} - x^{(2)}|^2|x^{(3)} - x^{(4)}|^2)^{\omega-1}}.$$

This result is finite when $x(\tau)$ is a smooth curve.

5.3 Insertion of one loop corrections to propagators

Using the one loop self-energies of the vector and scalar fields that we found in section 2.2, we can find the corrections to the expression (16) resulting from insertion of one loop into the vector and scalar field propagators.

We begin with the correction to the scalar propagator in momentum space. From equation (18) it is
$$-\delta^{ab} g^4 N \frac{\Gamma(2-\omega)\Gamma(\omega)\Gamma(\omega-1)}{(4\pi)^\omega \Gamma(2\omega)} 4(2\omega-1) \frac{1}{[p^2]^{3-\omega}}.$$

The Fourier transform of this expression, computed using (7), is
$$-\delta^{ab} g^4 N \frac{\Gamma^2(\omega-1)}{2^5 \pi^{2\omega}(2-\omega)(2\omega-3)} \frac{1}{[x^2]^{2\omega-3}}.$$

If we now combine this with the analogous expression for the vector field propagator and compute the correction to the circular loop the result is
$$\Sigma_2 = -g^4 N^2 \frac{\Gamma^2(\omega-1)}{2^7 \pi^{2\omega}(2-\omega)(2\omega-3)} \oint d\tau_1 \, d\tau_2 \frac{|\dot{x}^{(1)}||\dot{x}^{(2)}| - \dot{x}^{(1)} \cdot \dot{x}^{(2)}}{[(x^{(1)} - x^{(2)})^2]^{2\omega-3}}, \tag{19}$$

where a factor of $N^2/2$ came from taking the trace over gauge group generators, a factor of $1/N$ came from the normalization of the Wilson loop, and an additional factor of $1/2$ came from the combinatorics of expanding the Wilson loop operator to second order. We see that the integrand is identical to (16) with a correction to the coefficient
$$\frac{g^2 N}{4\pi^2} \mapsto \frac{g^2 N}{4\pi^2} - \frac{g^4 N^2 \Gamma^2(\omega-1)}{128 \, \pi^{2\omega}(2-\omega)(2\omega-3)}.$$

The coefficient diverges in four dimensions (at $\omega = 2$).

5.4 Diagrams with one internal vertex

The order $g^4 N^2$ contributions with one internal vertex come when we Taylor expand $W(C)$ to third order in A and Φ and Wick contract it with the relevant vertices to obtain the quantities:
$$\frac{i^3}{3!} \int d\tau_1 \, d\tau_2 \, d\tau_3 \Big\langle \text{Tr}\, \mathcal{P}[A(\tau_1)A(\tau_2)A(\tau_3)] \Big(-\int d^4 y \, f^{abc} \partial_\mu \phi_i^a(y) A_\mu^b(y) \phi_i^c(y)\Big) \Big\rangle,$$
$$\frac{i}{2!1!} \int d\tau_1 \, d\tau_2 \, d\tau_3 \Big\langle \text{Tr}\, \mathcal{P}[\Phi(\tau_1)A(\tau_2)\Phi(\tau_3)] \Big(-\int d^4 y \, f^{abc} \partial_\mu A_\nu^a(y) A_\mu^b(y) A_\nu^c(y)\Big) \Big\rangle, \tag{20}$$

where $A(\tau) = A_\mu^a(x) \dot{x}^\mu(\tau) T^a$ and $\Phi(\tau) = \Phi^a(x)|\dot{x}|T^a$. The minus signs in both vertices come from the expansion of e^{-S} (we work in Euclidean space). The sum of the two diagrams is
$$\Sigma_3 = -\frac{g^4 N^2}{4} \oint d\tau_1 \, d\tau_2 \, d\tau_3 \, \epsilon(\tau_1 \, \tau_2 \, \tau_3)(|\dot{x}^{(1)}||\dot{x}^{(3)}| - \dot{x}^{(1)} \cdot \dot{x}^{(3)})$$
$$\times \dot{x}^{(2)} \cdot \frac{\partial}{\partial x^{(1)}} \int d^{2\omega} w \, \Delta(x^{(1)} - w) \Delta(x^{(2)} - w) \Delta(x^{(3)} - w). \tag{21}$$

Here, ϵ is the antisymmetric path ordering symbol: we define $\epsilon(\tau_1 \, \tau_2 \, \tau_3) = 1$ for $\tau_1 > \tau_2 > \tau_3$ and let ϵ be antisymmetric under any transposition of τ_i. It is straightforward to introduce Feynman parameters and do the integral over w to obtain
$$\Sigma_3 = g^4 N^2 \frac{\Gamma(2\omega-2)}{2^7 \pi^{2\omega}} \int_0^1 d\alpha \, d\beta \, d\gamma \, (\alpha\beta\gamma)^{\omega-2} \delta(1 - \alpha - \beta - \gamma) \oint d\tau_1 \, d\tau_2 \, d\tau_3 \, \epsilon(\tau_1 \, \tau_2 \, \tau_3)$$
$$\times \frac{\big(|\dot{x}^{(1)}||\dot{x}^{(3)}| - \dot{x}^{(1)} \cdot \dot{x}^{(3)}\big)\big(\alpha(1-\alpha)\dot{x}^{(2)} \cdot x^{(1)} - \alpha\gamma \dot{x}^{(2)} \cdot x^{(3)} - \alpha\beta \dot{x}^{(2)} \cdot x^{(2)}\big)}{[\alpha\beta|x^{(1)} - x^{(2)}|^2 + \alpha\gamma|x^{(1)} - x^{(3)}|^2 + \beta\gamma|x^{(3)} - x^{(2)}|^2]^{2\omega-2}}. \tag{22}$$

6 Cancellation of divergences to order $g^4 N^2$

A logarithmic divergence arises in the integral (21) from where τ_1 is coincident with τ_2. This divergence should cancel with the divergence in the coefficient of Σ_2 in (19) so that the order $g^4 N^2$ contribution

$$\Sigma_1 + \Sigma_2 + \Sigma_3$$

is finite. In extracting the divergences from (21), we must consider the integral

$$G(\tau_i) = \int d^4w \, \Delta(w - x^{(1)}) \Delta(w - x^{(2)}) \Delta(w - x^{(3)}), \tag{23}$$

in detail. This integral is singular in the limit $x^{(2)} \to x^{(1)}$. The divergent contribution comes from the integration over w close to $x^{(1)}$ and $x^{(2)}$. We can approximate $|w - x^{(3)}| \approx |x^{(1)} - x^{(3)}|$ introducing simultaneously an infrared cutoff δ, so that the divergent part of (23) takes the form

$$G(\tau_i) \sim \Delta(x^{(1)} - x^{(3)}) \int^\delta d^4w \, \Delta(w - x^{(1)}) \Delta(w - x^{(2)}). \tag{24}$$

As follows from dimensional counting, this integral depends only on the dimensionless ratio $|x^{(1)} - x^{(2)}|/\delta$. The limit where $x^{(1)}$ and $x^{(2)}$ are coincident is then equivalent to the limit of infinite δ. Without the infrared cutoff the integral would diverge logarithmically, so up to terms regular in the limit $\delta \to \infty$, we get

$$G(\tau_i) \sim \frac{1}{64\pi^6} \frac{1}{|x^{(1)} - x^{(3)}|^2} \int \frac{d^4w}{w^4} = -\frac{1}{64\pi^4} \frac{\log |x^{(1)} - x^{(2)}|^2/\delta^2}{|x^{(1)} - x^{(3)}|^2}. \tag{25}$$

Since G only goes as a logarithm as points approach each other, (21) receives divergent contributions only from $x^{(1)}$ near $x^{(2)}$; the parts of the integral with $x^{(1)}$ or $x^{(2)}$ near $x^{(3)}$ are finite. This divergence is regularized by cutting off the integral over τ_1. Since the overall divergence is logarithmic, the result is independent of the method of regularization. Writing $\tau = \tau^{(1)} - \tau^{(2)}$ and Taylor expanding $x^{(1)} = x^{(2)} + \dot{x}^{(2)} \tau + \cdots$, we see that the divergent part of (21) is

$$\Sigma_3 \sim -\frac{g^4 N^2}{128\pi^4} \oint d\tau_2 \oint d\tau_3 \frac{|\dot{x}^{(2)}||\dot{x}^{(3)}| - \dot{x}^{(2)} \cdot \dot{x}^{(3)}}{|x^{(2)} - x^{(3)}|^2} \int d\tau \, \text{sign}(\tau) \frac{\dot{x}^{(2)} \cdot (x^{(1)} - x^{(2)})}{|x^{(1)} - x^{(2)}|^2}$$

$$= -\frac{g^4 N^2}{64\pi^4} \oint d\tau_2 \oint d\tau_3 \frac{|\dot{x}^{(2)}||\dot{x}^{(3)}| - \dot{x}^{(2)} \cdot \dot{x}^{(3)}}{|x^{(2)} - x^{(3)}|^2} \log \epsilon. \tag{26}$$

This cancels exactly against (19), for one should replace the pole $1/(2-\omega)$ at $\omega = 2$ by $-2 \log \epsilon$.

7 The circular loop

In this section, we consider a circular Wilson loop, whose radius we can assume to be unity. A convenient parameterization of this loop is

$$x(\tau) = (\cos \tau, \sin \tau, 0, 0). \tag{27}$$

7.1 Summing the planar ladder graphs

First, we will sum all planar diagrams which have no internal vertices. These include all rainbow diagrams. Our strategy is the following: the large N limit with $g^2 N$ held fixed is given by planar diagrams. It is well known that each planar graph will contain the same group theoretical factors. We observe that, in fact, each planar diagram without internal vertices gives an identical contribution to the loop expectation value. We choose a fixed ordering of the times and compute a particular, convenient diagram. Then we multiply by the number of diagrams that occur to that order and sum over all orders.

First, consider the $2n$-th order term in the Taylor expansion of the loop

$$\frac{1}{N}\int_0^{2\pi}d\tau_1\int_0^{\tau_1}d\tau_2\cdots\int_0^{\tau_{2n-1}}d\tau_{2n}\operatorname{Tr}\langle(iA(\tau_1)+\Phi(\tau_1))\cdots(iA(t_{2n})+\Phi(t_{2n}))\rangle.$$

Here we have chosen a particular time ordering, which cancels the factor of $1/(2n)!$ which would come from the Taylor expansion of the exponential. We are interested in all Wick contractions which represent planar diagrams. Note that for the circular loop

$$\langle(iA^a(\tau_1)+\Phi^a(\tau_1))\left(iA^b(\tau_2)+\Phi^b(\tau_2)\right)\rangle_0 = \frac{g^2\delta^{ab}}{4\pi^2}\frac{|\dot{x}^{(1)}||\dot{x}^{(2)}| - \dot{x}^{(1)}\cdot\dot{x}^{(2)}}{|x^{(1)} - x^{(2)}|^2} = \frac{g^2\delta^{ab}}{8\pi^2}. \qquad (28)$$

Thus, the contributions of all (free-field) Wick contractions giving planar diagrams are identical (if we sum over all ways of choosing scalar or gluon lines). The color factor can be computed by repeated application of the identity:

$$T^aT^a = \frac{N}{2}\mathbf{1}.$$

Thus, for the sum of ladder-like diagrams with n propagators we obtain

$$\frac{(g^2N/4)^n}{(2n)!} \times (\text{number of planar graphs with n internal lines}), \qquad (29)$$

where the factor $1/(2n)!$ accounts for the integral over the $\tau^{(i)}$.

We now have the task of counting the number of planar graphs with n internal lines. If we define A_{n+1} as the number of such diagrams n+1 propagators then A_{n+1} satisfies the recursion relation

$$A_{n+1} = \sum_{k=0}^{n}A_{n-k}A_k,$$

with $A_0 = 1$. If we define a generating function f by

$$f(z) = \sum_{n=0}^{\infty}A_nz^n,$$

then f satisfies $zf^2(z) = f(z) - 1$. So,

$$f(z) = \frac{1-\sqrt{1-4z}}{2z} = \sum_{n=0}^{\infty}\frac{(2n)!}{(n+1)!n!}z^n. \qquad (30)$$

The sign of the square root is chosen by requiring that f be finite at $z = 0$. Hence

$$A_n = \frac{(2n)!}{(n+1)!n!}, \qquad (31)$$

so the sum of all planar diagrams without vertices on the loop is, from (29) and (31)

$$\langle W(C)\rangle_{\text{ladders}} = \sum_{n=0}^{\infty}\frac{(g^2N/4)^n}{(n+1)!n!} = -\frac{2}{\sqrt{g^2N}}I_1(\sqrt{g^2N}). \qquad (32)$$

Thus, the large g^2N behavior is

$$\langle W(C)\rangle_{\text{ladders}} \sim \frac{e^{\sqrt{g^2N}}}{(\pi/2)^{1/2}(g^2N)^{3/4}}. \qquad (33)$$

The supergravity prediction is that
$$\langle W(C)\rangle_{\text{AdS/CFT}} \sim e^{\sqrt{g^2 N}}, \tag{34}$$
so the ladder diagrams have the same leading behavior as the prediction of the AdS/CFT correspondence.

It is worth mentioning that the cancellation of coordinate dependence in the Wick contraction (28) maps the problem of summing the ladder-like diagrams to the zero dimensional theory. In particular, the number of planar graphs with no vertices and n propagators can be calculated from the infinite N limit of the matrix integral

$$A_n = \left\langle \frac{1}{N}\text{Tr}\, M^{2n} \right\rangle = \frac{1}{Z}\int [dM]\, \frac{1}{N}\text{Tr}\, M^{2n}\, \exp\left\{-\frac{N}{2}\text{Tr}\, M^2\right\},$$

where
$$Z = \int [dM]\, \exp\left\{-\frac{N}{2}\text{Tr}\, M^2\right\}.$$

This can be evaluated by extracting the term of order z^{2n} in the Taylor expansion of the zero-dimensional Wilson loop:

$$\Omega(z) = \left\langle \frac{1}{N}\text{Tr}\,\frac{1}{1-zM}\right\rangle. \tag{35}$$

Using the identity
$$\frac{1}{(2n)!} = \oint_C \frac{dz}{2\pi i}\,\frac{e^z}{z^{2n+1}}$$
for a positively oriented contour C containing the origin, we can represent the sum of the ladder diagrams as

$$\langle W(C)\rangle_{\text{ladders}} = \oint_C \frac{dz}{2\pi i}\,\frac{e^z}{z}\,\Omega(g^2 N/4z), \tag{36}$$

where C must be chosen large enough to encircle all singularities of the integrand. The function $\Omega(z)$ satisfies a zero-dimensional loop equation that follows from the identity

$$0 = \int [dM]\,\frac{\partial}{\partial M_{ij}}\left\{\left(\frac{1}{1-zM}\right)_{ij}\exp(-2\text{Tr}\, M^2)\right\}$$

and large N factorization. In the infinite N limit, this equality reduces to the algebraic equation for $\Omega(z)$:

$$z\Omega^2(z) - \frac{1}{z}\Omega(z) + \frac{1}{z} = 0.$$

This has solution
$$\Omega(z) = \frac{1-\sqrt{1-4z^2}}{2z^2}.$$

Substituting $z^2 \to z$, this is the same as (30). Shrinking the contour of integration in (36) to the branch cut of $\Omega(z)$, we obtain

$$\langle W(C)\rangle_{\text{ladders}} = 4\int_{-1}^{1} \frac{dx}{2\pi}\, e^{\sqrt{g^2 N}\, x}\,\sqrt{1-x^2},$$

which is just the Bessel function (32).

7.2 Diagrams with vertices

The sum of the diagrams with one internal vertex attaching to three points on the Wilson loop is given by the expression (22). It is convenient to abbreviate $\tau_{ij} = \tau_i - \tau_j$. For a circular loop, $|x^{(i)}|^2 = 1$, $|x^{(i)} - x^{(j)}|^2 = 2(1 - \cos \tau_{ij})$, $x^{(i)} \cdot \dot{x}^{(j)} = \sin \tau_{ij}$, and $\dot{x}^{(i)} \cdot \dot{x}^{(j)} = \cos \tau_{ij}$. Thus, from (21)

$$\Sigma_3 = g^4 N^2 \frac{\Gamma(2\omega - 2)}{2^{2\omega+5} \pi^{2\omega}} \int_0^1 d\alpha \, d\beta \, d\gamma \, (\alpha\beta\gamma)^{\omega-2} \delta(1 - \alpha - \beta - \gamma)$$

$$\times \oint d\tau_1 \, d\tau_2 \, d\tau_3 \, \frac{\epsilon(\tau_1 \tau_2 \tau_3)(1 - \cos \tau_{13})(\alpha(1 - \alpha) \sin \tau_{12} + \alpha\gamma \sin \tau_{23})}{[\alpha\beta(1 - \cos \tau_{12}) + \beta\gamma(1 - \cos \tau_{23}) + \gamma\alpha(1 - \cos \tau_{13})]^{2\omega-2}}. \qquad (37)$$

We are going to use integration by parts to rewrite (37) as a sum of a term which will cancel with the order $g^4 N^2$ diagrams with internal vertices in (19) and a term which vanishes when $\omega = 2$. For compactness, write the denominator

$$\Delta = \alpha\beta(1 - \cos \tau_{12}) + \beta\gamma(1 - \cos \tau_{23}) + \gamma\alpha(1 - \cos \tau_{13}).$$

Consider the identity

$$\oint d\tau_1 \, d\tau_2 \, d\tau_3 \, \frac{\partial}{\partial \tau_1} \frac{\epsilon(\tau_1 \tau_2 \tau_3)(1 - \cos \tau_{13})}{\Delta^{2\omega-3}} = 0. \qquad (38)$$

Using

$$\frac{\partial}{\partial \tau_1} \epsilon(\tau_1 \tau_2 \tau_3) = 2\delta(\tau_{12}) - 2\delta(\tau_{13}), \qquad (39)$$

that $\alpha + \beta + \gamma = 1$, and the fact that the integrand in vanishes when $\tau_1 = \tau_3$, we get

$$\oint d\tau_1 \, d\tau_2 \, d\tau_3 \, \Bigg\{ -\frac{\sin \tau_{13} \, (\alpha\beta(1 - \cos \tau_{12}) + \beta\gamma(1 - \cos \tau_{23}) + \gamma\alpha(1 - \cos \tau_{13}))}{\Delta^{2\omega-2}}$$

$$+ (2\omega - 3) \frac{(1 - \cos \tau_{13})(\alpha\beta \sin \tau_{12} + \gamma\alpha \sin \tau_{13})}{\Delta^{2\omega-2}} \Bigg\} \epsilon(\tau_1 \tau_2 \tau_3)$$

$$= 2 \oint d\tau_1 \, d\tau_2 \, \frac{1}{[\gamma(1-\gamma)]^{2\omega-3}} \frac{1}{[1 - \cos \tau_{12}]^{2\omega-4}}. \qquad (40)$$

Now, change variables in the first term on the left hand side so that the cosines in the denominator all have argument τ_{13} (note that this involves permuting α, β, and γ as well, so the following holds only after inserting $\delta(1 - \alpha - \beta - \gamma)$ and integrating these parameters over the unit cube). Then add part of the second term, so that the remaining part is proportional to $(2\omega - 4)$ to obtain

$$\oint d\tau_1 \, d\tau_2 \, d\tau_3 \, \epsilon(\tau_1 \tau_2 \tau_3) \Bigg\{ \frac{(1 - \cos \tau_{13})(\alpha(1 - \alpha) \sin \tau_{12} + \alpha\gamma \sin \tau_{23})}{\Delta^{2\omega-2}}$$

$$+ (2\omega - 4) \frac{(1 - \cos \tau_{13})(\alpha\beta \sin \tau_{12} + \gamma\alpha \sin \tau_{13})}{\Delta^{2\omega-2}} \Bigg\}$$

$$= 2 \oint d\tau_1 \, d\tau_2 \, \frac{1}{[\gamma(1-\gamma)]^{2\omega-3}} \frac{1}{[1 - \cos \tau_{12}]^{2\omega-4}}. \qquad (41)$$

The first term on the left hand side is precisely the term occurring in (37). Note that the second term can be rewritten as

$$-\frac{2\omega - 4}{2\omega - 3} \oint d\tau_1 \, d\tau_2 \, d\tau_3 \, \epsilon(\tau_1 \tau_2 \tau_3)(1 - \cos \tau_{13}) \frac{\partial}{\partial \tau_1} \frac{1}{\Delta^{2\omega-3}}$$

$$= \frac{2\omega - 4}{2\omega - 3} \Bigg\{ \oint d\tau_1 \, d\tau_2 \, d\tau_3 \, \epsilon(\tau_1 \tau_2 \tau_3) \frac{\sin \tau_{13}}{\Delta^{2\omega-3}}$$

$$+ 2 \oint d\tau_1 \, d\tau_2 \, \frac{1}{[\gamma(1-\gamma)]^{2\omega-3}} \frac{1}{[1 - \cos \tau_{12}]^{2\omega-4}} \Bigg\}, \qquad (42)$$

using integration by parts. Finally, we have

$$\oint d\tau_1\, d\tau_2\, d\tau_3\, \epsilon(\tau_1\, \tau_2\, \tau_3) \frac{(1-\cos\tau_{13})\left(\alpha(1-\alpha)\sin\tau_{12}+\alpha\gamma\sin\tau_{23}\right)}{\Delta^{2\omega-2}}$$

$$= -\frac{2\omega-4}{2\omega-3}\oint d\tau_1\, d\tau_2\, d\tau_3\, \epsilon(\tau_1\, \tau_2\, \tau_3)\frac{\sin\tau_{13}}{\Delta^{2\omega-3}}$$

$$+\frac{2}{2\omega-3}\oint d\tau_1\, d\tau_2\, \frac{1}{[\gamma(1-\gamma)]^{2\omega-3}}\frac{1}{[1-\cos\tau_{12}]^{2\omega-4}}. \quad (43)$$

If we symmeterize, the integral in the first term on the right hand side may be rewritten (setting $\omega = 2$) as

$$\frac{1}{3}\frac{2\omega-4}{2\omega-3}\oint d\tau_1\, d\tau_2\, d\tau_3\, \epsilon(\tau_1\, \tau_2\, \tau_3)\frac{\sin\tau_{12}+\sin\tau_{23}+\sin\tau_{31}}{\alpha\beta(1-\cos\tau_{12})+\beta\gamma(1-\cos\tau_{23})+\gamma\alpha(1-\cos\tau_{13})}.$$

This integration is completely finite. Since this term appears with a coefficient that vanishes when $\omega = 2$, it vanishes in four dimensions. Inserting this expression into (37), we have the result

$$\Sigma_3 = g^4 N^2 \frac{\Gamma^2(\omega-1)}{2^{2\omega+4}\pi^{2\omega}(2\omega-3)(2-\omega)}\oint d\tau_1\, d\tau_2\, \frac{1}{[1-\cos\tau_{12}]^{2\omega-4}} + \mathcal{O}(2\omega-4). \quad (44)$$

If we specialize (19) to the case of a circular loop, we obtain

$$\Sigma_2 = -g^4 N^2 \frac{\Gamma^2(\omega-1)}{2^{2\omega+4}\pi^{2\omega}(2-\omega)(2\omega-3)}\oint d\tau_1\, d\tau_2\, \frac{1}{[1-\cos\tau_{12}]^{2\omega-3}}. \quad (45)$$

The two contributions (44) and (45) cancel exactly when $2\omega = 4$:

$$\Sigma_2 + \Sigma_3 = 0.$$

8 Correlator of Wilson loop with chiral primary operators

When probed from a distance much larger than the size of the loop, the Wilson loop operator should behave effectively as a local operator. It is reasonable that, for this purpose, it can be expanded in a series of local operators with some coefficients [20],[15]:

$$W[C] = \sum_\Delta \mathcal{C}_A R^{\Delta_A} \mathcal{O}^A(0) \quad (46)$$

where $\mathcal{O}^A(0)$ is a local operator evaluated at the center of the loop, Δ_A is the conformal dimension of $\mathcal{O}^A(x)$ and R is the radius of the loop.

An interesting problem is to attempt to compute the coefficients $\mathcal{C}_A$ in this operator product expansion (OPE). These coefficients tell us about how the Wilson loop couples to other operators.

The contribution of the sum of all planar rainbow graphs to the coefficients for a particular class of chiral primary operators (CPO) was computed in [14]. It was also able to show that the leading order corrections to this sum, which come from diagrams with internal vertices, cancels identically. This leads us to conjecture that the radiative corrections cancel to all orders and the sum of planar rainbow graphs gives the exact result.

We find that the coefficients that we compute are non-trivial functions of the coupling constant. In the limit of large λ they coincide with results of the AdS/CFT correspondence [15]. This gives a large array of non-trivial functions of the coupling constant which could be compared with string computations of the strong coupling limit. There are various reasons why these computations could be simpler than the $1/\sqrt{\lambda}$ corrections to the expectation value of the Wilson loop itself.

For primary operators, one can choose a basis where

$$\langle \mathcal{O}^A(x)\mathcal{O}^B(y)\rangle = \frac{\delta^{AB}}{|x-y|^{\Delta_A+\Delta_B}} \tag{47}$$

Then their OPE coefficients can be extracted from the large distance behavior of connected two-point correlation functions,

$$\frac{\langle W(C)\,\mathcal{O}^A(L)\rangle_c}{\langle W(C)\rangle} = \mathcal{C}_A\,\frac{R^{\Delta_A}}{L^{2\Delta_A}} + \cdots \tag{48}$$

where $L \gg R$ and the omitted terms are of higher order in R^2/L^2.

The coefficient corresponding to the CPO of lowest conformal dimension, which in this case is $\Delta = 2$, is important as it determines the correlator of two Wilson loops with large separation

$$\frac{\langle W[C_1]W[C_2]\rangle_c}{\langle W[C_1]\rangle\langle W[C_2]\rangle} = \mathcal{C}_2^2\left(\frac{R}{L}\right)^4 + \cdots \tag{49}$$

The coefficients of various CPOs in the expansion of the circular Wilson loop were calculated in ref. [15] both perturbatively at $\lambda \sim 0$ and at strong coupling, $\lambda \sim \infty$, using the AdS/CFT correspondence. Evaluation of the correlators that define coefficients in the strong-coupling regime involves a hybrid of the supergravity and the string calculations.

8.1 The example of dimension two operators

Let us begin by considering the CPO with smallest conformal dimension, $\Delta = 2$. It is the symmetric traceless part of a gauge invariant product of scalar fields,

$$\mathcal{O}^{ij} = \frac{8\pi}{\sqrt{2}\,\lambda}\,\mathrm{Tr}\left(\Phi^i\Phi^j - \frac{1}{6}\delta^{ij}\Phi^2\right). \tag{50}$$

This operator is the lowest weight component of a short multiplet of $\mathcal{N}=4$ super-conformal algebra. Such chiral primary operators have very special properties. The super-conformal algebra guarantees that their conformal dimensions do not receive radiative corrections, so in this case the conformal dimension is exactly two. Furthermore, it is known that their two and three-point correlation functions are given by the free field values, that is, that they are independent of the coupling constant, g. It is known that their four-point functions are non-trivial, so they are not free fields in disguise [21]-[24].

In (50), the overall coefficient is chosen to give a canonical normalization of the two-point function:

$$\langle \mathcal{O}^{ij}(x)\mathcal{O}^{kl}(y)\rangle = \frac{1}{2}\left(\delta^{ik}\delta^{jl} + \delta^{il}\delta^{jk} - \frac{1}{3}\delta^{ij}\delta^{kl}\right)\frac{1}{|x-y|^4}. \tag{51}$$

The small coupling limit of the correlator of $\mathcal{O}^{ij}$ with the Wilson loop is straightforward to obtain. To leading order in perturbation theory:

$$\frac{\langle W(C)\,\mathcal{O}^{ij}\rangle}{\langle W(C)\rangle} = \frac{1}{2\sqrt{2}}\lambda\left(\theta^i\theta^j - \frac{1}{6}\delta^{ij}\right)\frac{R^2}{L^4} \tag{52}$$

The linear dependence on λ is an obvious consequence of the fact that the correlator contains two propagators and one power of λ is cancelled by the normalization.

The AdS dual of the dimension two operator is the negative mass scalar which is a linear combination of the trace of the metric and the Ramond-Ramond four-form field. Its contribution to the OPE of the circular Wilson loop was calculated in [15]:

$$\frac{\langle W(C)\,\mathcal{O}^{ij}\rangle}{\langle W(C)\rangle} = \sqrt{2\lambda}\left(\theta^i\theta^j - \frac{1}{6}\delta^{ij}\right)\frac{R^2}{L^4} \quad (\lambda \to \infty). \tag{53}$$

Comparing OPE coefficients at strong and at weak coupling, we see that the scaling with λ is different. The OPE coefficients are clearly renormalized by radiative corrections. We shall conjecture that, in the large N limit, this renormalization is entirely due to planar rainbow diagrams. We shall also obtain the sum of planar rainbows as

$$\frac{\langle W(C)\,\mathcal{O}^{ij}\rangle}{\langle W(C)\rangle} = \sqrt{2\lambda}\,\frac{I_2\left(\sqrt{\lambda}\right)}{I_1\left(\sqrt{\lambda}\right)}\left(\theta^i\theta^j - \frac{1}{6}\delta^{ij}\right)\frac{R^2}{L^4}, \tag{54}$$

where I_2 and I_1 are modified Bessel functions. By construction, this expression reduces to (52) at small λ. Since

$$\lim_{\lambda\to\infty}\frac{I_k\left(\sqrt{\lambda}\right)}{I_1\left(\sqrt{\lambda}\right)} = 1 \tag{55}$$

for any k, the AdS/CFT prediction (53) is also exactly reproduced at large λ. The sum of rainbow diagrams thus interpolates between perturbative and strong coupling limits of the OPE coefficient.

8.2 Chiral primary operators

The $\mathcal{O}^{ij}$ is the first in an infinite sequence of CPOs. The operator of dimension k in this sequence is a symmetrized trace of k scalar fields:

$$\mathcal{O}_k^I = \frac{(8\pi^2)^{k/2}}{\sqrt{k}\lambda^{k/2}}\,C_{i_1\ldots i_k}^I\,\mathrm{Tr}\Phi^{i_1}\ldots\Phi^{i_k}, \tag{56}$$

where $C_{i_1\ldots i_k}^I$ are totally symmetric traceless tensors normalized as

$$C_{i_1\ldots i_k}^I C_{i_1\ldots i_k}^J = \Delta^{IJ}. \tag{57}$$

The AdS duals of CPOs are Kaluza Klein modes of the AdS_5 tachyonic scalar on S^5 and each CPO is associated with a spherical harmonic:

$$Y^I(\theta) = C_{i_1\ldots i_k}^I\,\theta^{i_1}\ldots\theta^{i_k}. \tag{58}$$

Here, we are following the notation of refs. [21],[15].

The OPE coefficients depend on how the operators are normalized. When comparing perturbative calculations with the AdS/CFT predictions, we need to use the same normalization. For operators in short multiplets of $\mathcal{N}=4$ supersymmetry, this is easy to achieve, since the two point correlation functions of such operators do not receive radiative corrections and can be used to fix normalization. The coefficient in (56) is chosen to unit normalize the two point function:

$$\langle \mathcal{O}_k^I(x)\mathcal{O}_k^J(y)\rangle = \frac{\Delta^{IJ}}{|x-y|^{2k}}. \tag{59}$$

The same conventions were used in the supergravity calculations of ref. [15].

At weak coupling, the OPE coefficient of the circular Wilson loop is proportional to $\lambda^{k/2}$, where k is the dimension of the CPO:

$$\frac{\langle W(C)\,\mathcal{O}_k^I\rangle}{\langle W(C)\rangle} = 2^{-k/2}\,\frac{\sqrt{k}}{k!}\,\lambda^{k/2}\,\frac{R^k}{L^{2k}}\,Y^I(\theta) \tag{60}$$

It turns out that AdS/CFT correspondence predicts a universal scaling of the OPE coefficients with λ at strong coupling: all of them are proportional to $\sqrt{\lambda}$ independently of k. This can be easily understood by considering a pair correlator of the Wilson loops, which is quadratic in OPE coefficients. The Wilson loop correlator is described by an annulus string amplitude and therefore is proportional to the string coupling g_s. According to the AdS/CFT dictionary,

$$g_s = \frac{g^2}{4\pi} = \frac{\lambda}{4\pi N}.$$

Hence, OPE coefficients must scale as $\sqrt{\lambda}$. An explicit calculation gives [15]:

$$\frac{\langle W(C)\,\mathcal{O}_k^I\rangle}{\langle W(C)\rangle} = 2^{k/2-1}\sqrt{k\lambda}\,\frac{R^k}{L^{2k}}\,Y^I(\theta) \quad (\lambda\to\infty). \tag{61}$$

Our main result is an expression for correlators of the circular Wilson loop with CPOs:

$$\frac{\langle W(C)\,\mathcal{O}_k^I\rangle}{\langle W(C)\rangle} = 2^{k/2-1}\sqrt{k\lambda}\,\frac{I_k\left(\sqrt{\lambda}\right)}{I_1\left(\sqrt{\lambda}\right)}\,\frac{R^k}{L^{2k}}\,Y^I(\theta) \tag{62}$$

which we expect is exact in the large N limit. Its expansion in λ reproduces (60). The strong-coupling limit exactly coincides with the AdS/CFT prediction (using eq. (55)).

9 Rainbow diagrams

Our calculation of the OPE coefficients begins with summing all planar rainbow diagrams of the kind shown in fig. 1. They contain k scalar propagators connecting the point L to the Wilson loop and propagators of scalars and gauge fields connecting points in segments of the loop.

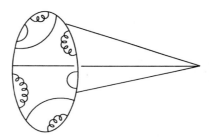

Figure 1: A typical diagram that contributes to the correlator of the circular Wilson loop with the dimension k CPO.

If L lies on the axis of symmetry of the circle, the scalar propagators are constants equal to

$$\frac{1}{8\pi^2(L^2+R^2)}.$$

If the origin is displaced from the axis of symmetry, the propagators will depend on positions of their endpoints on the circle. In any case, we will be interested in the large-distance asymptotics, and the propagators can be set to $1/8\pi^2 L^2$, up to corrections of higher order in $1/L$.

The problem thus reduces to re-summation of rainbow diagrams for each of the segments of the circle. This problem was solved in ref. [25]. In the following we will review the salient points involved in finding the solution. We start with the dimension two operator (50), when there are only two segments.

9.1 Computations for the example of dimension two operators

In this case we have:

$$\langle W(C)\,\mathcal{O}^{ij}\rangle = \frac{8\pi}{\sqrt{2}\,\lambda}\frac{\lambda^2}{(8\pi^2 L^2)^2}\left(\theta^i\theta^j - \frac{1}{6}\Delta^{ij}\right)R^2\,2\pi\int_0^{2\pi}d\varphi\,W(\varphi)W(2\pi-\varphi), \qquad (63)$$

where $W(\varphi' - \varphi)$ denotes the sum of rainbow graphs for a segment of the circle between polar angles φ and φ'. To compute this sum, we notice that the sum of the scalar and the gluon propagators between any two points on a circle does not depend on the positions of these points:

$$\langle (iA(x)\cdot x + \Phi^i(x)\theta^i|\dot{x}|)_{ab}\,(iA(y)\cdot y + \Phi^i(y)\theta^i|\dot{y}|)_{cd}\rangle_0 = \frac{\lambda}{N}\delta_{ad}\delta_{bc}\,(|\dot{x}||\dot{y}| - \dot{x}\cdot\dot{y})\frac{1}{8\pi^2|x-y|^2}$$

$$= \frac{\lambda}{16\pi^2 N}\delta_{ad}\delta_{bc}\,.$$

This observation allows us to replace the field-theory Wick contraction by the matrix-model average defined by the partition function

$$Z = \int dM\,\exp\left(-\frac{8\pi^2}{\lambda}N\mathrm{Tr}M^2\right). \qquad (64)$$

Upon the replacement of $iA_\mu(x)\dot{x}_\mu + \Phi^i(x)\theta^i|\dot{x}|$ by M, the sum of rainbow diagrams in the segment of the length φ reduces the matrix-model counterpart of the Wilson loop:

$$W(\varphi) = \langle\frac{1}{N}\mathrm{Tr}e^{\varphi M}\rangle. \qquad (65)$$

The matrix model can be viewed as a combinatorial tool which simply counts the number of planar graphs [12].

The Wilson loop in the matrix model satisfies Schwinger-Dyson identity in the large-N limit (the loop equation [27]):

$$W'(\theta) = \frac{\lambda}{16\pi^2}\int_0^\theta d\varphi\,W(\varphi)W(\theta-\varphi). \qquad (66)$$

The solution [12],[26],[25] of the loop equation is

$$W(\varphi) = \frac{4\pi}{\sqrt{\lambda}\,\varphi}I_1\left(\frac{\sqrt{\lambda}\,\varphi}{2\pi}\right). \qquad (67)$$

The integral we need to compute in order to calculate the correlation function (63) is the right-hand-side of the loop equation at $\theta = 2\pi$. Using properties of the modified Bessel functions,

$$I'_k(z) = \frac{1}{2}\left(I_{k-1}(z) + I_{k+1}(z)\right),$$

$$kI_k(z) = \frac{z}{2}\left(I_{k-1}(z) - I_{k+1}(z)\right), \qquad (68)$$

we get:
$$\int_0^\theta d\varphi\, W(\varphi) W(\theta - \varphi) = \frac{32\pi^2}{\lambda \varphi} I_2\left(\frac{\sqrt{\lambda}\,\varphi}{2\pi}\right). \tag{69}$$

Setting $\theta = 2\pi$ and substituting into (63), we obtain:
$$\langle W(C)\, \mathcal{O}^{ij}\rangle = 2\sqrt{2}\, I_2 \,(\text{Tr})\left(\theta^i \theta^j - \frac{1}{6}\delta^{ij}\right) \frac{R^2}{L^4}. \tag{70}$$

Dividing by the vacuum expectation value of the Wilson loop,
$$\langle W(C)\rangle = W(2\pi) = \frac{2}{\sqrt{\lambda}} I_1\left(\sqrt{\lambda}\right), \tag{71}$$

we arrive at the result (54) which we quoted earlier.

9.2 Computation for general chiral primary operators

The correlator of the Wilson loop with the CPO of dimension k contains an integral over $k-1$ endpoints of the scalar propagators (one integration yields an overall factor of 2π):

$$\begin{aligned}\langle W(C)\, \mathcal{O}_k^I\rangle &= \frac{(8\pi^2)^{k/2}}{\sqrt{k}\,\lambda^{k/2}} \frac{\lambda^k}{(8\pi^2 L^2)^k} C_{i_1\ldots i_k}^I \theta^{i_1}\ldots \theta^{i_k}\, R^k \\ &\quad \times 2\pi \int_0^{2\pi} d\varphi_1 \ldots \int_0^{\varphi_{k-2}} d\varphi_{k-1}\, W(\varphi_{k-1}) W(\varphi_{k-2} - \varphi_{k-1})\ldots W(2\pi - \varphi_1)\end{aligned} \tag{72}$$

It is useful to introduce
$$F_k(\varphi) = \int_0^\varphi d\varphi_1 \ldots \int_0^{\varphi_{k-2}} d\varphi_{k-1}\, W(\varphi_{k-1}) W(\varphi_{k-2} - \varphi_{k-1})\ldots W(\varphi - \varphi_1). \tag{73}$$

The correlator is expressed in terms of $F_k(2\pi)$ as
$$\langle W(C)\, \mathcal{O}_k^I\rangle = \frac{2\pi}{\sqrt{k}\,(8\pi^2)^{k/2}} \lambda^{k/2} F_k(2\pi) \frac{R^k}{L^{2k}} Y^I(\theta). \tag{74}$$

To find the functions $F_k(\varphi)$, we again use the loop equation. Differentiating $F_k(\varphi)$ and using (66), we get the recurrence relations:
$$F_k'(\varphi) = F_{k-1}(\varphi) + \frac{\lambda}{16\pi^2} F_{k+1}(\varphi), \tag{75}$$

which are supplemented by initial conditions
$$F_1(\varphi) = W(\varphi), \quad F_0(\varphi) = 0. \tag{76}$$

These unambiguously determine all F_k. A systematic way to solve these recurrence relations is to introduce a generating function and then use a Laplace transform to convert differential equations into algebraic equations. The result is
$$F_k(\varphi) = \frac{k}{\varphi}\left(\frac{4\pi}{\sqrt{\lambda}}\right)^k I_k\left(\frac{\sqrt{\lambda}\,\varphi}{2\pi}\right). \tag{77}$$

It is straightforward to check that this expression solves the recurrence relations with the help of (68).

Substituting (77) into (74), we obtain
$$\langle W(C)\, \mathcal{O}_k^I\rangle = 2^{k/2} \sqrt{k}\, I_k\left(\sqrt{\lambda}\right) \frac{R^k}{L^{2k}} Y^I(\theta). \tag{78}$$

Normalizing by the Wilson loop expectation value (71), we get (62).

10 Cancellation of radiative corrections

The leading radiative corrections come form the Feynman diagrams which are shown in fig.2.

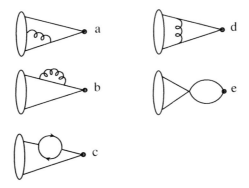

Figure 2: Leading radiative corrections to the correlator of the Wilson loop operator with the $k = 2$ CPO. These are not taken into account by the sum over planar rainbow graphs which were computed in Sec.3.

Each of these diagrams is separately divergent and regularization is required to define them properly. We use a regularization by dimensional reduction which was previously used in ref. [12]. The essential observation is that $\mathcal{N} = 4$ SYM is obtained by dimensional reduction of ten dimensional SYM. This dimensional reduction retains sixteen supersymmetries in any dimension. Thus, a supersymmetric dimensional regularization of $\mathcal{N} = 4$ SYM theory is obtained by dimensionally reducing ten dimensions to $4 - \epsilon$ dimensions.

In this dimensional regularization, the diagram in fig.2a is of higher order than the relevant leading power, R^2/L^4, and therefore does not contribute to $\mathcal{C}_2$.

In the limit $L \gg R$, using dimensional regularization, the leading, R^2/L^4, contributions of the remaining diagrams in fig.2 can be seen to be identical to results of computing the diagrams which are displayed in fig.3. This sum of diagrams is known to vanish when the

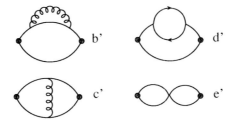

Figure 3: The contribution to the leading term in R^2/L^4 of the diagrams in fig.2b-e are given by these 2-loop diagrams. This combination of diagrams is known to vanish due to the non-renormalization theorem of the 2-point function of the CPO.

dimension is exactly four, due to the non-renormalization theorem for the two-point function of the CPO. This non-renormalization results from super-conformal invariance.

Similar arguments apply to the higher CPO's for which similar non-renormalization theorems can be applied [28].

This is by no means a proof that all radiative corrections vanish. But the excellent agreement with strong coupling AdS/CFT results gives optimism that it is indeed the case.

11 Remarks

The AdS/CFT correspondence has been inspiring since it appears to allow for analytic calculations in strongly coupled $SU(N)$ gauge theory, at least when N is infinite and $\mathcal{N}$ (the number of supersymmetries) is large ($\mathcal{N} = 4$). It will be interesting to eventually work out all of the consequences of this correspondence for gauge theory. This will of course involve quantitative, non-kinematical analytic tests of the kind that have been discussed in the present paper..

The current status of this subject leaves many questions unanswered. The most obvious ones fall into two categories:

• There should be a more rigorous proof that radiative corrections to the results in this paper actually cancel. One approach which was suggested in [13] is to show that the result for the circle comes from a conformal anomaly. Establishing this at a rigorous level would be an important step in the right direction. It also leads to many other interesting speculations about Wilson loops [13].

• It should be possible to study other kinds of Wilson loops [29] [30]. One promising type are those which deviate infinitesimally from a straight line, or a so-called wavy line.

• Most desirable would be to obtain some results for non-conformally invariant gauge theories. At this point this appears to be very difficult as most of the analytic computations depend heavily on conformal invariance.

Acknowledgments

This work was supported by NSERC of Canada and NATO Collaborative Linkage Grant SA(PST.CLG.977361)5941.

References

[1] G. 't Hooft, "A Planar Diagram Theory For Strong Interactions," Nucl. Phys. B **72**, 461 (1974).

[2] S. Coleman, in *C79-07-31.7* SLAC-PUB-2484 *Presented at 1979 Int. School of Subnuclear Physics, Pointlike Structures Inside and Outside Hadrons, Erice, Italy, Jul 31-Aug 10, 1979.*

[3] J. Maldacena, "The large N limit of super-conformal field theories and supergravity," Adv. Theor. Math. Phys. **2**, 231 (1998) [Int. J. Theor. Phys. **38**, 1113 (1998)] [hep-th/9711200].

[4] J. Polchinski and M. J. Strassler, "The string dual of a confining four-dimensional gauge theory," hep-th/0003136.

[5] I. R. Klebanov and M. J. Strassler, "Supergravity and a confining gauge theory: Duality cascades and chiSB-resolution of naked singularities," JHEP **0008**, 052 (2000) [hep-th/0007191].

[6] E. Witten, "Anti-de Sitter space and holography," Adv. Theor. Math. Phys. **2**, 253 (1998) [hep-th/9802150].

[7] J. L. Petersen, "Introduction to the Maldacena conjecture on AdS/CFT," Int. J. Mod. Phys. A **14**, 3597 (1999) [hep-th/9902131].

[8] P. Di Vecchia, "Large N gauge theories and AdS/CFT correspondence," hep-th/9908148.

[9] E. T. Akhmedov, "Introduction to the AdS/CFT correspondence," hep-th/9911095.

[10] O. Aharony, S. S. Gubser, J. Maldacena, H. Ooguri and Y. Oz, "Large N field theories, string theory and gravity," Phys. Rept. **323**, 183 (2000) [hep-th/9905111].

[11] D. Z. Freedman and P. Henry-Labordere, "Field theory insight from the AdS/CFT correspondence," hep-th/0011086.

[12] J. K. Erickson, G. W. Semenoff and K. Zarembo, "Wilson loops in N = 4 supersymmetric Yang-Mills theory," Nucl. Phys. B **582**, 155 (2000) [hep-th/0003055].

[13] N. Drukker and D. J. Gross, "An exact prediction of N = 4 SUSYM theory for string theory," hep-th/0010274.

[14] G. W. Semenoff and K. Zarembo, "More exact predictions of SUSYM for string theory," hep-th/0106015.

[15] D. Berenstein, R. Corrado, W. Fischler and J. Maldacena, "The operator product expansion for Wilson loops and surfaces in the large N limit," Phys. Rev. D **59**, 105023 (1999) [hep-th/9809188].

[16] J. Maldacena, "Wilson loops in large N field theories," Phys. Rev. Lett. **80**, 4859 (1998) [hep-th/9803002].

[17] S. Rey and J. Yee, "Macroscopic strings as heavy quarks in large N gauge theory and anti-de Sitter supergravity," hep-th/9803001.

[18] N. Drukker, D. J. Gross and H. Ooguri, "Wilson loops and minimal surfaces," Phys. Rev. D **60**, 125006 (1999) [hep-th/9904191].

[19] P. Ramond, "Field Theory: A Modern Primer," *REDWOOD CITY, USA: ADDISON-WESLEY (1989) 329 P. (FRONTIERS IN PHYSICS, 74).*

[20] M. A. Shifman, "Wilson Loop In Vacuum Fields," Nucl. Phys. B **173**, 13 (1980).

[21] S. Lee, S. Minwalla, M. Rangamani and N. Seiberg, "Three-point functions of chiral operators in D = 4, N = 4 SYM at large N," Adv. Theor. Math. Phys. **2**, 697 (1998) [hep-th/9806074].

[22] G. Arutyunov, B. Eden, A. C. Petkou and E. Sokatchev, "Exceptional non-renormalization properties and OPE analysis of chiral four-point functions in N = 4 SYM(4)," hep-th/0103230.

[23] L. Hoffmann, A. C. Petkou and W. Ruhl, "Aspects of the conformal operator product expansion in AdS/CFT correspondence," hep-th/0002154.

[24] M. Bianchi, S. Kovacs, G. Rossi and Y. S. Stanev, "On the logarithmic behavior in N = 4 SYM theory," JHEP **9908**, 020 (1999) [hep-th/9906188].

[25] K. Zarembo, "String breaking from ladder diagrams in SYM theory," JHEP **0103**, 042 (2001) [hep-th/0103058].

[26] G. Akemann and P. H. Damgaard, "Wilson loops in N = 4 supersymmetric Yang-Mills theory from random matrix theory," hep-th/0101225.

[27] A. A. Migdal, "Loop Equations And 1/N Expansion," Phys. Rept. **102**, 199 (1983).

[28] E. D'Hoker, D. Z. Freedman and W. Skiba, "Field theory tests for correlators in the AdS/CFT correspondence," Phys. Rev. D **59**, 045008 (1999) [hep-th/9807098].

[29] J. K. Erickson, G. W. Semenoff, R. J. Szabo and K. Zarembo, "Static potential in N = 4 supersymmetric Yang-Mills theory," Phys. Rev. D **61**, 105006 (2000) [hep-th/9911088].

[30] J. Erickson, G. W. Semenoff and K. Zarembo "BPS vs. non-BPS Wilson loops in N = 4 supersymmetric Yang-Mills theory," Phys. Lett. B **466**, 239 (1999) [hep-th/9906211].

Production of Heavy Quarks in Deep-Inelastic Lepton-Hadron Scattering

W.L.van Neerven[1]

Instituut-Lorentz, Universiteit Leiden, P.O. Box 9506, 2300 RA Leiden, The Netherlands.

Abstract. We will give a review of the computation of exact next-to-leading order corrections to heavy quark production in deep inelastic lepton-hadron scattering and discuss the progress made in this field over the past ten years. In this approach, hereafter called EXACT, where the heavy quark mass is taken to be of the same order of magnitude as the other large scales in the process, one can apply perturbation theory in all orders of the strong coupling constant α_s. The results are compared with another approach, called the *variable flavor number scheme* (VFNS), where the heavy quark is also treated as a massless quark. It turns out that the differences between the two approaches are very small provided both of them are carried out up to next-to-next-to-leading order.

INTRODUCTION

In the last ten years one has made much progress on the theoretical and experimental level in the study of heavy flavor production in deep inelastic lepton-hadron scattering. Computations of the cross section in the Born approximation, where the heavy quark is treated as a massive particle, were finished by the end of the seventies [1] whereas the first experimental results came from the EMC-collaboration [2] in the early eighties. In the nineties the $\mathcal{O}(\alpha_s)$ corrections to the Born process were computed in [3]. Moreover other methods to study heavy flavor production were advocated like the intrinsic quark approach [4] and the variable flavor number scheme [5] where the heavy quark is also treated as a massless particle. From the experimental side the electron (positron)-proton collider HERA has given us a wealth of information about charm quark production (for recent experimental results see [6], [7], [8], [9]). We will report about this progress below.

The semi-inclusive process describing heavy quark production in deep inelastic electron-proton scattering is given by (see Fig. 1)

$$e^-(k_1) + P(p) \to e^-(k_2) + Q(p_1) +' X', \tag{1}$$

---

[1] Work supported by the EC network 'QCD and Particle Structure' under contract No. FMRX-CT98-0194.

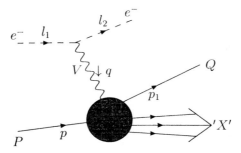

FIGURE 1. Kinematics of heavy quark Q-production in deep inelastic electron-proton scattering

where V represents the intermediate vector boson Z or γ carrying the momentum q and Q denotes the heavy (anti-) quark. The scaling variables are given by

$$x = \frac{Q^2}{2p \cdot q}, \qquad y = \frac{p \cdot q}{p \cdot k_1},$$

$$q^2 = (k_1 - k_2)^2 \equiv -Q^2 < 0, \qquad 0 < x \leq 1, \qquad 0 < y < 1. \qquad (2)$$

In the ongoing experiments carried out at HERA [6], [7], [8], [9] and in the fixed target experiments carried out in the past [2], $Q^2 \ll M_Z^2$ so that the intermediate Z-boson can be neglected and the reaction is dominated by photon exchange. In this case the inclusive cross section simplifies considerably and when the incoming particles are unpolarised it can be written as

$$\frac{d^2\sigma}{dx\, dQ^2} = \frac{2\pi\alpha^2}{xQ^4} \left[\{1 + (1-y)^2\} F_{2,Q}(x, Q^2, m^2) - y^2 F_{L,Q}(x, Q^2, m^2) \right], \qquad (3)$$

where $F_{k,Q}$ ($k = 2, L$) denote the heavy quark contibutions to the structure functions. Notice that an analogous formula exists for the semi-inclusive cross section

$$\frac{d^4\sigma}{dx\, dy\, dp_{T,Q}\, d\eta_Q}, \qquad (4)$$

where $p_{T,Q}$ and η_Q denote the transverse momentum and rapidity of the heavy quark Q respectively. Both of them are considered in the center of mass of the photon-proton system. For neutral current reactions one has proposed two different production mechanisms for heavy quark production. They are distinguished as follows

I Intrinsic Heavy Quark Production

Here one assumes that, besides light quarks u, d, s and gluons g, the wave function of the proton also consists of heavy quarks like c, b, t [4]. In the context of the QCD improved parton model this means that the production mechanism is described as indicated in Fig. 2. In this picture the heavy quark emerges directly from the proton and interacts with the virtual photon γ^*. The consequence is that it is described by a heavy flavor density $f_Q(z, \mu^2)$ with $p_Q = z\,p$ and μ denotes the factorization scale. For this mechanism the heavy quark structure function has the following representation

$$F_{k,Q}(x, Q^2, m^2) = e_Q^2 \int_x^1 \frac{dz}{z}\, f_Q(x/z, \mu^2)\, H_{k,Q}(z, Q^2, m^2, \mu^2)$$

$$\equiv e_Q^2\, f_Q(\mu^2) \otimes H_{k,Q}(Q^2, m^2, \mu^2)\,. \qquad (5)$$

Here e_Q denotes the charge of the heavy quark and $\otimes$ stands for the convolution

$$f \otimes g(x) = \int_x^1 \frac{dz}{z}\, f(x/z)\, g(z)\,. \qquad (6)$$

Further the heavy quark coefficient function $H_{k,Q}$ can be expanded in the strong coupling constant $\alpha_s(\mu^2)$ as follows

$$H_{k,Q} = \sum_{n=0}^{\infty} a_s^n\, H_{k,Q}^{(n)}\,, \quad \text{with} \quad a_s \equiv \frac{\alpha_s}{4\pi}\,. \qquad (7)$$

Some of the contributions to $H_{k,Q}$ are given by the diagrams in Fig. 2a,b.

II Extrinsic Heavy Quark Production

In this case the proton wave function does not contain the heavy quark components. In lowest order of perturbation theory the heavy quark and heavy anti-quark appear in pairs and are produced via photon-gluon fusion [1] as presented in Fig. 3. Here the gluon emerges from the proton in a similar way as the heavy quark in Fig. 2. In this approach the heavy quark structure function reads in lowest order

$$F_{k,Q}(x, Q^2, m^2) = a_s\, e_Q^2\, f_g(\mu^2) \otimes H_{k,g}^{(1)}(Q^2, m^2, \mu^2)\,. \qquad (8)$$

The main difference between the two production mechanisms can be attributed to the fact that for extrinsic heavy quark production two heavy particles are produced in the final state instead of one as in the case of the intrinsic heavy quark approach. This reveals itself in the transverse momentum p_T-distribution where for mechanism II the quark and anti-quark appear back to back. The experiments carried out HERA [9] confirm the p_T-spectrum predicted by the latter mechanism. However in the past the EMC-collaboration [2] found a discrepancy at large x ($x = 0.237$)

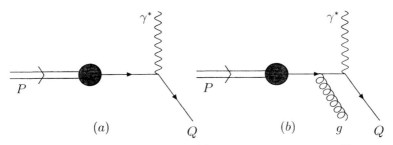

FIGURE 2. (a) $\gamma^* + Q \to Q$ ($H_{k,Q}^{(0)}$), (b) $\gamma^* + Q \to Q + g$ ($H_{k,Q}^{(1)}$).

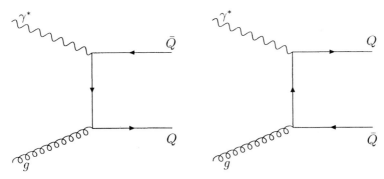

FIGURE 3. Feynman diagrams for the lowest-order photon-gluon fusion process contributing to the coefficient functions $H_{k,g}^{(1)}$.

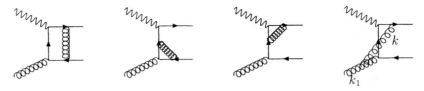

FIGURE 4. Virtual gluon corrections to the process $\gamma^* + g \to Q + \bar{Q}$ contributing to the coefficient functions $H_{i,g}^{(2)}$.

between the structure function $F_{2,Q}$, predicted by mechanism II, and the charm quark data. This difference can be explained by also invoking mechanism I [4] (see also [10]). Nevertheless because of the success of extrinsic heavy quark production revealed by the HERA charm quark data we will continue with this approach in our calculations below.

HEAVY QUARK STRUCTURE FUNCTIONS UP TO $\mathcal{O}(\alpha_S^2)$

In the extrinsic heavy quark approach there are two different production mechanisms [11], [12] which are of importance for the derivation of the variable flavor number scheme (VFNS) treated in the next section. In the case of electro-production the virtual photon either interacts with the heavy quark appearing in the final state or it is attached to the light (anti-) quark in the initial state. This distinction is revealed by the Feynman graphs. Some of them will be shown as illustration below.

A The photon interacts with the heavy quark

Here the lowest order (LO) process is given by gluon fusion as presented in Fig. 3. It is given by the partonic reaction

$$\gamma^* + g \to Q + \bar{Q}, \tag{9}$$

which is calculated in [1]. In next-to-leading order (NLO) one has to compute the virtual corrections to this process. Some of the Feynman graphs are shown in Fig. 4. Besides the virtual corrections one also has to calculate all processes which contain an extra particle in the final state. Adding a gluon to reaction (9) we observe gluon bremsstrahlung given by

$$\gamma^* + g \to Q + \bar{Q} + g. \tag{10}$$

Some of the Feynman graphs corresponding to the process above are presented in Fig. 5. Besides this reaction we have another one which appears for the

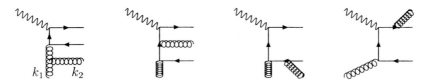

FIGURE 5. The bremsstrahlung process $\gamma^* + g \to Q + \bar{Q} + g$ contributing to the coefficient functions $H_{i,g}^{(2)}$.

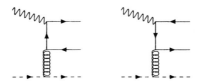

FIGURE 6. The Bethe-Heitler process $\gamma^* + Q \to Q + \bar{Q} + q$ contributing to the coefficient functions $H_{i,q}^{(2)}$.

first time if the computations are extended to NLO. It is represented by the Bethe-Heitler process

$$\gamma^* + q(\bar{q}) \to Q + \bar{Q} + q(\bar{q}), \tag{11}$$

where the Feynman diagrams are shown in Fig. 6. The contributions to the heavy quark structure functions due to the reactions above can be written as

$$F_{k,Q}(Q^2, m^2) = e_Q^2 \bigg[f_g(3, \mu^2) \otimes \big\{ a_s(\mu^2) H_{k,g}^{(1)}(Q^2, m^2) + a_s^2(\mu^2) H_{k,g}^{(2)}(Q^2, m^2, \mu^2) \big\}$$

$$+ a_s^2(\mu^2) f_q^S(3, \mu^2) \otimes H_{k,q}^{(2)}(Q^2, m^2, \mu^2) \bigg]. \tag{12}$$

Here e_Q denotes the charge of Q indicating that the photon couples to the heavy quark and $H_{k,i}$ ($i = q, g$), characteristic of these type of processes, represent the heavy quark coefficient functions which emerge from the calculation of the Feynman graphs after renormalization and mass factorization is carried out. The former procedure leads to a dependence on a renormalization scale μ of the coupling constant a_s, the coefficient function $H_{k,i}$ and the parton densities f_i. Moreover the latter two also depend on a factorization scale which for convenience is set equal to the parameter μ defined above. The factorization scale, which can be attributed to mass factorization, appears in all coefficient functions except for the lowest order one i.e. $H_{k,i}^{(1)}$. In the structure function

FIGURE 7. The Compton process $\gamma^* + Q \to Q + \bar{Q} + q$ contributing to the coefficient functions $L_{i,q}^{(2)}$.

of Eq. (12) there appear two different types of flavor singlet parton densities i.e. the gluon density f_g and the quark singlet density which in a three flavor number scheme reads as

$$f_q^S(3, \mu^2) =$$
$$f_u(3, \mu^2) + f_{\bar{u}}(3, \mu^2) + f_d(3, \mu^2) + f_{\bar{d}}(3, \mu^2) + f_s(3, \mu^2) + f_{\bar{s}}(3, \mu^2). \quad (13)$$

Hence Eq. (12) represents the singlet part of the heavy quark structure function.

B <u>The photon interacts with the light quark.</u>

The second production mechanism is represented by the Feynman diagrams where the photon couples to the light quark or the anti-quark. This happens for the first time in NLO where one observes the Compton process

$$\gamma^* + q(\bar{q}) \to Q + \bar{Q} + q(\bar{q}), \quad (14)$$

which is depicted in Fig. 7. The coefficient functions corresponding to this type of process are denoted by $L_{k,i}$ ($i = q, g$) and the contribution to the heavy quark structure functions is characterized by the expression

$$F_{k,Q}(Q^2, m^2) = \sum_{i=u,d,s} e_i^2 \, a_s^2(\mu^2) \left(f_i(3, \mu^2) + f_{\bar{i}}(3, \mu^2) \right) \otimes L_{k,q}^{(2)}(Q^2, m^2), \quad (15)$$

where e_i denotes the charge of the light quark represented by $i = u, d, s$ in a three flavor number scheme. Since this process appears for the first time in second order mass factorization is not needed which explains the independence of $L_{k,q}^{(2)}$ on the parameter μ.

The computation of the second order contributions to the heavy quark coefficient functions $H_{k,i}$, $L_{k,i}$ has been carried out in [3]. While calculating the Feynman graphs in Figs. 4- 6 one encounters several type of singularities which have to be regularized and subsequently to be subtracted off before one obtains a finite result. The singularities are of the following nature i.e. infrared (IR), ultraviolet (UV) and collinear (C). Sometimes the latter are also called mass singularities. The IR divergences cancel between the virtual and the bremsstrahlung corrections to reaction (9). The UV divergences, regularized by n-dimensional regularization, are removed by mass and coupling constant renormalization. For the mass renormalization we choose the on-shell scheme. In this case the UV divergence will be removed by replacing the bare mass $\hat{m}$ by the renormalized mass m via

$$\hat{m} = m \left[1 + \hat{a}_s \, \delta_0 \, \frac{2}{\varepsilon} + \cdots \right], \tag{16}$$

where the UV pole term is indicated by $1/\varepsilon$ with $\varepsilon = n - 4$. If we choose for example process (10) together with the virtual corrections to the Born reaction (9) (see Figs. 4,5) the unrenormalized coefficient function takes the form

$$\hat{H}_{k,g}^{(2)} = \hat{a}_s^2 \left[\left\{ \frac{1}{\varepsilon_C} + \frac{1}{2} \ln \left(\frac{m^2}{\mu^2} \right) \right\} P_{gg}^{(0)} \otimes H_{k,g}^{(1)} - \beta_0 \left\{ \frac{2}{\varepsilon_{UV}} + \ln \left(\frac{m^2}{\mu^2} \right) \right\} H_{k,g}^{(1)} \right]$$

$$+ H_{k,g}^{(2)}|_{\mu=m}, \tag{17}$$

where $H_{k,g}^{(2)}|_{\mu=m}$ is finite and $\hat{a}_s$ denotes the bare coupling constant. Further we have also regularized the collinear divergences by n-dimensional regularization. In order to distinguish between ultraviolet and collinear divergences we have indicated them by $1/\varepsilon_{UV}$ and $1/\varepsilon_C$ respectively. The residues of the collinear divergences are represented by the so called splitting functions denoted by P_{ij} ($i,j = q,g$). The origin of the collinear divergences is explained by the first diagram in Fig. 5. The propagator carrying the momentum $k_1 - k_2$ behaves as

$$\frac{1}{(k_1 - k_2)^2} = \frac{1}{2\,\omega_1\,\omega_2\,(1 - \cos\theta)}, \tag{18}$$

which diverges for $\theta \to 0$. This propagator only shows up if three massless particles are coupled to each other like three gluons or when a gluon is attached to a quark line provided the quark is massless. If the gluon is attached to a heavy quark the mass of the latter, which is unequal to zero, prevents that the denominator in Eq. (18) vanishes when $\theta \to 0$. The UV-divergence in Eq. (17) is removed by coupling constant renormalization which is achieved by adding $\hat{a}_s \, H_{k,g}^{(1)}$ to Eq. (17) and replacing the bare coupling constant $\hat{a}_s$ by the renormalized one represented by $a_s(\mu^2)$

$$\hat{a}_s = a_s(\mu^2) \left[1 + a_s(\mu^2) \, \beta_0 \, \frac{2}{\varepsilon} + \cdots \right]. \tag{19}$$

Finally one has to remove the collinear divergences. This is achieved by mass factorization. It proceeds in a similar way as multiplicative renormalization so that one can write

$$\hat{H}_{k,i}\left(\frac{1}{\varepsilon_C},Q^2,m^2\right) = \Gamma_{ji}\left(\frac{1}{\varepsilon_C},\mu^2\right) \otimes H_{k,j}\left(Q^2,m^2,\mu^2\right), \tag{20}$$

where Γ_{ji} represents the transition function which removes all collinear divergences from the bare heavy quark coefficient functions. Further we have introduced the notion of bare parton density $\hat{f}_i$ so that the heavy quark structure function can be written as

$$F_{k,Q}(Q^2,m^2) = e_Q^2\, \hat{f}_i \otimes \hat{H}_{k,i}\left(\frac{1}{\varepsilon_C},Q^2,m^2\right). \tag{21}$$

Substitution of $\hat{H}_{k,i}$ (see Eq. (20)) into the expression above we can derive

$$F_{k,Q}(Q^2,m^2) = e_Q^2\, f_j(\mu^2) \otimes H_{k,j}\left(Q^2,m^2,\mu^2\right), \tag{22}$$

where $f_j(\mu^2)$ is the renormalized parton density defined by

$$f_j(\mu^2) = \Gamma_{ji}\left(\frac{1}{\varepsilon_C},\mu^2\right) \otimes \hat{f}_i. \tag{23}$$

In the case of the example presented in Eq. (18) the collinear divergence is removed by adding $H_{k,g}^{(1)}$ to $\hat{H}_{k,g}^{(2)}$ and choosing the following transition function

$$\Gamma_{gg}\left(\frac{1}{\varepsilon_C},\mu^2\right) = \delta(1-z) + a_s\, N \left[\frac{1}{\varepsilon_C} P_{gg}^{(0)}\right], \tag{24}$$

where N denotes the number of colors.

A comparison of the next-to-leading order (NLO) heavy quark structure function $F_{2,c}$ with the data for charm quark production measured at HERA reveals a fairly good agreement between theory and experiment. The data cover the range $1 < Q^2 < 1350$ GeV^2 and $5 \times 10^{-5} < x < 5.6 \times 10^{-2}$. In [13] one has made a comparison with the data obtained by the ZEUS-collaboration [6], [8]. From the cross section in Eq. (2) one can derive the integrated quantities

$$\frac{d\sigma}{dQ^2} = \int_{x_{min}}^{x_{max}} dx\, \frac{d^2\sigma}{dx\, dQ^2}, \qquad \frac{d\sigma}{dx} = \int_{Q^2_{min}}^{Q^2_{max}} dQ^2\, \frac{d^2\sigma}{dx\, dQ^2}. \tag{25}$$

Notice that the quantities above represent D_c^*-meson production rather than charm quark production. The meson appears as a fragmentation product of the quark and the cross sections in Eq. (25) are obtained by convoluting Eq. (2) with fragmentation functions. Furthermore one has to impose experimental cuts on the kinematics which are indicated by *max* and *min*. The results are presented in Figs. 8, 9 which originate from [13] where one can also find the maximal and minimal values for x and Q^2. The figures show that there is a good agreement between the exact NLO result (called EXACT in the figure) and the data except for $x \sim 10^{-3}$ where there is a small discrepancy. Furthermore in [14] one has also made

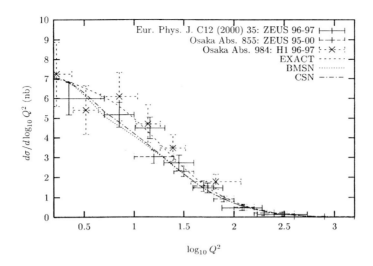

FIGURE 8. The combined Osaka and published Zeus-data for $d\sigma/d\log_{10}Q^2$ in nb for deep inelastic production of $D^{*\pm}$-mesons. The dashed line is the exact NLO result from the program HVQDIS which follows from $F_{2,c}^{\text{EXACT}}$. The dotted line (BMSN-scheme) and dashed-dotted line (CSN-scheme) is based on $F_{2,c}^{\text{VFNS}}$.

a comparison between the program HQVDIS [15] based on the NLO computations above and the experimental differential distributions for D_c^* production where the following cross sections are studied.

$$\frac{d\sigma}{dp_{T,D_c}}, \quad \frac{d\sigma}{d\eta_{D_c}}, \quad \frac{d\sigma}{dW}, \qquad (26)$$

where W ($W^2 = (p+q)^2$) is the center of mass energy of the photon-proton system. Also the differential distributions agree with the NLO predictions except for the rapidity η-distribution. Here it appears that for $\eta_D > 0$ the experimental cross section is larger than the one computed in NLO (see also [9]).

Summarizing our findings for the exact NLO calculations we conclude

1. There exist a fairly good agreement between the data and the NLO calculations.

2. The theoretical curves are insensitive to the choice of the renormalization/factorization scale μ occurring in the parton densities and the coefficient

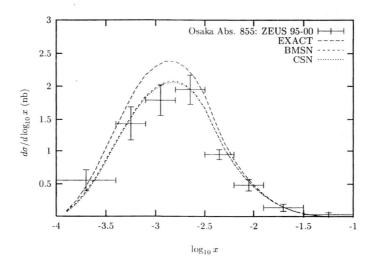

FIGURE 9. The combined Osaka and published Zeus-data for $d\sigma/dlog_{10}x$ in nb for deep inelastic production of $D^{*\pm}$-mesons. The dashed line is the exact NLO result from the program HVQDIS which follows from $F_{2,c}^{\text{EXACT}}$. The dotted line (BMSN-scheme) and dashed-dotted line (CSN-scheme) is based on $F_{2,c}^{\text{VFNS}}$.

functions [16].

3. The theoretical curves are sensitive to the choice of the charm mass m_c which is in the range $1.3 < m_c < 1.7$ GeV/c^2.

4. Processes with a gluon in the initial state (Eqs. (9),(10)) constitute the bulk of the contribution to the heavy quark structure function at $x < 10^{-2}$. Hence they are an excellent probe to measure the gluon density $f_g(z, \mu^2)$.

ASYMPTOTIC HEAVY QUARK STRUCTURE FUNCTIONS

In this section we want to study the heavy quark coefficient functions $H_{k.i}$, $L_{k,i}$ in the asymptotic region $Q^2 \gg m^2$. In this region they have the following form

$$H_{k,i}^{\text{ASYMP}}(z,Q^2,m^2,\mu^2) \sim \sum_{l=1}^{\infty} a_s^l \sum_{n+m\leq l} a_{nm}(z) \ln^n\left(\frac{\mu^2}{m^2}\right) \ln^m\left(\frac{Q^2}{m^2}\right). \tag{27}$$

A similar expression exists for $L_{k,i}^{\text{ASYMP}}$. An example is the asymptotic expression for the Born reaction in Eq. (9) (Fig. 3) which can be written as

$$H_{2,g}^{\text{ASYMP}}(z,Q^2,m^2) = a_s \left[\frac{1}{2}P_{qg}(z)\ln\left(\frac{Q^2}{m^2}\right) + a_{Qg}(z) + c_{2,g}(z)\right]. \tag{28}$$

The origin of this asymptotic behaviour can be attributed to the property that in the limit $m \to 0$ the heavy quark coefficient functions become collinear divergent which is revealed by the logarithmic singularities $\ln Q^2/m^2$ and $\ln \mu^2/m^2$. The reason why this behaviour is of interest can be summarized as follows

1. The results obtained for the exact coefficient functions in the previous section are semi-analytic. In [3] one has obtained exact results for the virtual corrections to process (9) (Fig. 4) but for the reactions with three particles in the final state like Eqs. (10),(11), (14) a full analytical expression could only be presented for the Compton process in Eq. (14) (see [11]). In the other cases only the integration over the angles could be carried out but the integration over the final state energies are so tedious that they have to be done in a numerical way. However the latter integration becomes more amenable when $m^2 \ll Q^2$ so that terms of the order m^2/Q^2 can be neglected. Therefore an analytical result for the asymptotic heavy quark coefficient functions provides us with a check of the exact expressions computed for arbitrary m.

2. The asymptotic coefficient functions play an important role in the derivation of the variable flavor number scheme [5], [17] discussed at the end of this section. This scheme is only useful if the following questions are answered. They are:

    a. Do the logarithmic terms of the type $\ln^n Q^2/m^2$, occurring in the coefficient functions, really dominate the heavy quark structure functions or the heavy quark cross sections?

    b. Are the logarithmic terms $\ln^n Q^2/m^2$ so large that they bedevil the perturbation series so that they have to be re-summed?

There are two ways to compute the asymptotic heavy quark coefficient functions.

1. One can follow the procedure for the derivation of the exact coefficient functions in [3] but since the mass m can be neglected one can now carry out the additional integrations over the energies of the final state particles in an analytical way.

2. One can use the operator product expansion (OPE) techniques which however are only applicable for inclusive quantities like the structure functions $F_{k,Q}(x,Q^2,m^2)$.

In [11] one has adopted the latter approach which will be outlined below. In the derivation of the cross section in Eq. (3) one encounters the hadronic tensor leading to the definition of the structure functions. It is defined by

$$W_{\mu\nu}(p,q) = \frac{1}{4\pi}\int d^4z\, e^{iq\cdot z}\, \langle P(p)|\big[J_\mu(z), J_\nu(0)\big]|P(p)\rangle$$

$$= \left(p_\mu p_\nu - \frac{p\cdot q}{q^2}(p_\mu q_\nu + p_\nu q_\mu) + g_{\mu\nu}\frac{(p\cdot q)^2}{q^2}\right)\frac{F_2(x,Q^2,M^2)}{p\cdot q}$$

$$+ \left(g_{\mu\nu} - \frac{q_\mu q_\nu}{q^2}\right)\frac{F_L(x,Q^2,M^2)}{2\,x},$$

$$p^2 = M^2, \qquad q^2 = -Q^2, \qquad x = \frac{Q^2}{2p\cdot q}. \tag{29}$$

In the limit $Q^2 \gg M^2$ the current-current correlation function appearing in the integrand of Eq. (29) is dominated by the light cone. Hence one can make an operator product expansion near the light cone and write

$$\big[J(z), J(0)\big] \underset{z^2\to 0}{\sim} \sum_i \sum_m c_i^{(m)}(z^2\mu^2)\, z_{\mu_1}\cdots z_{\mu_m}\, O_i^{\mu_1\cdots\mu_m}(0,\mu^2), \tag{30}$$

where for convenience we have dropped the Lorentz indices of the currents. Here $c_i^{(m)}$ denote the coefficient functions which are distributions and $O_i^{\mu_1\cdots\mu_m}$ are local operators. Both are renormalized which is indicated by the renormalization scale μ. When dropping all terms of the order M^2/Q^2 we can limit ourselves to leading twist operators. In QCD they are given by

non-singlet operators

$$O_{q,r}^{\mu_1\cdots\mu_m}(z) = \frac{1}{2}i^{m-1}\mathcal{S}\Big[\bar\psi(z)\gamma^{\mu_1}D^{\mu_2}\cdots D^{\mu_m}\frac{\lambda_r}{2}\psi(z)\Big] + \text{trace terms},$$

singlet operators

$$O_q^{\mu_1\cdots\mu_m}(z) = \frac{1}{2}i^{m-1}\mathcal{S}\Big[\bar\psi(z)\gamma^{\mu_1}D^{\mu_2}\cdots D^{\mu_m}\psi(z)\Big] + \text{trace terms},$$

$$O_g^{\mu_1\cdots\mu_m}(z) = \frac{1}{2}i^{m-2}\mathcal{S}\big[F_\alpha^{a,\mu_1}(z)D^{\mu_2}\cdots D^{\mu_{m-1}}F_\alpha^{a,\alpha\mu_m}(z)\big] + \text{trace terms}. \tag{31}$$

The symbol $\mathcal{S}$ indicates that one has to symmetrize over all Lorentz indices. The covariant derivative is given by $D^\mu = \partial^\mu - ig\, T_a\, A_a^\mu$ and λ_r is the generator of the flavor group $SU(n_f)_F$. Insertion of the OPE (Eq. (30)) into the hadronic tensor of Eq. (29) leads to the following result

$$F_k^{(m)}(Q^2, M^2) \equiv \int_0^1 dx \, x^{m-1} \, F(x, Q^2) = \sum_{i=q,g} A_i^{(m)}\left(\frac{\mu^2}{M^2}\right) C_i^{(m)}\left(\frac{Q^2}{\mu^2}\right), \quad (32)$$

where the operator matrix element and the coefficient function are defined by

$$A_i^{(m)}\left(\frac{\mu^2}{M^2}\right) = \langle P(p)|O_i^m(0,\mu^2)|P(p)\rangle, \quad C_i^{(m)}\left(\frac{Q^2}{\mu^2}\right) = \int d^4z \, e^{iq\cdot z} \, c_i^{(m)}(z^2\mu^2). \quad (33)$$

The OPE techniques can also be applied when the the proton state $|P(p)\rangle$ in Eq. (29) is replaced by a light quark state $|q(p)\rangle$ or a gluon state $|g(p)\rangle$. However when the proton is replaced by massless quarks and gluons the external momentum satisfies the relation $p^2 = 0$ so that the partonic structure functions and the partonic operator matrix elements become collinearly divergent. One can show (see [17]) that instead of Eq. (32) one obtains more complicate expressions which are given by

$$\hat{\mathcal{F}}_{k,q}^{\mathrm{NS}}\left(\frac{Q^2}{\mu^2}, \frac{1}{\varepsilon_C}\right) + \hat{L}_{k,q}^{\mathrm{ASYMP}}\left(\frac{Q^2}{m^2}, \frac{m^2}{\mu^2}, \frac{1}{\varepsilon_C}\right) = \hat{A}_{qq}^{\mathrm{NS}}\left(\frac{m^2}{\mu^2}, \frac{1}{\varepsilon_C}\right) \otimes \mathcal{C}_{k,q}^{\mathrm{NS}}\left(\frac{Q^2}{\mu^2}\right),$$

$$\hat{\mathcal{F}}_{k,i}^{\mathrm{S}}\left(\frac{Q^2}{\mu^2}, \frac{1}{\varepsilon_C}\right) + \hat{L}_{k,i}^{\mathrm{ASYMP}}\left(\frac{Q^2}{m^2}, \frac{m^2}{\mu^2}, \frac{1}{\varepsilon_C}\right) + \hat{H}_{k,i}^{\mathrm{ASYMP}}\left(\frac{Q^2}{m^2}, \frac{m^2}{\mu^2}, \frac{1}{\varepsilon_C}\right)$$

$$= \hat{A}_{ji}^{\mathrm{S}}\left(\frac{m^2}{\mu^2}, \frac{1}{\varepsilon_C}\right) \otimes \mathcal{C}_{k,j}^{\mathrm{NS}}\left(\frac{Q^2}{\mu^2}\right),$$

$$i, j = q, g. \quad (34)$$

Here $\hat{\mathcal{F}}_{k,i}$ represent the partonic structure functions which are given by Feynman graphs containing massless particles only and therefore contain collinear singularities indicated by ε_C. These singularities also appear in the asymptotic heavy quark coefficient functions $\hat{H}_{k,i}^{\mathrm{ASYMP}}$, $\hat{L}_{k,i}^{\mathrm{ASYMP}}$ which are determined in the asymptotic regime $Q^2 \gg m^2$ before mass factorization is carried out. These collinear divergences can be traced back to the massless quarks and gluons appearing in Figs. 3-7. Finally $\hat{A}_{ji}$ represent the operator matrix elements on the partonic level and are defined by

$$\hat{A}_{ji}\left(\frac{m^2}{\mu^2}, \frac{1}{\varepsilon_C}\right) = \langle i|O_j(\mu^2, 0)|i\rangle, \quad i = q, g, \quad j = q, g, Q, \quad p_i^2 = 0. \quad (35)$$

The operator matrix elements which depend on the heavy flavor mass consist of two classes. The first class is given by heavy quark operators sandwiched between gluon or light quark states. The second class contains light quark or gluon operators which contain a heavy flavor loop. Examples of the first class together with the corresponding process are given in Figs. 10 and 11. An example of the second class

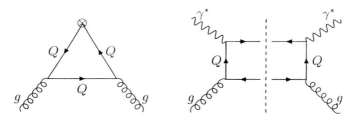

FIGURE 10. The operator matrix element $A_{Qg}^{(1)}$ and the corresponding Feynman graph for the process $\gamma^* + g \to Q + \bar{Q}$ ($H_{k,g}^{(1)}$).

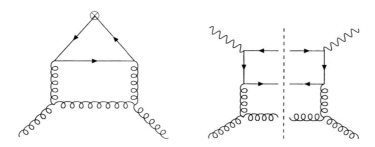

FIGURE 11. The operator matrix element $A_{Qg}^{(2)}$ and the corresponding Feynman graph for the process $\gamma^* + g \to Q + \bar{Q} + g$ ($H_{k,g}^{(2)}$).

is shown in 12. The light partonic coefficient functions defined by $\mathcal{C}_{k,j}$ are derived from the light partonic structure functions via mass factorization

$$\hat{\mathcal{F}}_{k,i}\left(\frac{Q^2}{\mu^2}, \frac{1}{\varepsilon_C}\right) = \Gamma_{ji}\left(\frac{1}{\varepsilon_C}, \mu^2\right) \otimes \mathcal{C}_{k,j}\left(\frac{Q^2}{\mu^2}\right), \qquad (36)$$

where Γ_{ji} denote the transition functions which are discussed below Eq. (20). The quantities $\hat{\mathcal{F}}_{k,i}$ and $\mathcal{C}_{k,i}$ have been calculated up to second order in α_s in [18]. The operator matrix elements $\hat{A}_{ij}$ are also known up to second order and the calculation of them is presented in [11], [17]. Like in the case of the coefficient functions n-dimensional regularization has been used to regularize the ultraviolet and collinear divergences and one has chosen the same renormalization conditions for the mass and the strong coupling constant. Hence from Eq. (34) one infers the asymptotic heavy quark coefficient functions $\hat{H}_{k,i}^{\text{ASYMP}}$, $\hat{L}_{k,i}^{\text{ASYMP}}$ up to the same order (see [17]). Finally one can remove the remaining collinear divergences via the same mass factorization as is done for the exact heavy quark coefficient functions in Eq. (20). After this outline of the calculation of the asymptotic heavy quark

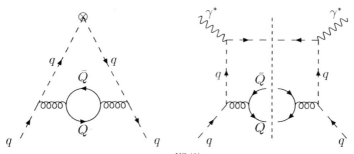

FIGURE 12. The operator matrix element $A_{qq}^{\text{NS},(2)}$ and the corresponding Feynman graph for the process $\gamma^* + q \to Q + \bar{Q} + q$ ($L_{k,q}^{\text{NS},(2)}$).

coefficient function one might ask the question whether it is not easier to compute them in a more direct way. The main problem of radiative corrections is the computation of the phase space integrals in particular if one has massive particles in the final state even if one takes $m^2 \ll Q^2$. Since this work was already done for the light partonic structure functions $\hat{\mathcal{F}}_{k,i}$ in [18] it was not needed to repeat this procedure anymore. On the contrary it is much easier to compute two-loop operator matrix elements because of the zero momentum flowing into the operator vertex indicated by the symbol $\otimes$ in Figs. 10-12. The difference between the exact and asymptotic coefficient functions can be attributed to the power corrections of the type $(m^2/Q^2)^l$, with $l \geq 1$, which occur in the former but are absent in the latter. The asymptotic coefficient functions only contain the logarithms $\ln^m Q^2/m^2$ and $\ln^n \mu^2/m^2$ and terms which survive in the limit $Q^2 \to \infty$. We will now first answer the question whether these logarithmic terms dominate the heavy quark structure function $F_{k,Q}(x, Q^2, m^2)$. For that purpose we have studied the structure functions for charm production i.e. $Q = c$ in [17], [19]. Here one has computed the ratio

$$R_{k,c} = \frac{F_{k,c}^{\text{ASYMP}}}{F_{k,c}^{\text{EXACT}}}, \qquad (37)$$

where $F_{k,c}^{\text{EXACT}}$ and $F_{k,c}^{\text{ASYMP}}$ are represented in the three flavor number scheme

$$F_{k,c}^{\text{EXACT}} = \frac{4}{9} \sum_{i=q,g} f_i^S(3, \mu^2) \otimes H_{k,i}^{\text{EXACT}} + \sum_{j=u,d,s} e_j^2 \left(f_j(3, \mu^2) + f_{\bar{j}}(3, \mu^2) \right) \otimes L_{k,q}^{\text{EXACT}}, \qquad (38)$$

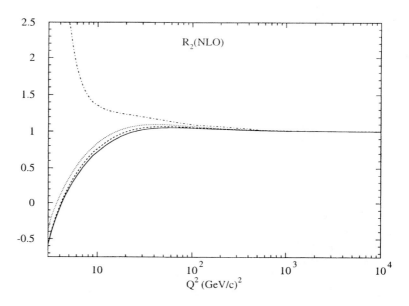

FIGURE 13. $R_{2,c}$ in NLO (Eq. (37)) plotted as function of Q^2 at fixed x; $x = 10^{-1}$ (dashed-dotted line), $x = 10^{-2}$ (dotted line), $x = 10^{-3}$ (dashed line), $x = 10^{-4}$ (solid line)

$$F_{k,c}^{\text{ASYMP}} = \frac{4}{9} \sum_{i=q,g} f_i^S(3,\mu^2) \otimes H_{k,i}^{\text{ASYMP}} + \sum_{j=u,d,s} e_j^2 \left(f_j(3,\mu^2) + f_{\bar{j}}(3,\mu^2) \right) \otimes L_{k,q}^{\text{ASYMP}}. \tag{39}$$

In Fig. 13 we have plotted $R_{2,c}$. Here one observes that this quantity becomes very close to one for $x < 10^{-2}$ and $Q^2 > 20$ GeV^2 which belongs to the region explored by the experiments carried out at HERA. This shows that the logarithms mentioned above dominate the structure function except in the threshold region given by $x \sim 1$ and small Q^2 which is characteristic of the EMC experiment [2].

In order to answer the second question whether these logarithmic terms vitiate the perturbation series one has to resum them in all orders of perturbation theory and show that the resummed structure function differs from the one which is computed exactly in fixed order of perturbation theory. This resummation procedure is provided by the variable flavor number scheme (VFNS) [5]. An example of a resummed structure function has been derived in [17]. In the case of charm quark production one obtains

$$F_{k,c}^{\text{VFNS}} = \sum_{j=u,d,s,c} e_j^2 \left[(f_j(4,\mu^2) + f_{\bar{j}}(4,\mu^2)) \otimes \mathcal{C}_{k,j}^{\text{VFNS}} + f_g(4,\mu^2) \otimes \mathcal{C}_{k,g}^{\text{VFNS}} \right], \quad (40)$$

with the conditions

$$\lim_{m^2 \to 0} \mathcal{C}_{k,i}^{\text{VFNS}} \left(\frac{Q^2}{m^2}, \frac{\mu^2}{m^2} \right) = \mathcal{C}_{k,i} \left(\frac{Q^2}{\mu^2} \right), \quad (41)$$

$$\lim_{s \to 4m^2} F_{k,c}^{\text{VFNS}}(x, Q^2, m^2) = F_{k,c}^{\text{EXACT}}(x, Q^2, m^2), \quad s = \frac{1-x}{x} Q^2. \quad (42)$$

The consequence of condition (41) is that in the asymptotic regime $Q^2 \gg m^2$, $F_{k,c}^{\text{VFNS}}$ turns into the structure function represented in the four flavor number scheme which contains the contribution of light flavors only including the charm quark. The mass singular logarithms, occurring in the heavy quark coefficient functions in the three flavor number scheme (see Eqs. (12), (15)), are shifted to the parton densities defined in the four flavor number scheme appearing in Eq. (40). The most conspicuous feature is the appearance of the charm quark density which is absent in a three flavor number scheme but shows up in the four flavor number scheme. It is given by

$$f_c(4,\mu^2) + f_{\bar{c}}(4,\mu^2) = f_q^S(3,\mu^2) \otimes A_{Qq}^S \left(\frac{\mu^2}{m^2} \right) + f_g(3,\mu^2) \otimes A_{Qg}^S \left(\frac{\mu^2}{m^2} \right). \quad (43)$$

The operator matrix elements satisfy renormalization group equations which enable us to resum all logarithmic terms of the type $\ln \mu^2/m^2$. In order to get the boundary condition (42) one needs matching conditions for which one can make various choices (see e.g. [5], [17], [20]). Two of them are proposed in [17], [21] (BMSN scheme) and in [22] (CSN scheme). In the former one equates

$$\mathcal{C}_{k,c}^{\text{VFNS}} \left(\frac{Q^2}{m^2}, \frac{\mu^2}{m^2} \right) = \mathcal{C}_{k,q} \left(\frac{Q^2}{\mu^2} \right), \quad q = u, d, s. \quad (44)$$

Both schemes have been calculated up to next-to-next-to-leading order (NNLO) and are compared in [13] with $F_{k,c}^{\text{EXACT}}$ (NLO) in Eq. (38). The results are shown in Figs. 8, 9 from which one infers that there is hardly any difference between the two schemes representing VFNS. This shows that one can neglect the power contributions $\mathcal{O}(m^2/Q^2)$ in $\mathcal{C}_{k,c}^{\text{VFNS}}$ which are absent in $\mathcal{C}_{k,q}$. Also the difference between the two versions of VFNS on one hand and the exact NLO approach on the other hand is hardly noticeable except in Fig. 9 where in the vicinity of $x = 10^{-3}$ it seems that the data are better described by the BMSN and CSN schemes than by the exact NLO result. The main conclusion that one can draw from these figures is that the resummation effect is very small which means that the so called large logarithms of the type $\ln Q^2/m^2$ and also $\ln \mu^2/m^2$, when $\mu^2 \sim Q^2$, do not vitiate the perturbation series.

Summarizing our results we conclude

1. The past ten years have shown much progress in the computation of higher order corrections to heavy flavor production. In particular the results obtained in electro-production agree well with the data obtained by the experiments carried out at HERA.

2. The asymptotic heavy quark coefficient functions can be calculated using operator product expansion techniques. The results obtained for the operator matrix elements can be also used for processes where the light cone does not dominate the reaction which e.g. holds for $e^+ e^- \to \mu^+ \mu^-$ [23].

3. The heavy quark structure function is dominated by the logarithmic terms $\ln Q^2/m^2$ and $\ln \mu^2/m^2$ provided x and Q^2 are chosen in such a way that they are outside the threshold region of the production process i.e. $s = (1-x)Q^2/x \gg 4m^2$.

4. In spite of the fact that the logarithms above dominate the structure function they do not bedevil the convergence of the perturbation series so that a resummation is in principle not necessary. Therefore one can use fixed order (exact) perturbation theory which is simple to apply and to interpret in particular if one studies the differential distributions presented in Eqs. (4), (26).

REFERENCES

1. Witten E., *Nucl. Phys.* **B104**, 445 (1976);
   Babcock J. and Sivers D., *Phys. Rev.* **D18**, 2301 (1978);
   Shifman M.A., Vainstein A.I. and Zakharov V.J., *Nucl. Phys.* **B136**, 157 (1978);
   Glück M. and Reya E., *Phys. Lett.* **B83** 98 (1978);
   Leveille J.V. and Weiler T., *Nucl. Phys.* **B147**, 147 (1979).
2. Aubert J.J. et al., (EMC-collaboration), *Nucl. Phys.* **B213**, 31 (1983).
3. Laenen E, Riemersma S., Smith J. and van Neerven W.L., *Nucl. Phys.* **B392**, 162 (1993); ibid. *Nucl. Phys.* **B392** 229 (1993);
   Riemersma S., Smith J and van Neerven W.L., *Phys. Lett.* **B347**, 43 (1995).
4. Brodsky S.J., Hoyer P., Mueller A.H., Tang W.-K., *Nucl. Phys.* **B369** 519 (1992);
   Vogt R., Brodsky S.J., *Nucl. Phys.* **B438** 261 (1995).
5. Aivazis M.A.G., Collins J.C., Olness F.I. and W. -K. Tung, *Phys. Rev.* **D50**, 3102 (1994).
6. Breitweg J. et al. (ZEUS-collaboration), *Eur. Phys. J.* **C12**, 35 (2000).
7. Adloff C. et al. (H1-collaboration), *Nucl. Phys.* **B545**, 21 (1999).
8. ZEUS-collaboration, Abstract 855, submitted to the *XXXth International Conference on High Energy Physics*, July 27-August 2, 2000, Osaka, Japan.

9. H1-collaboration, Abstract 984, submitted to the *XXXth International Conference on High Energy Physics*, July 27-August 2, 2000, Osaka, Japan.
10. Harris B.W., Smith J, Vogt R., *Nucl. Phys.* **B461**, 181 (1996).
11. Buza M., Matiounine Y., Smith J., Migneron R., van Neerven W.L., *Nucl. Phys.* **B472**, 611 (1996); ibid. *Nucl. Phys. Proc. Suppl.* **51C**, 183 (1996).
12. van Neerven W.L., *Acta Phys. Polon.* **B28**, 2715 (1997), hep-ph/9708452.
13. Chuvakin A, Smith J. and Harris B.W., *Eur. Phys. J.* **C18**, 547 (2001).
14. Harris B.W., Smith J., *Phys. Rev.* **D57**, 2806 (1998).
15. Harris B.W. and Smith J., *Nucl. Phys.* **B452** 109 (1995); ibid. *Nucl. Phys. Proc. Suppl* **51C** 188-194 (1996), hep-ph/9605358.
16. Glück M, Reya E., Stratmann M., *Nucl. Phys.* **B222** 37 (1994);
    Vogt A., hep-ph/9601352, Proceedings of *Deep Inelastic Scattering and Related Phenomena "DIS96"*. Editors 'Agostini G.D. and Nigro A., (World Scientific), p. 254.
17. Buza. M, Matiounine Y., Smith J. and van Neerven W.L., *Eur. Phys. J.* **C1**, 301 (1998); ibid. *Phys. Lett.* **B411**, 211 (1997).
18. van Neerven W.L. and Zijlstra E.B., *Phys. Lett.* **B272**, 127 (1991);
    Zijlstra E.B. and van Neerven W.L., *Phys. Lett.* **B273**, 476 (1991);
    Zijlstra E.B. and van Neerven W.L., *Nucl. Phys.* **B383**, 525 (1992).
19. Laenen E. et al., hep-ph/9609351, Published in the proceedings of the *Workshop on Future Physics at HERA*, Hamburg, Germany, 25-26 Sep 1995, p. 393-401.
20. Thorne R.S. and Roberts R.G., *Phys. Lett.* **B421**, 303 (1998); ibid. *Phys. Rev.* **D57**, 6871 (1998).
21. van Neerven W.L., hep-ph/9804445, Published in the Proceedings of the *6th International Workshop on Deep Inelastic Scattering and QCD "DIS98"*, Editors Coremans GH. and Roosen R., World Scientific, p. 162-166.
22. Chuvakin A, Smith J. and van Neerven W.L., *Phys. Rev.* **D61**, 096004 (1999), ibid. *Phys. Rev.* **D62**, 036004 (2000).
23. Berends F.A., van Neerven W.L., Burgers G.J.H., *Nucl. Phys.* **B297**, 429 (1988); Erratum: *Nucl. Phys.* **B304**, 921 (1988).

Spinor and Supersymmetry in Spaces of Various Dimensions and Signatures

D.G.C. McKeon*
T.N. Sherry[†]

University of Western Ontario
[†]*National University of Ireland, Galway*

Abstract. We consider the nature of spinors and of supersymmetry in flat spaces of three and four dimensions and spaces of constant curvature in two dimensions.

The analysis of spinors and supersymmetry (SUSY) in $3+1$ dimensions is quite standard. (See for example ref. [1].) In a representation in which Dirac matrices are given by

$$\gamma^\mu = \left[\begin{pmatrix} 0 & 1 \\ 1 & 0 \end{pmatrix}, \begin{pmatrix} 0 & \vec{\sigma} \\ -\vec{\sigma} & 0 \end{pmatrix} \right] \tag{1}$$

and a charge conjugation matrix C is defined by

$$C^{-1}\gamma^\mu C = -\gamma^{\mu T}, \tag{2}$$

we have spinors

$$\Psi = \begin{pmatrix} \psi_\alpha \\ \overline{\chi}^{\dot\alpha} \end{pmatrix} \quad \overline{\Psi} \equiv \Psi^\dagger \gamma^0 = \left(\chi^\alpha, \overline{\psi}_{\dot\alpha} \right) \tag{3a}$$

and

$$\Psi_C \equiv C\overline{\Psi}^T = \begin{pmatrix} \chi_\alpha \\ \overline{\psi}^{\dot\alpha} \end{pmatrix} \quad \overline{\Psi}_C = (\psi^\alpha, \overline{\chi}_{\dot\alpha}) \tag{3b}$$

forming representations of the Lorentz group. The two spinorial generators of $N=2$ SUSY extension of the Lorentz group are Majorana (ie, $Q_i = Q_{Ci}$ for $i=1,2$) and satisfy the algebra

$$\left\{ Q_\alpha^i, \overline{Q}_{\dot\beta j} \right\} = 2\delta^i_j \sigma_{\alpha\dot\beta} P_\mu \tag{4a}$$

$$\left\{ Q_\alpha^i, Q_\beta^j \right\} = \epsilon_{\alpha\beta}\epsilon^{ij} Z. \tag{4b}$$

A representation of this algebra can be found using Fermionic creation and annihilation operators

$$a_\alpha = \frac{1}{\sqrt{2}}\left(Q_\alpha^1 + \epsilon_{\alpha\beta}Q_\beta^{2\dagger}\right), \quad b_\alpha = \frac{1}{\sqrt{2}}\left(Q_\alpha^1 - \epsilon_{\alpha\beta}Q_\beta^{2\dagger}\right). \tag{5}$$

¿From (4) it follows that

$$\left\{a_\alpha, a_\beta^\dagger\right\} = \delta_{\alpha\beta}(2M + Z) \tag{6a}$$

$$\left\{b_\alpha, b_\beta^\dagger\right\} = \delta_{\alpha\beta}(2M - Z) \tag{6b}$$

in a frame in which $P_\mu = (M, \vec{0})$. By (6b) we obtain the "BPS" bound

$$2M \geq Z. \tag{7}$$

The situation in $4+0$ dimensions is quite different. In this case

$$\gamma^\mu = \left[\begin{pmatrix} 0 & 1 \\ 1 & 0 \end{pmatrix}, \begin{pmatrix} 0 & i\vec{\sigma} \\ -i\vec{\sigma} & 0 \end{pmatrix}\right] \tag{8}$$

so that

$$\Psi = \begin{pmatrix} \psi_\alpha \\ \chi^{\dot{\alpha}} \end{pmatrix} \quad \overline{\Psi} = \Psi^\dagger = \left(\overline{\psi}^\alpha, -\overline{\chi}_{\dot\alpha}\right) \tag{9a}$$

and

$$\Psi_C = C\overline{\Psi}^T = \begin{pmatrix} -\overline{\psi}_\alpha \\ \overline{\chi}^{\dot\alpha} \end{pmatrix} \quad \overline{\Psi}_C = (\psi^\alpha, \chi_{\dot\alpha}) \tag{9b}$$

for representations of $SO(4) = SU(2) \times SU(2)$. The two spinors ψ_α and χ_α now transform separately under each $SU(2)$ subgroup; it is also evident from (9) that as $\left(\overline{\Psi}_C\right)_C = -\Psi$, we cannot have Majorana spinors in $4+0$ dimensions. The simplest SUSY algebra is now [2]

$$\left\{G, G_C^\dagger\right\} = 0 \tag{10a}$$

$$\left\{G, G^\dagger\right\} = i\gamma_5\gamma^\mu P^\mu + Z + Z\gamma_5 \tag{10b}$$

or, equivalently, if

$$G = \begin{pmatrix} Q_a \\ R^{\dot a} \end{pmatrix} \implies \begin{array}{ll} S_{a1} = Q_a & S_{a2} = \overline{Q}_a \\ T_{\dot a 1} = -R_{\dot a} & T_{\dot a 2} = \overline{R}_{\dot a} \end{array} \tag{11}$$

we obtain an equivalent algebra which displays an $SU(2)$ structure

$$\left\{S_{ai}, S_{bj}\right\} = \epsilon_{ab}\epsilon_{ij} Z_{Q\overline{Q}} \tag{12a}$$

$$\left\{T_{\dot a i}, T_{\dot b j}\right\} = \epsilon_{\dot a \dot b}\epsilon_{ij} Z^{R\overline{R}} \tag{12b}$$

$$\{S_{ai}, T_{bj}\} = i\epsilon_{ij}\sigma^\mu_{ab}P^\mu. \tag{12c}$$

This is similar in form to (4) with the roles of Z and P^μ "reversed"; (12) in terms of Fermionic creation and annihilation operators becomes, in the frame where $P^\mu = (0,0,0,P)$,

$$\{A_a, A_b^\dagger\} = \delta_{ab}\left[1 + P\left(Z_{Q\overline{Q}}Z^{R\overline{R}}\right)^{-1/2}\right] \tag{13a}$$

$$\{B_a, B_b^\dagger\} = \delta_{ab}\left[1 - P\left(Z_{Q\overline{Q}}Z^{R\overline{R}}\right)^{-1/2}\right]. \tag{13b}$$

We hence see that P has an <u>upper</u> bound in $4+0$ dimensions

$$P \leq \left(Z_{Q\overline{Q}}Z^{R\overline{R}}\right)^{1/2}. \tag{14}$$

A similar upper bound arises when one has extended SUSY in $4+0$ dimensions with algebra [3]

$$\{Q_{ai}, \overline{Q}_{bj}\} = \epsilon_{ab}Z^Q_{ij} \tag{15a}$$

$$\{R_{\dot{a}i}, \overline{R}_{\dot{b}j}\} = \epsilon_{\dot{a}\dot{b}}Z^R_{ij} \tag{15b}$$

$$\{Q_{ai}, \overline{R}_{\dot{b}j}\} = i\sigma^\mu_{a\dot{b}}\epsilon_{ij}P^\mu. \tag{15c}$$

Just as $N = 2$ super Yang-Mills theory in $3+1$ dimensions can be obtained by dimensional reduction of the $N = 1$ gauge theory in $5+1$ dimensions, so also the supersymmetric gauge model of Zumino in $4+0$ dimensions can be generated; it has the action

$$S = \int d^4x_E \left[-\frac{1}{4}F^2_{\mu\nu} + \frac{1}{2}(D_\mu A)^2 - \frac{1}{2}(D_\mu B)^2 - \frac{i}{2}\left(\psi^\dagger \gamma \cdot \overleftrightarrow{D} \psi\right)\right.$$
$$\left. + ig\psi^\dagger(A - B\gamma_5)\psi + \frac{1}{2}g^2(A \times B)^2\right]. \tag{16}$$

One simply drops dependence on one space variable and the time variable in the $5+1$ dimensional model and has the corresponding components of the vector field identified with the scalar fields A and B. Explicit calculation [4] shows that the β-function in this model is the same as that in $N = 2$ gauge theory in $3+1$ dimensions despite the peculiar kinetic terms in (16) for the scalars A and B.

A model with extended SUSY invariance in $4+0$ dimensions can be obtained by dimensional reduction of $N = 1$ gauge theory in $9+1$ dimensions. It is expected that the β-function in this model vanishes, just as it does for $N = 4$ gauge theory in $3+1$ dimensions.

The $SU(2)$ structure of (12) allows one to define a Harmonic superspace in conjunction with $4+0$ dimensions [5].

We also note that in $4+0$ dimensions, one can define a model which is (a) Hermitian (b) gauge invariant under an axial $U(1)$ gauge transformation (c) anomaly free. Its action is

$$S = \int d^4x_E \left(\frac{1}{4}F_{\mu\nu}(A)F_{\mu\nu}(A) + \Psi_C^\dagger(\slashed{\partial} + \slashed{A}\gamma_5)\Psi + \Psi^\dagger(\slashed{\partial} - \slashed{A}\gamma_5)\Psi_C\right]. \tag{17}$$

No analogue of this model can be defined in $3+1$ dimensions.

In $2+2$ dimensions, spinors can be both Majorana and Weyl [6]. Spinors take the form

$$\Psi = \begin{pmatrix} \phi_a \\ \chi^{\dot{a}} \end{pmatrix} \quad \overline{\Psi} = \begin{pmatrix} i\epsilon^{ab}\phi_b^*, i\epsilon_{\dot{a}\dot{b}}\chi^{\dot{b}*} \end{pmatrix} \quad (18)$$

and the two simplest SUSY algebras are

$$\{q_a, r_{\dot{b}}\} = 2\left(\sigma_\mu\right)_{a\dot{b}} P^\mu \quad (19a)$$

and

$$\{Q, \overline{Q}\} = 2\gamma^\mu P_\mu + Z + Z_5\gamma_5 \quad (19b)$$

for Majorana and Dirac spinorial generators $Q = \begin{pmatrix} q_a \\ r^{\dot{a}} \end{pmatrix}$ respectively. In [6] it is shown that both of these algebras can be rewritten in terms of Fermionic creation and annihilation operators that generate a Hilbert space with negative norm states; this is taken to indicate that SUSY is incompatible with a $2+2$ dimensional space.

In $3+0$ dimensions, the simplest SUSY algebra is [7,8]

$$\{Q, Q^\dagger\} = \vec{\sigma} \cdot \vec{p} + Z \quad (20)$$

where Q is a two component spinorial generator, $\vec{\sigma}$ is a set of Pauli matrices and Z is a central charge operator. Forming a superspace with coordinates $(x^\mu, \zeta, \theta_i$ and $\theta_i^\dagger)$ allows one to make the identifications

$$Q_i = \frac{\partial}{\partial \theta_i^\dagger} - \frac{i}{2}\left(\vec{\sigma} \cdot \vec{\nabla}\theta\right)_i - \frac{i}{2}\left(\theta \frac{\partial}{\partial \zeta}\right)_i \quad (21a)$$

$$P_\mu = -i\frac{\partial}{\partial x^\mu}, \quad Z = -i\frac{\partial}{\partial \zeta}. \quad (21b,c)$$

This makes it possible to formulate supersymmetric models in $3+0$ dimensions which are analogous to both the Wess-Zumino and $N=1$ gauge models in $3+1$ dimensions. A similar analysis can be applied to $N=2$ supersymmetric models in $2+1$ dimensions. Dimensional reduction can be used to establish a relationship between supersymmetric models in $3+1$ dimensions and three dimensional supersymmetric models.

We note that just in (7) and (14), in $2+1$ dimensions the central charge bounds the momentum below while in $3+0$ dimensions, it bounds the momentum above.

An analysis of supersymmetry in five dimensions [7] reveals that much as in four dimensions, no time dimensions implies an upper bound on momentum; one time dimension implies a lower bound on momentum and two time dimensions implies that for all momentum, negative norm states occur.

The simplest SUSY algebra [9] associated with the two dimensional surface of a sphere embedded in three dimensions is

$$\{Q_i, Q_j\} = 0 \ , \ \{Q_i, Q_j^\dagger\} = Z\delta_{ij} - 2\vec{\sigma}_{ij} \cdot \vec{P} \tag{22}$$

$$[J^a, Q] = -\frac{1}{2}\sigma^a Q, \ [J^a, J^b] = i\epsilon^{abc} J^c, \ [Z, Q] = -Q$$

(Note that Z is no longer a "central charge" as it does not commute with Q.). To examine representations of this superalgebra, we define a state $|I>$ such that

$$J^2|I> = j(j+1)|I>, \ J_Z|I> = m|I>$$
$$Z|I> = \zeta|I>, \ Q|I> = 0. \tag{23}$$

Now if $|i> = Q_i^\dagger|I>$ and $|F> = Q_1^\dagger Q_2^\dagger|I>$, we find that $<1|1> = (\zeta + 2m)$, $<2|2> = \zeta - 2m$, $<F|F> = (\zeta - 2j)(\zeta + 2j + 2)$, showing that a positive definite Hilbert space occurs if

$$\zeta \geq 2j. \tag{24}$$

A model invariant under transformations generated by the SUSY algebra of (22) is

$$S = \int \frac{dA}{A^2} \left\{ \frac{1}{2}\Psi^\dagger(\sigma \cdot L + x)\Psi - \Phi^*\left(L^2 + x(1-x)\right)\Phi - \frac{1}{4}F^*F \right. \tag{25}$$
$$\left. +\lambda_N \left(2(1-2x)\Phi^*\Phi - (F^*\Phi + F\Phi^*) - \Psi^\dagger\Psi\right)^N \right\};$$

the transformations are

$$\delta\Phi = \xi^\dagger\Psi, \ \delta\Psi = 2(\sigma \cdot L + 1 - u)\Phi\xi - F\xi, \ \delta F = -2\xi^\dagger(\sigma \cdot L + x)\Psi \tag{26}$$

$$\delta_Z\Phi = [2(1-2x)\Phi - F], \ \delta_Z\Psi = [1 + 2\sigma \cdot L]\Psi, \ \delta_Z F = -4\left[L^2 + x(1-x)\right]\Phi + 2xF.$$

A superspace representation of the algebra of (22) is provided by

$$Q = (\sigma \cdot \bar{r} + \zeta)\frac{\partial}{\partial\theta^\dagger} - \left(\frac{\partial}{\partial\zeta} - \sigma \cdot \nabla\right)\theta \tag{27a}$$

$$Q^\dagger = \frac{\partial}{\partial\theta}(\sigma \cdot r + \zeta) + \theta^\dagger\left(\frac{\partial}{\partial\zeta} - \sigma \cdot \nabla\right) \tag{27b}$$

$$J^a = -i(r \times \nabla)^a + \frac{1}{2}\left(\theta^\dagger\sigma^a\frac{\partial}{\partial\theta^\dagger} + \frac{\partial}{\partial\theta}\sigma^a\theta\right) \tag{27c}$$

$$Z = -\theta^\dagger\frac{\partial}{\partial\theta^\dagger} + \theta\frac{\partial}{\partial\theta}. \tag{27d}$$

Currently we are attempting to formulate (25) in terms of superfields. It is difficult to do so as no operators that commute with Q and $Q^\dagger$ are possible. If these existed they would allow us to find irreducible representations of our algebra.

The role of ζ in (27) is not at all clear. However, it is necessary to introduce ζ in order for Q to be the "square root" of the non-Abelian operator J^a.

We have also examined the SUSY algebra on three and four dimensional spherical surfaces as well as two dimensional AdS space. A full superfield formalism for AdS_2 space can be defined [10].

ACKNOWLEDGEMENTS

NSERC provided financial support. R. and D. MacKenzie had helpful suggestions.

REFERENCES

1. D. Balin and A. Love, "Supersymmetric Gauge Field Theory and String Theory", IOP Publishing, Bristol 1994.
2. D.G.C. McKeon and T.N. Sherry, Ann. of Phys. 288 (2001) 2.
3. D.G.C. McKeon and T.N. Sherry, Ann. of Phys. 285 (2000) 221.
4. R. Clarkson and D.G.C. McKeon, Can. J. Phys. (to be published).
5. D.G.C. McKeon, Can. J. Phys. (in press).
6. F.T. Brandt, D.G.C. McKeon and T.N. Sherry, Mod. Phys. Lett. A 15 (2000) 1349.
7. D.G.C. McKeon, Nucl. Phys. B591 (2000) 591.
8. D.G.C. McKeon and T.N. Sherry, UWO/NUIG report (2001).
9. D.G.C. McKeon and T.N. Sherry, UWO/NUIG report (2001).
10. D.G.C. McKeon and T.N. Sherry, UWO/NUIG report (2001).

Dynamically generating the quark-level SU(2) linear sigma model

M.D. SCADRON

Physics Department, University of Arizona, Tucson, AZ, 85721 USA

First we study Nambu-type gap equations, $\delta f_\pi = f_\pi$ and $\delta m_q = m_q$. Then we exploit the dimensional regularization lemma, subtracting quadratic from log-divergent integrals. The nonperturbative quark loop LσM solution recovers the original Gell-Mann-Levy (tree level) equations along with $m_\sigma = 2m_q$ and meson-quark coupling $g = 2\pi/\sqrt{N_c}$. Next we use the Ben Lee null tadpole condition to reconfirm that $N_c = 3$ even through loop order. Lastly we show that this loop order LσM a) reproduces the (remarkably successful) VMD scheme in tree order, and b) could be suggested as the infrared limit of low energy QCD.

§1. Introduction

To begin, we give the original [1] tree-level chiral-broken SU(2) interacting LσM lagrangian density, but after the spontaneous symmetry breaking (SSB) shift:

$$\mathcal{L}^{int}_{L\sigma M} = g\bar{\psi}(\sigma' + i\gamma_5 \boldsymbol{\tau} \cdot \boldsymbol{\pi})\psi + g'\sigma'(\sigma'^2 + \boldsymbol{\pi}^2) - \lambda(\sigma'^2 + \boldsymbol{\pi}^2)^2/4. \quad (1\cdot 1)$$

In refs.1) the couplings g, g', λ in (1·1) satisfy the quark-level Goldberger-Treiman relation (GTR) for $f_\pi \approx 93$ MeV and $f_\pi \sim 90$ MeV in the chiral limit (CL):

$$g = m_q/f_\pi, \quad g' = m_\sigma^2/2f_\pi = \lambda f_\pi. \quad (1\cdot 2)$$

We work in loop order and dynamically generate mass terms in (1·1) via non-perturbative Nambu-type gap equations $\delta f_\pi = f_\pi$, $\delta m_q = m_q$. The CL $m_\pi = 0$, corresponds to $<0|\partial A|\pi> = 0$ for $<0|A_\mu^3|\pi^0> = if_\pi q_\mu$. The latter requires the GTR $m_q = f_\pi g$ to be valid in tree and loop order, fixing g, g', λ in loop order.

In §2, 3, this quark-level LσM is nonperturbatively solved via loop-order gap equations. In §4, the NGT is expressed in LσM language with charge radius $r_\pi = 1/m_q$ characterizing quark fusion for the tightly bound $q\bar{q}$ pion. In §5, the Lee null tadpole sum is shown to require $N_c = 3$ for the true vacuum. §6 discusses s-wave chiral cancellations in the LσM. §7 shows VMD follows directly from the LσM. Finally §8 suggests this LσM is the infrared limit of nonperturbative QCD. We give our conclusions in §9.

§2. Quark loop gap equations

First we compute $\delta f_\pi = f_\pi$ in the CL via the u and d quark loops of Fig.1a. Replacing f_π by m_q/g and taking the quark trace, giving $4m_q q_\mu$, the factors $m_q q_\mu$ cancel, requiring the CL log-divergent gap equation (LDGE) [2,3], $\bar{d}^4 p = d^4 p/(2\pi)^4$:

$$1 = -4iN_c g^2 \int (p^2 - m_q^2)^{-2} \bar{d}^4 p. \quad (2\cdot 1)$$

CP601, *Theoretical High Energy Physics: MRST 2001*, edited by V. Elias et al.
© 2001 American Institute of Physics 0-7354-0045-8/01/$18.00

Fig. 1. Quark loop for f_π(a), quark tadpole loop for m_q(b)

Anticipating $g \sim 320$ MeV/90MeV ~ 3.6 from the CL GTR, this LDGE (2·1) suggests an UV cutoff $\Lambda \sim 750$ MeV. Such a 750 MeV cutoff separates LσM elementary particle $\sigma(600) < \Lambda$ from bound states $\rho(770)$, $\omega(780)$, $a_1(1260) > \Lambda$. This is a $Z = 0$ compositeness condition [4], requiring $g = 2\pi/\sqrt{N_c}$. We later derive this from our dynamical symmetry breaking (DSB) loop order LσM.

Next we study $\delta m_q = m_q$ in the CL, with zero current quark mass; m_q is the nonstrange constituent quark mass. The needed mass gap is formed via the quadratically divergent quark tadpole loop of Fig.1b; additional quark π- and σ-mediated self-energy graphs then cancel [3], giving the quadratic divergent mass gap

$$1 = 8iN_c g^2/(-m_\sigma^2) \cdot \int (p^2 - m_q^2)^{-1} \bar{d}^4 p. \quad (2\cdot 2)$$

Here the $q^2 = 0$ tadpole σ propagator $(0 - m_\sigma^2)^{-1}$ means the right-hand side (rhs) of the integral in eq.(2·2) acts as a counterterm quadratic divergent NJL [5] mass gap.

References 3) first subtract the quadratic-from the log-divergent integrals of eqs. (2·1), (2·2) to form the dimensional regularization (dim. reg.) lemma for $2l = 4$:

$$\int \bar{d}^4 p \left[\frac{m_q^2}{(p^2 - m_q^2)^2} - \frac{1}{p^2 - m_q^2} \right] = \lim_{l \to 2} \frac{i m_q^{2l-2}}{(4\pi)^l} [\Gamma(2-l) + \Gamma(1-l)] = \frac{-im_q^2}{(4\pi)^2}. \quad (2\cdot 3)$$

This dim. reg. lemma (2·3) follows because $\Gamma(2-l) + \Gamma(1-l) \to -1$ as $l \to 2$ due to the gamma function defining identity $\Gamma(z+1) = z\Gamma(z)$. This lemma eq.(2·3) is more general than dim. reg.; (i) use partial fractions to write

$$\frac{m^2}{(p^2 - m^2)^2} - \frac{1}{p^2 - m^2} = \frac{1}{p^2} \left[\frac{m^4}{(p^2 - m^2)^2} - 1 \right], \quad (2\cdot 4)$$

(ii) integrate eq.(2·4) via $\bar{d}^4 p$ and neglect the latter massless tadpole $\int \bar{d}^4 p/p^2 = 0$ (as is also done in dim. reg., analytic, zeta function and Pauli-Villars regularization [3]) (iii) Wick rotate $d^4 p = i\pi^2 p_E^2 dp_E^2$ in the integral over eq.(2·4) to find

$$\int \bar{d}^4 p \left[\frac{m^2}{(p^2 - m^2)^2} - \frac{1}{p^2 - m^2} \right] = -\frac{im^4}{(4\pi)^2} \int_0^\infty \frac{dp_E^2}{(p_E^2 + m^2)^2} = \frac{-im^2}{(4\pi)^2}. \quad (2\cdot 5)$$

So (2·5) gives the dim.reg.lemma (2·3); both are *regularization scheme independent*.

Following refs.3) we combine eqs. (2·3) or (2·5) with the LDGE (2·1) to solve the quadratically divergent mass gap integral (2·2) as

$$m_\sigma^2 = 2m_q^2(1 + g^2 N_c/4\pi^2). \quad (2\cdot 6)$$

Also the Fig.2 quark bubble plus tadpole graphs dynamically generate the σ mass [3]:

Fig. 2. Quark bubble plus quark tadpole loop for m_σ^2.

Fig. 3. Quark bubble plus quark tadpole loop for m_π^2.

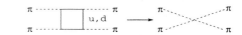

Fig. 4. Quark triangle shrinks to point for $m_\sigma \to \pi\pi$.

Fig. 5. Quark box shrinks to point contact for $\pi\pi \to \pi\pi$.

$$m_\sigma^2 = 16iN_c g^2 \int \bar{d}^4 p \left[\frac{m_q^2}{(p^2-m_q^2)^2} - \frac{1}{p^2-m_q^2} \right] = \frac{N_c g^2 m_q^2}{\pi^2}, \quad (2\cdot 7)$$

where we have deduced the rhs of eq.(2·7) by using (2·3) or (2·5). Finally solving the two equations (2·6) and (2·7) for the two unknowns m_σ^2/m_q^2 and $g^2 N_c$, one finds [3)]

$$m_\sigma = 2m_q, \qquad g = 2\pi/\sqrt{N_c}. \quad (2\cdot 8)$$

Not surprisingly, the lhs equation in (2·8) is the famous NJL four quark result [5)], earlier anticipated for the LσM in refs.6). The rhs equation in (2·8) is also the consequence of the Z=0 compositeness condition [4)], as noted earlier.

Finally we compute m_π^2 from the analog pion bubble plus tadpole graphs of Fig.3. Since both quark loops (ql) are quadratic divergent in the CL, one finds [2),3)]

$$m_{\pi,ql}^2 = 4iN_c[2g^2 - 4gg'm_q/m_\sigma^2] \int (p^2-m_q^2)^{-1} \bar{d}^4 p = 0; \quad g' = m_\sigma^2/2f_\pi, (2\cdot 9)$$

using the GTR. Not suprisingly, eq.(2·9) is the dynamical version of the SSB (1·2).

§3. Loop order three- and four-point functions

Having studied all 2-point functions in §2, we now look at 3- and 4-point functions. In the CL the u and d quark loops of Fig.4 generate $g_{\sigma\pi\pi}$ [2),3)] as

$$g_{\sigma\pi\pi} = -8ig^3 N_c m_q \int (p^2-m_q^2)^{-2} \bar{d}^4 p = 2gm_q \quad (3\cdot 1)$$

by virtue of the LDGE (2·1). Using the GTR and $m_\sigma = 2m_q$, eq.(3·1) reduces to

$$g_{\sigma\pi\pi} = 2gm_q = m_\sigma^2/2f_\pi = g'. \quad (3\cdot 2)$$

In effect, the $g_{\sigma\pi\pi}$ loop of Fig.4 "shrinks" to the LσM cubic meson coupling g' in the tree-level lagrangian eq.(1·1), but only when $m_\sigma = 2m_q$ and $g/m_q = 1/f_\pi$.

Next we study the 4-point $\pi\pi$ quark box of Fig.5, giving a CL log divergence [3)]:

$$\lambda_{box} = -8iN_c g^4 \int (p^2-m_q^2)^{-2} \bar{d}^4 p = 2g^2 = g'/f_\pi = \lambda_{tree}, \quad (3\cdot 3)$$

employing the LDGE (2·1) to reduce (3·3) to $2g^2$. Equation(3·3) shrinks to λ_{tree}, by virtue of eq.(1·2). Substituting (2·8) into (3·3), we find $\lambda = 8\pi^2/N_c$.

We have dynamically generated the entire LσM lagrangian (1·1), but using the DSB true vacuum, satisfying specific values of g, g', λ in eq.(1·1).

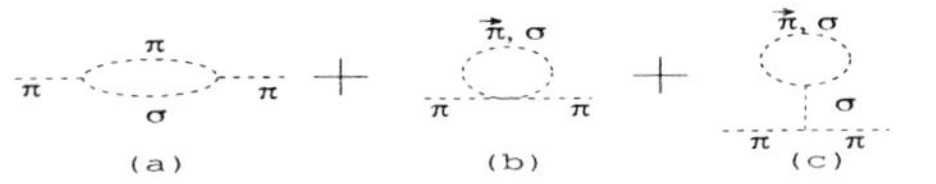

Fig. 6. Meson bubble(a) meson quartic (b) meson tadpole(c) graphs for m_π^2.

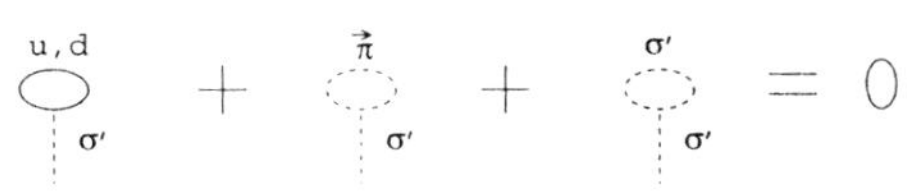

Fig. 7. Null tadpole sum for SU(2) LσM.

§4. Nambu-Goldstone theorem (NGT) in LσM loop order

Having dynamically generated the chiral pion and σ as elementary, we must add to Fig.3 the five meson loops of Fig.6. The first bubble graph in Fig.6 is log divergent, while the latter four quartic and tadpole graphs are quadratic divergent.

To proceed, first one uses a partial fraction identity to rewrite the log-divergent bubble graph as the difference of π and σ quadratic divergent integrals [2],[7]. Then the six meson loops (ml) of Fig.6 can be separated into three quadratic divergent π and three quadratic divergent σ integrals [7]:

$$m_{\pi,ml}^2 = (-2\lambda + 5\lambda - 3\lambda)i\int (p^2 - m_\pi^2)^{-1}\bar{d}^4p + (2\lambda + \lambda - 3\lambda)i\int (p^2 - m_\sigma^2)^{-1}\bar{d}^4p. \quad (4\cdot1)$$

Adding eq.(4·1) to eq.(2·9), the total m_π^2 in the CL is in loop order

$$m_\pi^2 = m_{\pi,ql}^2 + m_{\pi,\pi l}^2 + m_{\pi,\sigma l}^2 = 0 + 0 + 0 = 0. \quad (4\cdot2)$$

Moreover, eq.(4·2) is chirally regularized and renormalized because the tadpole graphs of Figs.3, 6c are already counterterm masses acting as subtraction constants.

A second aspect of the chiral pion concerns the pion charge radius r_π in the CL. First one computes the pion form factor $F_{\pi,ql}(q^2)$ due to quark loops (ql) and then differentiates it with respect to q^2 at $q^2 = 0$ to find $r_{\pi,ql}^2$ as

$$r_{\pi,ql}^2 = \frac{6dF_{\pi,ql}(q^2)}{dq^2}\bigg|_{q^2=0} = 8iN_cg^2\int_0^1 dx 6x(1-x)\int (p^2 - m_q^2)^{-3}\bar{d}^4p$$
$$= 8iN_c(4\pi^2/N_c) \cdot (-i\pi^2/2m_q^2 16\pi^4) = 1/m_q^2. \quad (4\cdot3)$$

Although r_π was originally expressed as $\sqrt{N_c}/2\pi f_\pi$ [8],[7], we prefer the result (4·3) or $r_\pi = 1/m_q$, as it requires the tightly bound $q\bar{q}$ pion to have the two quarks *fused* in the CL. Later we will show that $N_c = 3$; $m_q \approx 325$ MeV in the CL gives $r_\pi = 1/m_q \approx 0.6$ fm. The observed r_π is [9] (0.63±0.01) fm. The alternative ChPT requires $r_\pi \propto L_9$, a low energy constant (LEC)! However VMD successfully predicts

$$r_\pi^{VMD} = \sqrt{6}/m_\rho \approx 0.63 fm, \quad (4\cdot4)$$

not only accurate but r_π^{VMD} and $r_\pi^{L\sigma M}$ in (4·3) and (4·4) are clearly related [7].

§5. Lee null tadpole sum in SU(2) LσM finding $N_c = 3$

To characterize the true DSB (not the false SSB) vacuum, B. Lee [10] requires the *sum* of loop-order tadpoles to vanish, see *eg.* our Fig.7. This tadpole sum is [3]

$$<\sigma'> = 0 = -i8N_cgm_q\int (p^2 - m_q^2)^{-1}\bar{d}^4p + 3ig'\int (p^2 - m_\sigma^2)^{-1}\bar{d}^4p. \quad (5\cdot1)$$

Replacing g by m_q/f_π, g' by $m_\sigma^2/2f_\pi$ and scaling the quadratic divergent q(or σ) loop integrals by m_q^2 (or m_σ^2), eq.(5·1) requires [3] (neglecting the pion tadpole)

$$N_c(2m_q)^4 = 3m_\sigma^4. \tag{5·2}$$

But we know from eq.(2·8) that $2m_q = m_\sigma$, so the loop-order SU(2) LσM result (5·2) in turn *predicts* $N_c = 3$, a satisfying result ! Then the dynamically generated SU(2) loop-order LσM in §3 also predicts in the CL [3] $m_q \approx 325$ MeV, $m_\sigma \approx 650$ MeV and $g = 2\pi/\sqrt{3} = 3.6276$, $g' = 2gm_q \approx 2.36$ GeV, $\lambda = 8\pi^2/3 \approx 26.3$.

§6. Chiral s-wave cancellations in LσM

Away from the CL, the tree-order LσM requires the cubic meson coupling to be

$$g_{\sigma\pi\pi} = (m_\sigma^2 - m_\pi^2)/2f_\pi = \lambda f_\pi. \tag{6·1}$$

But at threshold $s = m_\pi^2$, so the net $\pi\pi$ amplitude then vanishes using (6·1):

$$M_{\pi\pi} = M_{\pi\pi}^{contact} + M_{\pi\pi}^{\sigma pole} \to \lambda + 2g_{\sigma\pi\pi}^2(m_\pi^2 - m_\sigma^2)^{-1} = 0. \tag{6·2}$$

In effect, the contact λ " chirally eats" the σ pole at the $\pi\pi$ threshold at tree level. Then σ poles from the cross channels predict a LσM Weinberg PCAC form [11], [12]

$$M_{\pi\pi}^{abcd} = A\delta^{ab}\delta^{cd} + B\delta^{ac}\delta^{bd} + C\delta^{ad}\delta^{bc},$$

$$A^{L\sigma M} = -2\lambda\left[1 - \frac{2\lambda f_\pi^2}{m_\sigma^2 - s}\right] = \left(\frac{m_\sigma^2 - m_\pi^2}{m_\sigma^2 - s}\right)\left(\frac{s - m_\pi^2}{f_\pi^2}\right). \tag{6·3}$$

So the I=0 s-channel amplitude $3A+B+C$ at threshold predicts a 23% enhancement of the Weinberg s-wave I=0 scattering length at $s = 4m_\pi^2$, $t = u = 0$ for $m_\sigma \approx 650$ MeV with $\epsilon = m_\pi^2/m_\sigma^2 \approx 0.045$ and [12] (using only eq.(6·3))

$$a_{\pi\pi}^{(0)}|_{L\sigma M} = \left(\frac{7+\epsilon}{1-4\epsilon}\right)\frac{m_\pi}{32\pi f_\pi^2} \approx (1.23)\frac{7m_\pi}{32\pi f_\pi^2} \approx 0.20 m_\pi^{-1}. \tag{6·4}$$

For a $\sigma(550)$ and $\epsilon \approx 0.063$ this LσM scattering length (6·4) increases to $0.22 m_\pi^{-1}$. Compare this simple LσM tree order result (6·4) with the analogue ChPT 0.22 m_π^{-1} scattering length requiring a 2-loop calculation involving about 100 LECs ! These $\pi\pi$ scattering length problems should be sorted out soon by R. Kamiński, et. al. [13].

In LσM loop order the analog cancellation is due to a Dirac matrix *identity* [14]:

$$(\gamma \cdot p - m)^{-1} 2m\gamma_5 (\gamma \cdot p - m)^{-1} = -\gamma_5(\gamma \cdot p - m)^{-1} - (\gamma \cdot p - m)^{-1}\gamma_5. \tag{6·5}$$

At a soft pion momentum, eq.(6·5) requires a σ meson to be "eaten" via a quark box-quark triangle cancellation for $a_1 \to \pi(\pi\pi)$ s wave, $\gamma\gamma \to 2\pi^0$, $\pi^- p \to \pi\pi n$ as suggested in each case by low energy data [14], [15]. Also a soft pion scalar kappa $\kappa(800-900)$ is "eaten" in $K^- p \to K^- \pi^+ n$ peripheral scattering [15].

Fig. 8. Quark triangle graphs contributing to $\rho^0 \to \pi\pi$.

§7. VMD and the LσM

Given the implicit LDGE (2·1) UV cutoff $\Lambda \approx 750$ MeV, the $\rho(770)$ can be taken as an external field (bound state $\bar{q}q$ vector meson). Accordingly the quark loop graphs of Fig.8 generate the loop order $\rho\pi\pi$ coupling [2),3)]

$$g_{\rho\pi\pi} = g_\rho[-i4N_c g^2 \int (p^2 - m_q^2)^{-2} \bar{d}^4 p] = g_\rho \tag{7·1}$$

via the LDGE (2·1). While the individual udu and dud quark graphs of Figs.8 are both linearly divergent, when added together with vertices $g_{\rho^0 uu} = -g_{\rho^0 dd}$, the net $g_{\rho\pi\pi}$ loop in Fig.8 is log divergent. Equation(7·1) is Sakurai's VMD universality condition. Also a $\pi^+\sigma\pi^+$ meson loop added to the quark loops of Fig.8 gives [7)]

$$g_{\rho\pi\pi} = g_\rho + g_{\rho\pi\pi}/6 \text{ or } g_{\rho\pi\pi}/g_\rho = 6/5. \tag{7·2}$$

If one first gauges the LσM lagrangian, the inverted squared gauge coupling is related to the $q^2 = 0$ polarization amplitude as [3)]

$$(g_\rho^{-2}) = \pi(0, m_q^2) = -8iN_c/6 \cdot \int (p^2 - m_q^2)^{-2} \bar{d}^4 p = (3g^2)^{-1} \tag{7·3}$$

by virtue of the LDGE(2·1). But since we know $g = 2\pi/\sqrt{3}$, eq.(7·3) requires $g_\rho = \sqrt{3}g = 2\pi$, reasonably near the observed values $g_{\rho\pi\pi} \approx 6.05$ and $g_\rho \approx 5.03$.

The chiral KSRF relation for the ρ mass [18)] $m_\rho^2 = 2g_{\rho\pi\pi} g_\rho f_\pi^2$ coupled with this LσM implies $m_\rho^2 = 2(2\pi)^2 f_\pi^2 5/6 \approx (754 \text{ MeV})^2$, close to the observed ρ mass. Also, the dynamically generated LσM extension to SU(3) by the authors of ref. 3.

§8. LσM as infrared limit of nonperturbative QCD

We suggest five links between the LσM and the infrared limit of QCD.

i) Quark mass: the LσM has $m_q = f_\pi \frac{2\pi}{\sqrt{3}} \approx 325$ MeV, while QCD has [19)] $m_{dyn} = (\frac{4\pi\alpha_s}{3} < -\bar{\Psi}\Psi >_{1\text{GeV}})^{1/3} \approx 320$ MeV at a 1 GeV near-infrared cutoff.

ii) Quark condensate: the LσM condensate is at infrared cutoff m_q [20)]:

$$< -\bar{\Psi}\Psi >_{m_q} = i4N_c m_q \int \frac{\bar{d}^4 p}{p^2 - m_q^2} = \frac{3m_q^3}{4\pi^2}\left[\frac{\Lambda^2}{m_q^2} - \ln\left(\frac{\Lambda^2}{m_q^2} + 1\right)\right] \approx (209 \text{MeV})^3,$$

while the condensate in QCD is $< -\bar{\Psi}\Psi >_{m_q} = 3m_{dyn}^3/\pi^2 \approx (215 \text{MeV})^3$.

iii) Frozen coupling strength: the LσM coupling is for $g = 2\pi/\sqrt{3}$ or $\alpha_{L\sigma M} = \frac{g^2}{4\pi} = \frac{\pi}{3}$, while in QCD $\alpha_s = \frac{\pi}{4}$ at infrared freezeout [21)] leads to $\alpha_s^{eff} = (4/3)\alpha_s = \pi/3$.

iv) σ mass: the LσM requires $m_\sigma = 2m_q$, while the QCD condensate gives [22] $m_{dyn} = \frac{g_{\sigma qq}}{m_\sigma^2} < -\bar{\Psi}\Psi >_{m_\sigma}$ for $\alpha_s(m_\sigma) \approx \pi/4$, or $m_\sigma^2/m_{dyn}^2 = \pi/\alpha_s(m_\sigma^2) \approx 4$.

v) Chiral restoration temperature T_c: the LσM requires [23] $T_c = 2f_\pi \approx 180$ MeV, while QCD computer lattice simulations find [24] $T_c = (150 \pm 30)$ MeV.

§9. Conclusion

In §2, 3, the SU(2) LσM lagrangian was dynamically generated in all (chiral) regularization schemes, via loop gap equations, predicting the NJL σ mass $m_\sigma = 2m_q$ along with meson-quark coupling $g = 2\pi/\sqrt{N_c}$. Then the three-and four-point quark loops were shown to "shrink" to tree graphs, giving the meson cubic and quartic couplings $g' = m_\sigma^2/2f_\pi$, $\lambda = 8\pi^2/N_c$. Next in §4, the Nambu-Goldstone theorem (NGT) was shown to hold in LσM loop order with the pion charge radius $r_\pi = 1/m_q$. In §5, the SU(2) LσM requires color number $N_c=3$ in loop order, then predicting $m_q \approx 325$ MeV, $m_\sigma \approx 650$ MeV, $g \approx 3.63$, $\lambda \approx 26$, $r_\pi \approx 0.6$ fm in the CL.

In §6, we considered LσM chiral cancellations, both in tree and in loop order. Next, in §7, Sakurai's vector meson dominance (VMD) empirically accurate scheme follows from the LσM, the latter further predicting $g_{\rho\pi\pi} = 2\pi$ and $g_{\rho\pi\pi}/g_\rho = 6/5$ along with the KSRF relation. Finally in §8, we suggested that the LσM is the infrared limit of nonperturbative QCD.

References

1) M. Gell-Mann and M. Levy, Nuovo Cimento **16** (1960), 705; V. de Alfaro, S. Fubini, G. Furlan and C. Rossetti, Currents in Hadron Physics (North Holland, 1973), chap. 5.
2) T. Hakioglu and M.D. Scadron, Phys. Rev. **D42** (1990), 941; **D43** (1991), 2439.
3) R. Delbourgo and M.D. Scadron, Mod. Phys. Lett. **A10** (1995), 251; R. Delbourgo, A. Rawlinson, and M.D. Scadron, ibid., **A13** (1998) 1893.
4) A. Salam, Nuovo Cimento **25** (1962), 224; S. Weinberg, Phys. Rev. **130** (1963), 776; M.D. Scadron, Phys. Rev. **D57** (1998), 5307.
5) Y. Nambu and G. Jona-Lasinio, Phys.Rev. **122** (1961), 345 (NJL); also see Y. Nambu, Phys.Rev.Lett. **4** (1960), 380.
6) T. Eguchi, Phys.Rev. **D14** (1976), 2755; **D17** (1978), 611.
7) A. Bramon, Riazuddin and M.D. Scadron, J.Phys.G **24** (1998), 1.
8) R. Tarrach, ZPhys.**C2** (1979), 221; S.B.Gerasimov, Sov.J.Nucl.Phys. **29** (1979) 259.
9) A.F. Grashin and M.V. Lepeshkin, Phys.Lett.**B146** (1984), 11.
10) B.W. Lee, Chiral Dynamics, (Gordon and Breach, NY, 1972), p. 12.
11) S. Weinberg, Phys.Rev.Lett. **17** (1966), 616.
12) M.D. Scadron, Eur.Phys.J. **C6** (1999), 141.
13) R. Kamiński, et. al. Z Phys.**C 74** (1997), 79.
14) A.N. Ivanov, M. Nagy, and M.D. Scadron, Phys.Lett.B **273** (1991), 137.
15) L.R. Babukhadia, et. al., Phys.Rev.**D 62** (2000) in press.
16) S. Ishida, et. al., Prog. Theor.Phys. **95** (1996), 745; **98** (1997), 621.
17) See eg. P.Ko and S. Rudaz, Phys.Rev.**D 50** (1994), 6877.
18) K. Kawarabayashi and M. Suzuki, Phys.Rev.Lett.**16** (1966), 255; Riazuddin and Fayyazudin, Phys.Rev.**147** (1966), 1071 (KSRF).
19) V. Elias and M.D. Scadron, Phys. Rev. **D30** (1984), 647.
20) L.R. Babukhadia, V. Elias and M.D. Scadron, J.Phys.G **23** (1997), 1065.
21) A. Mattingly and P. Stevenson, Phys.Rev.Lett. **69** (1992), 1320.
22) V. Elias and M.D. Scadron, Phys.Rev.Lett. **53** (1984), 1129.
23) N. Bilic, J. Cleymans and M.D. Scadron, Intern. J. Mod. Phys. **A 10** (1995), 1160.
24) Particle Data Group (PDG), C. Caso, et.al., Eur.J.**C 3** (1998), 1.

GRAVITY/GEOMETRY

Oscillating metrics and the cosmological constant

B. Holdom

Department of Physics, University of Toronto, Toronto, Ontario M5S 1A7, Canada

Abstract. Oscillating metric solutions are found in theories with higher powers of curvature in the action. Their consideration is motivated by the cosmological constant problem.

INTRODUCTION

A class of generic solutions to general gravitation actions will be described, namely metrics with an oscillating dependence on one or more coordinates [1]. The action will contain higher powers of the curvature tensor, as are expected to arise in a derivative expansion of some underlying theory. We shall proceed to describe the following cases.

1. time-dependent oscillations in 4D
2. time-dependent oscillations in 5D
3. y-dependent oscillations in 5D
4. the combination of (2) and (3).

Our earlier work [2] has been concerned with the third case, where y is the coordinate of a compact fifth dimension, and which leads to a "warped Kaluza-Klein" picture. Here we point out that time-dependent oscillations can arise in a similar way, and that they can arise most naturally in the five dimensional context.

It appears that the fourth case, a combination of t-dependent and y-dependent oscillations, is of the most interest for the cosmological constant problem. This picture has the following features.

1. There is a family of singularity free solutions parameterized by an effective 4D cosmological constant $\Lambda > 0$.
2. The t-dependent oscillations have Planck scale frequencies and an amplitude of order $\sqrt{\Lambda}/M_{\text{Pl}}$.
3. The source of the oscillating gravitational field involves, besides Λ, the positive kinetic energy contribution of a massless scalar field.
4. The decay of the oscillating gravitational field through particle production causes a relaxation of Λ towards zero.

The relaxation of the amplitude of rapid metric oscillations appears distinctly different from various other attempts at relaxation mechanisms involving slowly evolving and nearly constant scalar fields. Weinberg's no-go theorem [3] shows that some fine tuning is required if the extremum of an action involving constant fields is to give flat space.

This is not relevant for our discussion since our picture involves a family of solutions and a decay process, and derivatives of the oscillating fields are not in any way suppressed.

One of the features of our picture is that the size of the compact space is fixed; the inclusion of higher derivative terms makes it clear how Planck scale physics can resolve the radius stabilization problem of the original Kaluza-Klein picture. This allows standard cosmological evolution to emerge in a more natural way.

It is interesting to note that the decay process should produce particles with roughly Planck scale energies. The colored particles will fragment into jets of particles with a wide range of energies, and this leads to a model for the origin of ultra-high energy cosmic rays above the GZK cutoff. For more details on this and other issues see [1, 6].

TIME-DEPENDENT OSCILLATIONS IN 4D

We first introduce t-dependent metric oscillations in a purely four dimensional context. We consider the following metric,

$$ds^2 = -dt^2 + e^{B(t)} g_{ij}\, dx^i dx^j, \qquad (1)$$

and the following action,

$$S = M_{\rm Pl}^2 \int d^4x\, \sqrt{-g}\left(-2\Lambda + R + a\,R^2 + b\,R_{\mu\nu}R^{\mu\nu} + c\,R_{\mu\nu\lambda\kappa}R^{\mu\nu\lambda\kappa} + \cdots + \mathcal{L}\right), \qquad (2)$$

with arbitrary powers of the curvature. We assume that $\mathcal{L}$ contains a free massless scalar field and in particular contributes to $T_{\mu\nu}$ as $\rho = p = \frac{1}{4}\dot\phi^2 + C$. C is some additional constant contribution, perhaps arising as a self-consistent Casimir effect. We seek solutions with periodic and smooth $B(t)$.

We shall assume that the amplitude ε of the $B(t)$ oscillations is small, and thus we approach the problem by expanding in powers of this amplitude. The solutions we find have the property that the sources of the gravitational field, Λ, $\frac{1}{4}\dot\phi^2 M_{\rm Pl}^{-2}$, and $CM_{\rm Pl}^{-2}$, are all of order $\varepsilon^2 M_{\rm Pl}^2$. (In the following, $M_{\rm Pl}^{-2}$ factors will be absorbed into $\dot\phi^2$, C, ρ, p, etc.) Then at leading order in ε the field equations reduce to one equation,

$$\frac{d^2 B}{dt^2} + 2\nu \frac{d^4 B}{dt^4} + \sum_{k=3}^{\infty}\sum_i \nu_{k,i}\frac{d^{2k} B}{dt^{2k}} = 0, \qquad (3)$$

where $\nu = 3a + b + c$ and the $\nu_{k,i}$ are various combinations of coefficients of terms higher order in R. This clearly has sinusoidal solutions $B(t) = \varepsilon\cos(\omega t)$ as long as the polynomial

$$-\omega^2 + 2\nu\omega^4 + \sum_{k=3}^{\infty}\sum_i (-1)^k \nu_{k,i}\omega^{2k} = 0 \qquad (4)$$

has at least one real root. Thus for a range of parameters in the original action (no fine tuning) it appears that such solutions exist. We shall see how the amplitude ε is determined at next order in the expansion.

We are therefore suggesting that there are generic solutions of this type to arbitrary order in a derivative expansion. We view the derivative expansion and its associated generic solutions as indicative for what can happen in a true theory of Planck scale physics.

To obtain more explicit results we shall henceforth drop the terms in the action at order R^3 and higher. It should be clear that the existence of the solutions does not depend on this truncation. Our quantitative results certainly are sensitive to the truncation, but we expect that the qualitative picture is not. We emphasise that we are not expanding in powers of small derivatives; and the higher derivative terms in the equations are just as important as lower derivative terms.[1]

After the truncation we have the constraint $\nu > 0$ and

$$\omega = \frac{1}{\sqrt{2\nu}}. \tag{5}$$

At order ε^2 we introduce the next Fourier mode

$$B(t) = \varepsilon \cos(\omega t) + b_2 \varepsilon^2 \cos(2\omega t), \tag{6}$$

and the field equations at this order are satisfied by

$$\varepsilon^2 = -\frac{16}{3}\Lambda\nu, \quad b_2 = -\frac{3}{16}, \quad \dot{\phi}^2 + 4C = 4\Lambda. \tag{7}$$

Thus we see how ε is determined. Since $\nu > 0$ we also see that $\Lambda < 0$. Now expanding to third order we have

$$B(t) = \varepsilon \cos(\omega t) - \frac{3}{16}\varepsilon^2 \cos(2\omega t) + b_3 \varepsilon^3 \cos(3\omega t), \tag{8}$$

$$\omega = \frac{1}{\sqrt{2\nu}}(1 + h_3 \varepsilon^2), \tag{9}$$

$$\dot{\phi}^2 + 4C = 4\Lambda(1 + h'_3 \varepsilon \cos(\omega t)), \tag{10}$$

$$b_3 = \frac{13}{256}, \quad h_3 = -\frac{45}{64}, \quad h'_3 = -3. \tag{11}$$

Only at this order does a time dependence of ϕ appear, thus showing the necessity of a dynamical field in addition to gravity. Note that the correction to the frequency is of order ε^2 rather than ε. A numerical analysis of the full equations shows that there really is an exact solution that is being approximated here, and such a solution exists as long as $0 < -\nu\Lambda < \frac{1}{24}$.

We see the first instance of a generic feature of oscillating metric solutions, namely that the square of the oscillation amplitude ε is proportional to the cosmological constant. But the C parameter introduces a puzzle, as can be seen by considering the leading

---

[1] The usual procedure of using equations of motion at low order in a derivative expansion to simplify the analysis at higher order does not apply here.

order ε^2 contributions to $T_{\mu\nu}$.

$$\rho = \Lambda + \frac{1}{4}\dot{\phi}^2 + C = 2\Lambda < 0 \tag{12}$$

$$p = -\Lambda + \frac{1}{4}\dot{\phi}^2 + C = 0. \tag{13}$$

Thus the required $T_{\mu\nu}$ violates various positive energy conditions [5]; in other words it is difficult to see how the required negative C can arise. Perhaps this apparent difficulty is an artifact of our truncation of the action beyond the R^2 order, but we shall not pursue this possibility here.

TIME-DEPENDENT OSCILLATIONS IN 5D

We shall now consider the addition of a compact fifth dimension, and consider a metric of the form

$$ds^2 = -dt^2 + e^{B(t)}\delta_{ij}dx^i dx^j + e^{C(t)}dy^2. \tag{14}$$

At first order in ε_t (we attach a subscript to distinguish ε_t from ε_y in the next section),

$$B(t) = \varepsilon_t \cos(\omega_t t), \quad C(t) = \varepsilon_t \eta \cos(\omega_t t). \tag{15}$$

There are two solutions to the field equations,

$$\eta = 1, \; \omega_t = \sqrt{\frac{3}{\mu}} \quad \text{and} \quad \eta = -3, \; \omega_t = \frac{1}{\sqrt{3\mu - 16\nu}}. \tag{16}$$

Now the relevant combinations of R^2 terms in the five dimensional action are

$$\mu = 16a + 5b + 4c, \tag{17}$$
$$\nu = 3a + b + c, \tag{18}$$
$$\lambda = 5a + b + \tfrac{1}{2}c. \tag{19}$$

ν corresponds to the Weyl-squared term while λ, corresponding to the Gauss-Bonnet term, appears at order ε_t^3.

The first solution is completely analogous to the previous case and it will lead to the same problem. Thus we consider the second solution and continue the expansion to order ε_t^3.

$$B(t) = \varepsilon_t \cos(\omega_t t) + b_2 \varepsilon_t^2 \cos(2\omega_t t) + b_3 \varepsilon_t^3 \cos(3\omega_t t) \tag{20}$$
$$C(t) = \varepsilon_t \eta \cos(\omega_t t) + c_2 \varepsilon_t^2 \cos(2\omega_t t) + c_3 \varepsilon_t^3 \cos(3\omega_t t) \tag{21}$$

$$\omega_t = \frac{1}{\sqrt{3\mu - 16\nu}}(1 + d_3 \varepsilon_t^2) \quad \eta = -3(1 + d'_3 \varepsilon_t^2) \tag{22}$$

$$b_2 = c_2 = -\frac{3}{4}\frac{8\nu - \mu}{48\nu - 5\mu} \tag{23}$$

$$\varepsilon_t^2 = \frac{4}{3}\Lambda(3\mu - 16\nu) \quad \frac{1}{4}\dot{\phi}^2 = \Lambda \tag{24}$$

Here Λ denotes the 5D cosmological constant and the explicit results for the constants b_3, c_3, d_3 and d_3' are given in [1]. We see that $\dot{\phi}^2$ remains constant at this order, unlike the previous case, but this is not expected to hold at higher orders. More importantly, from the results for ω_t and ε_t^2 we see that $3\mu - 16\nu$ and Λ must both be positive, and thus we get a consistent result for $\dot{\phi}^2$ without any additional *ad hoc* contribution to $T_{\mu\nu}$. The various leading contributions to $T_{\mu\nu}$ give $\rho = 2\Lambda > 0$ and $p = p_y = 0$. The welcome change of sign has occurred through the introduction of an oscillating extra dimension.

We are now led to speculate about some kind of relaxation mechanism for the cosmological constant, given that the latter is tied to the amplitude of an oscillating field. If the latter could decay away through particle production, then the cosmological constant may be taken to zero with it. But this can't happen in the present picture because we only have a 5D Λ, and it is fixed. We need some way to make an effective 4D Λ adjustable, and to have that tied to t-dependent oscillations.

Y-DEPENDENT OSCILLATIONS IN 5D

With this in mind we turn to the y-dependent oscillation, with a metric of the form

$$ds^2 = e^{A(y)}\eta_{\mu\nu}dx^\mu dx^\nu + dy^2. \tag{25}$$

We search for a solution in which $A(y)$ is periodic, in which case we can identify the y coordinate after one period to form a compact space. In this way the size of the compact space, the period, is dynamically determined, and there is no radion stabilization problem. In other words, a scale factor introduced as $dy^2 \to \kappa^2 dy^2$ is without physical meaning, since it would result in the period of $A(y)$ being scaled by κ^{-1}. The physical size of the compact space, the product of scale factor and range of y, remains the same.

With a scalar field $\phi(y)$ we consider a contribution to $T_{\mu\nu}$ of the form $\rho = -p = p_y = \frac{1}{4}\phi'^2 + C$. We again introduce a constant C as a possible Casimir effect, although its presence is not strictly necessary [1]. It will turn out that the scalar field $\phi(y)$ is also compact, with the solution determining the range over which it varies. Thus we have a dynamical generation of a compact internal space, in parallel with the dynamical generation of a compact fifth dimension. No fine-tuning of parameters in the action is required for either.

We have the following results at order ε_y^3.

$$A(y) = \varepsilon_y \cos(\omega_y y) + a_2 \varepsilon_y^2 \cos(2\omega_y y) + a_3 \varepsilon_y^3 \cos(3\omega_y y) \tag{26}$$

$$\omega_y = \sqrt{\frac{-3}{\mu}}(1 + \bar{d}_3 \varepsilon_y^2) \tag{27}$$

$$\phi'^2 + 4C = -4\Lambda(1 + \bar{d}_3' \varepsilon_y \cos(\omega_y y)) \tag{28}$$

$$a_2 = -\frac{1}{4} \quad a_3 = \frac{1}{144}\frac{13\mu - \lambda}{\mu} \quad \bar{d}_3 = \frac{1}{4}\frac{\lambda - 5\mu}{\mu} \quad \bar{d}_3' = -4 \tag{29}$$

$$\varepsilon_y^2 = -\frac{4}{9}\Lambda\mu. \tag{30}$$

The expressions for ω_y and ε_y^2 indicate that μ must be negative and Λ must be positive. The leading contributions to $T_{\mu\nu}$ give $\rho = p = 0$, $p_y = -2\Lambda$. Thus we start with a positive 5D Λ, but the effective 4D Λ_{eff} vanishes.

Not surprisingly, it is also true that there are other solutions in which Λ_{eff} is nonvanishing.[2] In other words, we now have a situation in which Λ_{eff} is adjustable, by choosing solutions with different warpings of the fifth dimension. The problem here is that we don't know why or how the system would relax to the vanishing Λ_{eff} solution.

Y AND T-DEPENDENT OSCILLATIONS IN 5D

Now we are clearly motivated to combine the two previous cases for t and y oscillations, which both required a positive 5D Λ. The hope is that the adjustable Λ_{eff} can be tied to the amplitude of t-dependent oscillations, thus providing a relaxation mechanism. The metric at first order in ε_y and ε_t takes the form

$$ds^2 = -e^{A(y)}\,dt^2 + e^{A(y)+B(t)}\,\delta_{ij}\,dx^i dx^j + e^{C(t)}\,dy^2, \tag{31}$$

$$A(y) = \varepsilon_y \cos(\omega_y y) \quad B(t) = \varepsilon_t \cos(\omega_t t) \quad C(t) = -3\varepsilon_t \cos(\omega_t t), \tag{32}$$

where we have already seen that

$$\omega_y = \sqrt{\frac{-3}{\mu}} \quad \text{and} \quad \omega_t = \frac{1}{\sqrt{3\mu - 16\nu}}. \tag{33}$$

The range of y is determined by ω_y, but the size of the compact space oscillates in time with frequency ω_t.

At next order in ε_t and ε_y the nonlinear field equations induce a mixing between the t and y dependent terms, and thus the metric must take a more complicated form,

$$ds^2 = -e^{A(y)+E(y,t)}\,dt^2 + e^{A(y)+B(t)+F(y,t)}\,\delta_{ij}\,dx^i dx^j + e^{C(t)+G(y,t)}\,dy^2. \tag{34}$$

At second order we find

$$\begin{aligned}
A(y) &= \varepsilon_y \cos(\omega_y y) + a_2 \varepsilon_y^2 \cos(2\omega_y y) & (35)\\
B(t) &= \varepsilon_t \cos(\omega_t t) + b_2 \varepsilon_t^2 \cos(2\omega_t t) & (36)\\
C(t) &= -3\varepsilon_t \cos(\omega_t t) + c_2 \varepsilon_t^2 \cos(2\omega_t t) & (37)\\
E(y,t) &= e_2 \varepsilon_y \varepsilon_t \cos(\omega_y y)\cos(\omega_t t) & (38)\\
F(y,t) &= f_2 \varepsilon_y \varepsilon_t \cos(\omega_y y)\cos(\omega_t t) & (39)\\
G(y,t) &= g_2 \varepsilon_y \varepsilon_t \cos(\omega_y y)\cos(\omega_t t). & (40)
\end{aligned}$$

---

[2] We have numerically studied exact solutions of this type [4].

a_2, b_2, and c_2 are as before, e_2, f_2 and g_2 are given in [1]. We find that

$$\varepsilon_y^2 = \frac{4}{3}\frac{\Lambda}{\omega_y^2} - \frac{\omega_t^2}{\omega_y^2}\varepsilon_t^2, \tag{41}$$

$$\phi'^2 + 4C = -4\Lambda + 3\omega_t^2\varepsilon_t^2, \tag{42}$$

$$\dot{\phi}^2 = 3\omega_t^2\varepsilon_t^2, \tag{43}$$

where ω_y and ω_t are the first order values in (33).

The main point here is that solutions exist for a range of ε_t. For each such solution there is a positive effective 4D cosmological constant, which we now denote by Λ_{osc} to emphasize that it is driving t-dependent oscillations.

$$\Lambda_{\text{osc}} = \Lambda + C + \frac{1}{4}\phi'^2 = \frac{3}{4}\omega_t^2\varepsilon_t^2 \tag{44}$$

We have seen such a relation before, but now Λ_{osc} parameterizes a family of solutions rather than being a fundamental constant. The resulting $T_{\mu\nu}$ is

$$\rho = \Lambda_{\text{osc}} + \frac{1}{4}\dot{\phi}^2 = 2\Lambda_{\text{osc}} > 0, \tag{45}$$

$$p = -\Lambda_{\text{osc}} + \frac{1}{4}\dot{\phi}^2 = 0. \tag{46}$$

The 4D Λ_{osc} appears here rather than the much larger 5D Λ; the latter drives the warping of the fifth dimension, and it appears in $p_y = -2\Lambda + 2\Lambda_{\text{osc}}$. The positive Λ_{osc} and the associated metric oscillations are susceptible to decay through conversion to energetic particles.

ACKNOWLEDGMENTS

This research was done in collaboration with Hael Collins and was supported in part by the Natural Sciences and Engineering Research Council of Canada.

REFERENCES

1. H. Collins and B. Holdom, hep-th/0107042.
2. H. Collins and B. Holdom, Phys. Rev. D **63**, 084020 (2001) [hep-th/0009127].
3. S. Weinberg, Rev. Mod. Phys. **61**, 1 (1989).
4. H. Collins and B. Holdom, hep-ph/0103103.
5. R. M. Wald, *General Relativity*, Chicago University Press, Chicago (1984).
6. H. Collins and B. Holdom, work in progress.

Hiding a cosmological constant in a warped extra dimension

Hael Collins

Department of Physics, University of Toronto, Toronto, Ontario M5S 1A7, Canada

Abstract. We present a scenario in which extra dimensions are used to address the cosmological constant problem. For a theory of gravity in $4+1$ dimensions whose dynamics are governed by an effective action that includes quadratic terms in the curvature and a compact scalar field, the field equations admit solutions that are compact in one direction and Poincaré invariant in the remaining directions. These solutions do not require any fine-tuning of the parameters in the action—including the cosmological constant—only that they should satisfy some mild inequalities. We further discuss several features of this picture, including an example of a metric that localizes gravity to a hypersurface without including a brane as well as how to combine this approach with the Randall-Sundrum model.[1]

The old idea that the universe might contain more than the observed four space-time dimensions has re-emerged recently in novel attempts to explain the weakness of gravity compared to the other forces [1] and the hierarchy problem [2], but it was realized earlier [3] that such theories might be able to address the cosmological constant problem [4]. The hope is that with extra dimensions, the metric might be able, through a non-trivial dependence on the extra coordinates, both to accommodate an arbitrary value for the cosmological constant and to maintain Poincaré invariance in $3+1$ of the directions. If this warping is accomplished with a metric that is both smooth and periodic in the extra dimensions, the period provides a natural compactifaction size for the extra dimensions.

An explicit realization of this idea occurs in $4+1$ dimensions [5] for an action with a generic set of curvature invariants with up to four derivatives of the metric and a compact scalar field. The 4D cosmological constant is determined by both the 5D cosmological constant and the geometry of the extra dimension. Therefore, we can achieve $3+1$ dimensional Poincaré invariance even when the 5D cosmological constant is not zero by choosing the solution to the field equations with the appropriate behavior in the extra dimension. Yet while no fine-tuning of the action is required, some further mechanism is still required to explain why this particular solution should be preferred.

This approach can also be adapted to eliminate the fine-tuning present in the Randall-Sundrum scenario [6]. Starting with an effective action for gravity in six dimensions as well as two parallel 4-branes, with some mild bounds on the parameters in the action, upon integrating out the compact sixth dimension we recover the action originally considered by Randall and Sundrum [2].

---

[1] This work was done in collaboration with Bob Holdom and was supported in part by the Natural Sciences and Engineering Research Council of Canada.

A WARPED KALUZA-KLEIN MODEL

At energies approaching the Planck scale, corrections to the standard Einstein-Hilbert action can play an important role in determining the geometry of space-time. Treating gravity as an effective theory by expanded in powers of derivatives, an action that includes a generic set of terms with up to four derivatives of the metric is

$$S_{\text{gravity}} = M_5^3 \int d^4x dy \sqrt{-g} \left(-2\Lambda + R - aR^2 - bR_{ab}R^{ab} - cR_{abcd}R^{abcd} + \cdots \right), \quad (1)$$

We also include a scalar field whose dynamics are determined by[2]

$$S_\phi = M_5^3 \int d^4x dy \sqrt{-g} \left(-\tfrac{1}{2}\nabla_a\phi\nabla^a\phi - \tfrac{1}{4}k(\nabla_a\phi\nabla^a\phi)^2 + \cdots \right). \quad (2)$$

Here Λ and M_5 are respectively the cosmological constant and the five dimensional Planck constant. g_{ab} is the metric for the space-time. We denote the coordinates that correspond to the usual space-time dimensions by x^μ, where $\mu, \nu, \cdots = 0, 1, 2, 3$, and the fifth coordinate by y, with $a, b, c, \ldots = 0, 1, 2, 3, y$. We shall often set $M_5 = 1$.

To produce a universe that resembles a flat, $3+1$ dimensional universe at lengths scales that have been observed, we consider a space-time metric with a warped Kaluza-Klein form,

$$ds^2 = g_{ab} dx^a dx^b = e^{A(y)} \eta_{\mu\nu} dx^\mu dx^\nu + dy^2. \quad (3)$$

Since the extra dimension is to be small, we search for solutions in which $A(y)$ is smooth, periodic and non-singular. Unlike the usual Kaluza-Klein compactification, the metric depends strongly on the fifth coordinate y. We find [5]–[7] that the parameters of the higher-derivative terms in (1) determine a unique period for the extra dimension and this picture, provided such periodic functions $A(y)$ exist, does not suffer from any radius stabilization problem. The metric (3) is conformally flat so we can parameterize the effects of the R^2 terms by

$$\mu \equiv 16a + 5b + 4c \qquad \lambda \equiv 5a + b + \tfrac{1}{2}c. \quad (4)$$

The parameter λ, in particular, represents the coefficient of the Gauss-Bonnet term.

Varying the action, we obtain for a warped Kaluza-Klein geometry

$$3(A')^2 + 3A'' + \lambda \left[2A''(A')^2 + (A')^4 \right] \quad (5)$$
$$+ \mu \left[A'''' + 4A'A''' + 3(A'')^2 + 4A''(A')^2 \right] = -2\Lambda - \tfrac{1}{2}(\phi')^2 - \tfrac{1}{4}k(\phi')^4$$
$$3(A')^2 + \lambda(A')^4 + \mu \left[2A'A''' - (A'')^2 + 4A''(A')^2 \right] = -2\Lambda + \tfrac{1}{2}(\phi')^2 + \tfrac{3}{4}k(\phi')^4.$$

Here the prime denotes a y-derivative. We can solve for

$$(\phi'(y))^2 = (3k)^{-1}\Big[-1 \pm \{ 1 + 12k \left[2\Lambda + 3(A')^2 + \lambda(A')^4 \right] \quad (6)$$
$$+ 12k\mu \left(2A'A''' - (A'')^2 + 4A''(A')^2 \right) \}^{1/2} \Big]$$

---

[2] This choice for the scalar action is not necessary for the existence of periodic solutions. We also found periodic metrics when we include a free scalar field and the effects of asymmetric Casimir effect [6].

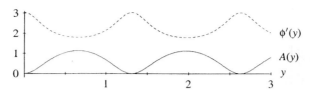

FIGURE 1. A periodic warp function $A(y)$ (solid line) and $\phi'(y)$ (dashed line) for $\Lambda = 1, \lambda = 0, \mu = 0.1$, and $k = -0.25$. The initial condition is $A''(0) = 23.77364592$.

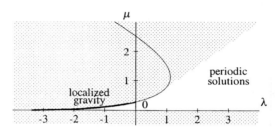

FIGURE 2. A plot of the parameter space $\{\lambda, \mu\}$ when $\Lambda = 1$. In the unshaded region, we have found periodic numerical solutions for $A(y)$ for arbitrarily chosen points. The curve depicts a set exact solutions discussed in [5] while the darker part of the curve shows the location of the solutions in (7).

and substitute the result into (5) to obtain a differential equation for $A(y)$.

A periodic solution for a generic set of values of Λ, μ, λ, and k is found by numerically integrating the resulting differential equation for $A(y)$. The coordinate y does not explicitly appear in the equations (5)–(6) which moreover only depend on the warp function $A(y)$ through its derivatives. Thus, we can choose $A(0) = A'(0) = 0$ without any loss of generality. We also chose $A'''(0) = 0$ for simplicity. The subsequent evolution of the warp function away from $y = 0$ then depended solely upon the initial value of the second derivative, $A''(0)$.

Numerically we find that there exists a precise value of $A''(0)$ that produces a periodic solution for each arbitrarily chosen set of parameters $\{\Lambda, \mu, \lambda, k\}$ within a region of the parameter space with a non-zero volume. This result demonstrates the existence of periodic solutions without finely tuning any of the parameters in the action. An example of such a solutions has been sketched in Fig. 1 for $\Lambda = 1, \lambda = 0, \mu = 0.1$, and $k = -0.25$ and choosing the minus root in (6). Many further examples appear in [5] and [6].

A slice of the region in parameter space for which periodic metrics exist is shown in Fig. 2, with $\Lambda = 1$ and $k \to 0$. The choice of the former is always possible by the rescaling, $y \to \sigma y$, $\Lambda \to \sigma^{-2}\Lambda$, $\mu \to \sigma^2\mu$, $\lambda \to \sigma^2\lambda$ and $k \to \sigma^2 k$, where σ is a real constant, which leaves (5) unchanged. We have found numerical solutions for arbitrarily chosen points throughout the unshaded portion of Fig. 2.

Using gravity to localize gravity

The purely gravitational part of the action (1) can also generate a warp function that is localized along a 3 + 1 dimensional hypersurface. The profile of the function $A(y)$ for these solutions superficially resembles that appearing in models in which a domain wall is used to localize gravity,

$$A(y) = -\frac{2}{\kappa l} \ln[2\cosh(\kappa y)]. \qquad [\phi(y) = 0] \qquad (7)$$

Here the width, κ^{-1}, and the asymptotic AdS$_5$ length, l, are respectively

$$\kappa = \left(\frac{3 - 4\sqrt{2\Lambda\mu}}{2\mu}\right)^{1/2} \qquad l = \left(\frac{3 - 4\sqrt{2\Lambda\mu}}{\Lambda}\right)^{1/2}, \qquad (8)$$

with $0 \leq \Lambda\mu \leq \frac{9}{32}$, $\Lambda > 0$ and $\mu \geq 0$. In this configuration the R^2 terms are in no sense negligible. This metric does require one fine-tuning among Λ, λ and μ, given in [5] by

$$\Lambda\lambda = -\left(3 - 4\sqrt{2\Lambda\mu}\right)\left(\tfrac{9}{8} - \tfrac{1}{2}\sqrt{2\Lambda\mu}\right), \qquad (9)$$

which can presumably be effected by adding a sixth dimension with an appropriately warped compactification.

THE RANDALL-SUNDRUM SCENARIO AS AN EFFECTIVE THEORY

Randall and Sundrum [2] proposed that if the universe were to consist of two 3-branes bounding a bulk region of five dimensional anti-de Sitter (AdS$_5$) space-time,

$$ds^2 = g_{ab}dx^a dx^b = e^{-2|r|/l}\eta_{\mu\nu}dx^\mu dx^\nu + dr^2, \qquad (10)$$

then the redshift induced by the bulk metric at one of the branes could generate an exponential hierarchy between the Planck scale and the scale of electroweak symmetry breaking. The bulk Einstein equations determine $\Lambda = -6l^{-2}$ and the specific choice of $\sigma = \pm 6l^{-1}$ for the brane tensions is necessary for the low energy four dimensional theory to be free of a cosmological constant.

As the cosmological constant and the surface tension appear in the action in [2], they represent fundamental parameters of the theory and we have no reason *a priori* that the fine-tuning condition is satisfied. If instead these quantities arise from some more fundamental theory, then it might be possible for a dynamical mechanism to exist that favors solutions in which the low energy, four dimensional theory is nearly flat.

We can adapt the picture developed above without branes to one which resembles the Randall-Sundrum construction but where the AdS$_5$ length, l, is not uniquely determined by the higher dimensional cosmological constant. The structure for such a model would include *two* extra dimensions—one small periodic dimension to avoid fine-tuning the

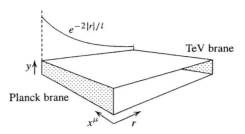

FIGURE 3. The geometry of a six dimensional model with two 4-branes. The small periodic coordinate is y. The direction orthogonal to the 4-branes, r, becomes the extra coordinate of the Randall-Sundrum model when we integrate out the y dimension. The model assumes an orbifold geometry about $r = 0$.

cosmological constant as before and a second to generate the electroweak-Planck hierarchy, as shown in Fig. 3.

By generalizing the four derivative action of (1)–(2) to six dimensions and adding two 4-branes at $r = 0$ and $r = r_c$, we shall show that after integrating out the y-dependence, we can recover the action of the Randall-Sundrum scenario [2]. The important new feature is that the 6D cosmological constant, $\tilde{\Lambda}$, no longer needs to be finely tuned with respect to the tensions on the branes, which we write as $\tilde{\sigma}^{(0)}$ and $\tilde{\sigma}^{(r_c)}$ respectively.

We begin with a metric of the form

$$ds_6^2 = \tilde{g}_{MN}(x^\lambda, r, y)\, dx^M dx^N = e^{A(y)} g_{ab}(x^\lambda, r)\, dx^a dx^b + dy^2 \qquad (11)$$

with the AdS$_5$ metric (10) for the (x^λ, r)-subspace. As in the 5D case earlier, when $A(y)$ is a periodic function of y, we can obtain a compact extra dimension with a very nontrivial y-dependence without any singularities. However, unlike the previous example, the (x^λ, r) subspace is not flat. The shape of $A(y)$ determines the effective cosmological constant of the g_{ab} metric. The metric (11) with an AdS$_5$ subspace is still conformally flat, but the definition of the coefficients of the linear combinations of $\tilde{a}$, $\tilde{b}$ and $\tilde{c}$ in the 6D version of (1) orthogonal to the Weyl squared term depend on the number of dimensions. In $5+1$ dimensions we have $\tilde{\mu} \equiv 20\tilde{a} + 6\tilde{b} + 4\tilde{c}$ and $\tilde{\lambda} \equiv 15\tilde{a} + \tfrac{5}{2}\tilde{b} + \tilde{c}$.

The 6D curvature tensors $\tilde{R}$, $\tilde{R}_{MN}$ and $\tilde{R}^L{}_{MNP}$ are related to their counterparts, R, R_{ab} and $R^a{}_{bcd}$ derived from the 5D metric g_{ab} in (11) and derivatives of the warp function. We thus derive a 5D effective action by integrating out the small y dimension in this background,

$$\begin{aligned} S_{\text{eff}} =\ & M_5^3 \int d^4x\, dr\, \sqrt{-g}\left(-2\Lambda + R - aR^2 - bR_{ab}R^{ab} - cR_{abcd}R^{abcd}\right) \\ & + M_5^3 \int_{r=0} d^4x\, \sqrt{-h}\left[-2\sigma^{(0)}\right] + M_5^3 \int_{r=r_c} d^4x\, \sqrt{-h}\left[-2\sigma^{(r_c)}\right] + \cdots. \quad (12)\end{aligned}$$

h_{ab} represents the metric induced on the branes by the metric g_{ab}. The new parameters that appear in this effective action depend partially upon the "fundamental" parameters of the original action but also upon the behavior of the warp function. Explicitly, the parameters in the low energy 5D theory are (M_6 is the 6D Planck mass)

$$M_5^3 \Lambda = M_6^4 \int_0^{y_c} dy\, e^{\frac{5}{2}A(y)} \left[\tilde{\Lambda} + \tfrac{1}{4}(\phi')^2 + \tfrac{1}{8}k(\phi')^4 - \tfrac{5}{2}(A')^2 + \tfrac{5}{8}\tilde{\mu}(A'')^2 - \tfrac{5}{24}\tilde{\lambda}(A')^4\right]$$

$$M_5^3 = M_6^4 \int_0^{y_c} dy\, e^{\frac{3}{2}A(y)} \left[1 + \tfrac{1}{8}(3\tilde{\mu} - 4\tilde{\lambda})(A')^2\right] \qquad (13)$$

$$M_5^3 a = M_6^4 \tilde{a} \int_0^{y_c} dy\, e^{\frac{1}{2}A(y)}$$

$$M_5^3 \sigma^{(0)} = M_6^4 \tilde{\sigma}^{(0)} \int_0^{y_c} dy\, e^{2A(y)}.$$

In the weak 5D gravity limit, $M_5 l \gg 1$, the R^2 terms become negligible and the leading behavior is governed by the Einstein-Hilbert terms in (12). Since $\Lambda \sim l^{-2}$, we require the effective 5D cosmological constant to be small which can easily occur when the contribution from the bulk cosmological constant is partially cancelled by effects from the warp function in (13). Thus, we can recover the Randall-Sundrum action.

The fine-tunings of the effective tensions on the two branes in [2] are

$$\sqrt{-6\Lambda} = \sigma^{(0)} = -\sigma^{(r_c)}. \qquad (14)$$

Numerically, we find solutions [6] periodic in the y-direction provided that the effective cosmological constant is of the same order or smaller than the full cosmological constant, $|\Lambda| \lesssim O(\tilde{\Lambda})$. Using the desired value of the 5D Λ from (14) and applying (13), we can thus find solutions that are periodic in y and satisfy (14) without finely tuning any of the fundamental parameters when $(\tilde{\sigma}^{(0)})^2 \lesssim O(\tilde{\Lambda})$.

CONCLUDING REMARKS

A theory with an extra compact dimension and an action with a generic set of R^2 terms and a compact scalar field contains sufficient freedom to admit periodic metrics with a $3+1$ dimensional Poincaré invariance without the need for finely tuning the action. Yet for each choice of parameters that allows such a solution, a family of other periodic solutions exists whose elements are specified by the value of the effective low energy 4D cosmological constant. We should further investigate whether a universe in a generic initial state can relax into one in which the effective 4D theory is nearly flat.

REFERENCES

1. N. Arkani-Hamed, S. Dimopoulos and G. Dvali, Phys. Lett. **B429**, 263 (1998) and I. Antoniadis, N. Arkani-Hamed, S. Dimopoulos and G. Dvali, Phys. Lett. **B436**, 257 (1998).
2. L. Randall and R. Sundrum, Phys. Rev. Lett. **83**, 3370 (1999).
3. V. A. Rubakov and M. E. Shaposhnikov, Phys. Lett. **B125**, 139 (1983).
4. S. Weinberg, Rev. Mod. Phys. **61**, 1 (1989).
5. H. Collins and B. Holdom, Phys. Rev. D **63**, 084020 (2001).
6. H. Collins and B. Holdom, *to appear in* Phys. Rev. D **64**, (2001) [hep-ph/0103103].
7. H. Collins and B. Holdom, hep-th/0107042.

5-Dimensional Warped Cosmological Solutions With Radius Stabilization by a Bulk Scalar

James M. Cline* and Hassan Firouzjahi*

Department of Physics, McGill University, Montreal, QC, Canada H3A 2T8

Abstract. We present the 5-dimensional cosmological solutions in the Randall-Sundrum warped compactification scenario, using the Goldberger-Wise mechanism to stabilize the size of the extra dimension. Matter on the Planck and TeV branes is treated perturbatively, to first order. The backreaction of the scalar field on the metric is taken into account. We identify the appropriate gauge-invariant degrees of freedom, and show that the perturbations in the bulk scalar can be gauged away. We confirm previous, less exact computations of the shift in the radius of the extra dimension induced by matter. We point out that the physical mass scales on the TeV brane may have changed significantly since the electroweak epoch due to cosmological expansion, independently of the details of radius stabilization.

INTRODUCTION

The Randall-Sundrum (RS) idea [1] for explaining the weak-scale hierarchy problem has garnered much attention from both the phenomenology and string-theory communities, providing a link between the two which is often absent. RS is a simple and elegant way of generating the TeV scale which characterizes the standard model from a set of fundamental scales which are of order the Planck mass (M_p). All that is needed is that the distance between a hidden and a visible sector brane be approximately $b = 35/M_p$ in a compact extra dimension, $y \in [0, 1]$. The warping of space in this extra dimension, by a factor e^{-kby}, translates the moderately large interbrane separation into the large hierarchy needed to explain the ratio TeV/M_p.

However the RS idea as originally proposed was incomplete due to the lack of any mechanism for stabilizing the brane separation, b. This was a modulus, corresponding to a massless particle, the radion, which would be ruled out because of its modification of gravity: the attractive force mediated by the radion would effectively increase Newton's constant at large distance scales. An attractive model for giving the radion a potential energy was proposed by Goldberger and Wise (GW) [2]; they introduced a bulk scalar field with different VEV's, v_0 and v_1, on the two branes. If the mass m of the scalar is small compared to the scale k which appears in the warp factor e^{-kby}, then it is possible to obtain the desired interbrane separation. One finds the relation $e^{-kb} \cong (v_1/v_0)^{4k^2/m^2}$.

An important benefit of stabilizing the radion is that cosmology is governed by the usual Friedmann equations, up to small corrections of order $\rho/(\text{TeV})^4$ [3]. Even with stabilization, there may be a problem with reaching a false minimum of the GW radion potential [4], but without stabilization, there is a worse problem: an unnatural tuning of the energy densities on the two branes is required for getting solutions where the extra dimension is static [5, 6], a result which can be derived using the (5,5) component of the Einstein equation $G_{mn} = \kappa^2 T_{mn}$. However when there is a nontrivial potential for the radius, $V(b)$, the (5,5) equation serves only to determine the shift δb in the radius due to the expansion, and there is no longer any constraint on the matter on the branes. Although this point is now well appreciated [7]-[10], it has not previously been explicitly demonstrated by solving the full 5-dimensional field equations using a concrete stabilization mechanism. Indeed, it has been claimed recently that such solutions are not possible with an arbitrary equation of state for the matter on the branes [12]-[13], and also that the rate of expansion does not reproduce normal cosmology on the negative tension brane despite stabilization [14]. Our purpose is to present the complete solutions, to leading order in an expansion in the energy densities on the branes, thus refuting these claims.

PRELIMINARIES

The action for 5-D gravity coupled to the stabilizing scalar field Φ and matter on the branes (located at $y = 0$ and $y = 1$, respectively) is

$$S = \int d^5 x \sqrt{g} \left(-\frac{1}{2\kappa^2} R - \Lambda + \tfrac{1}{2} \partial_\mu \Phi \partial^\mu \Phi - V(\Phi) \right)$$
$$+ \int d^4 x \sqrt{\tilde{g}} (\mathcal{L}_{m,0} - V_0(\Phi))|_{y=0} + \int d^4 x \sqrt{\tilde{g}} (\mathcal{L}_{m,1} - V_1(\Phi))|_{y=1}, \quad (1)$$

where κ^2 is related to the 5-D Planck scale M by $\kappa^2 = 1/(M^3)$. The negative bulk cosmological constant needed for the RS solution is parametrized as $\Lambda = -6k^2/\kappa^2$ and the scalar field potential is that of a free field, $V(\Phi) = \tfrac{1}{2} m^2 \Phi^2$. The brane potentials V_0 and V_1 can have any form that will insure nontrivial VEV's for the scalar field at the branes, for example $V_i(\Phi) = \lambda_i (\Phi^2 - v_i^2)^2$ [2]. In ref. [4] we pointed out that the choice $V_i(\Phi) = m_i (\Phi - v_i)^2$ is advantageous from the point of view of analytic calculability (see also [15]).

We will take the metric to have the form

$$ds^2 = n^2(t,y) dt^2 - a^2(t,y) \sum_i dx_i^2 - b^2(t,y) dy^2$$
$$= e^{-2N(t,y)} dt^2 - a_0(t)^2 e^{-2A(t,y)} \sum_i dx_i^2 - b(t,y)^2 dy^2, \quad (2)$$

where a perturbative expansion in the energy densities of the branes will be made around the static solution:

$$N(t,y) = A_0(y) + \delta N(t,y); \qquad A(t,y) = A_0(y) + \delta A(t,y)$$
$$b(t,y) = b_0 + \delta b(t,y); \qquad \Phi(t,y) = \Phi_0(y) + \delta \Phi(t,y). \quad (3)$$

The perturbations are taken to be linear in the energy densities ρ_* and ρ of matter on the Planck and TeV branes, located at $y = 0$ and $y = 1$, respectively.

This ansatz is to be substituted into the Einstein equations, $G_{mn} = \kappa^2 T_{mn}$, and the scalar field equation

$$\partial_t \left(\frac{1}{n}ba^3\dot{\Phi}\right) - \partial_y \left(\frac{1}{b}a^3 n\Phi'\right) + ba^3 n\left[V' + V_0'\delta(by) + V_1'\delta(b(y-1))\right] = 0. \quad (4)$$

Here and in the following, primes on functions of y denote $\frac{\partial}{\partial y}$, while primes on potentials of Φ will mean $\frac{\partial}{\partial \Phi}$. The nonvanishing components of the Einstein tensor are

$$\begin{aligned}
G_{00} &= 3\left[(\frac{\dot{a}}{a})^2 + \frac{\dot{a}\dot{b}}{ab} - \frac{n^2}{b^2}\left(\frac{a''}{a} + (\frac{a'}{a})^2 - \frac{a'b'}{ab}\right)\right] \\
G_{ii} &= \frac{a^2}{b^2}\left[(\frac{a'}{a})^2 + 2\frac{a'n'}{an} - \frac{b'n'}{bn} - 2\frac{b'a'}{ba} + 2\frac{a''}{a} + \frac{n''}{n}\right] \\
&+ \frac{a^2}{n^2}\left[-(\frac{\dot{a}}{a})^2 + 2\frac{\dot{a}\dot{n}}{an} - 2\frac{\ddot{a}}{a} + \frac{\dot{b}}{b}(-2\frac{\dot{a}}{a} + \frac{\dot{n}}{n}) - \frac{\ddot{b}}{b}\right] \\
G_{05} &= 3\left[\frac{n'\dot{a}}{na} + \frac{a'\dot{b}}{ab} - \frac{\dot{a}'}{a}\right] \\
G_{55} &= 3\left[\frac{a'}{a}\left(\frac{a'}{a} + \frac{n'}{n}\right) - \frac{b^2}{n^2}\left(\frac{\dot{a}}{a}\left(\frac{\dot{a}}{a} - \frac{\dot{n}}{n}\right) + \frac{\ddot{a}}{a}\right)\right] \quad (5)
\end{aligned}$$

and the stress energy tensor is $T_{mn} = g_{mn}(V(\Phi) + \Lambda) + \partial_m\Phi\partial_n\Phi - \frac{1}{2}\partial^l\Phi\partial_l\Phi g_{mn}$ in the bulk. On the branes, T_m^n is given by

$$\begin{aligned}
T_m^n &= \delta(by)\,\text{diag}(V_0 + \rho_*, V_0 - p_*, V_0 - p_*, V_0 - p_*, 0) \\
&+ \delta(b(y-1))\,\text{diag}(V_1 + \rho, V_1 - p, V_1 - p, V_1 - p, 0) \quad (6)
\end{aligned}$$

At zeroth order in the perturbations, the approximate solutions are

$$\Phi_0(y) \cong v_0 e^{-\varepsilon k b_0 y}; \qquad A_0(y) \cong k b_0 y + \frac{\kappa^2}{12}v_0^2(e^{-2\varepsilon k b_0 y} - 1) \quad (7)$$

where we have normalized $A_0(0) = 0$, and introduced $\varepsilon = \sqrt{4 + \frac{m^2}{k^2}} - 2 \cong \frac{m^2}{4k^2}$. For small ε, the GW solution coincides with an exact solution of the coupled equations that was presented in ref. [15].

PERTURBATION EQUATIONS

We can now write the equations for the perturbations of the metric, δA, δN, δb, and the scalar field, $\delta \Phi$. The equations take a simpler form when expressed in terms of the following combinations:

$$\Psi = \delta A' - A_0'\frac{\delta b}{b_0} - \frac{\kappa^2}{3}\Phi_0'\delta\Phi; \qquad \Upsilon = \delta N' - \delta A' \quad (8)$$

Further simplification comes from realizing that the perturbations will have the form, for example, $\Psi = \rho_*(t)g_0(y) + \rho(t)g_1(y)$, so that their time derivatives are proportional to $\dot\rho$ and $\dot\rho_*$. Below we will confirm that $\dot\rho = -3H(\rho+p)$, where $H \sim \sqrt{\rho}, \sqrt{\rho_*}$ is the Hubble parameter. Therefore time derivatives of the perturbations are higher order in ρ and ρ_* than are y derivatives, and can be neglected at leading order (except in the (05) Einstein equation, where $\rho^{3/2}$ *is* the leading order). Using this approximation, we can write the combinations (00), (00)−(ii), (05) and (55) of the Einstein equations as

$$4A_0'\Psi - \Psi' = \left(\frac{\dot a_0}{a_0}\right)^2 b_0^2 e^{2A_0} \quad (9)$$

$$-4A_0'\Upsilon + \Upsilon' = 2\left(\left(\frac{\dot a_0}{a_0}\right)^2 - \frac{\ddot a_0}{a_0}\right) b_0^2 e^{2A_0} \quad (10)$$

$$-\frac{\dot a_0}{a_0}\Upsilon + \dot\Psi = 0 \quad (11)$$

$$A_0'(4\Psi + \Upsilon) + \frac{\kappa^2}{3}\left(\Phi_0''\delta\Phi - \Phi_0'\delta\Phi' + \Phi_0'^2\frac{\delta b}{b_0}\right) = \left(\left(\frac{\dot a_0}{a_0}\right)^2 + \frac{\ddot a_0}{a_0}\right) b_0^2 e^{2A_0} \quad (12)$$

In addition, there is the scalar field equation,

$$\delta\Phi'' = (4\Psi+\Upsilon)\Phi_0' + \left(\frac{4\kappa^2}{3}\Phi_0'^2 + b_0^2 V''(\Phi_0)\right)\delta\Phi + 4A_0'\delta\Phi'$$
$$+ \left(2b_0^2 V'(\Phi_0) + 4A_0'\Phi_0'\right)\frac{\delta b}{b_0} + \Phi_0'\frac{\delta b'}{b_0} \quad (13)$$

Assuming Z_2 symmetry (all functions symmetric under $y \to -y$), the boundary conditions implied by the delta function sources at the branes are

$$\Psi(t,0) = +\frac{\kappa^2}{6}b_0\rho_*(t); \qquad \Psi(t,1) = -\frac{\kappa^2}{6}b_0\rho(t) \quad (14)$$

$$\Upsilon(t,0) = -\frac{\kappa^2}{2}b_0(\rho_*+p_*)(t); \qquad \Upsilon(t,1) = +\frac{\kappa^2}{2}b_0(\rho+p)(t) \quad (15)$$

$$\delta\Phi'(t,y) = \frac{\delta b(t,y)}{b_0}\Phi_0'(t,y) + (-1)^y\left(\frac{b}{2}\right)V_y''(\Phi_0(t,y)); \qquad y=0,1 \quad (16)$$

SOLUTIONS

Naively, it would appear that we have five equations for four unknown perturbations, but of course since gravity is a gauge theory, this is not the case. First, we have the relation $\frac{\partial}{\partial t}$[Eq. 9] $+ \frac{\dot a_0}{a_0}$[Eq. 10] = [Eq. 11]. Furthermore, the (55) Einstein equation and the scalar equation can be shown to be equivalent: [Eq. (12)]$' - 4A_0' \times$[Eq. (12)] $= \Phi_0' \times$[Eq. (13)]. So our system is actually underdetermined because of unfixed gauge degrees of freedom. To see this more directly, consider an infinitesimal diffeomorphism which leaves the coordinate positions of the branes unchanged: $y = \bar{y} + f(\bar{y})$, where $f(0) = f(1) = 0$. The metric and scalar perturbations transform as

$$\begin{aligned} \delta A &\to \delta A + A'_0 f; & \delta N &\to \delta N + A'_0 f \\ \delta b &\to \delta b + b_0 f'; & \delta \Phi &\to \delta \Phi + \Phi'_0 f \end{aligned} \quad (17)$$

If desired, one can form the gauge invariant combinations

$$\delta A' - A'_0 \frac{\delta b}{b_0} - \frac{\kappa^2}{3} \Phi'_0 \delta \Phi; \qquad \delta N' - \delta A'; \qquad \Phi''_0 \delta \Phi - \Phi'_0 \delta \Phi' + \Phi'^2_0 \frac{\delta b}{b_0} \quad (18)$$

the first two are precisely our variables Ψ and Υ and the last one appears in (55) equation. In terms of these gauge invariant variables, the system of equations closes.

It is now easy to verify the following solution from the (00) and (00)-(ii) equations, i.e., eqs. (9-10). Denoting the warp factor $\Omega = e^{-A_0(1)}$, we find

$$\Psi = \frac{\kappa^2 b_0}{6(1-\Omega^2)} e^{4A_0(y)} \left(F(y)(\Omega^4 \rho + \rho_*) - (\Omega^4 \rho + \Omega^2 \rho_*) \right) \quad (19)$$

$$\begin{aligned} \Upsilon &= \frac{\kappa^2 b_0}{2(1-\Omega^2)} e^{4A_0(y)} \left[-F(y)(\Omega^4(\rho + p) + \rho_* + p_*) \right. \\ &\quad \left. + (\Omega^4(\rho + p) + \Omega^2(\rho_* + p_*)) \right] \end{aligned} \quad (20)$$

where

$$F(y) = 1 - (1-\Omega^2) \frac{\int_0^y e^{-2A_0} dy}{\int_0^1 e^{-2A_0} dy} \cong e^{-2kb_0 y} \quad (21)$$

and the Friedmann equations are

$$\left(\frac{\dot{a}_0}{a_0} \right)^2 = \frac{8\pi G}{3} (\rho_* + \Omega^4 \rho) \quad (22)$$

$$\left(\frac{\dot{a}_0}{a_0} \right)^2 - \frac{\ddot{a}_0}{a_0} = 4\pi G \left(\rho_* + p_* + \Omega^4(\rho + p) \right) \quad (23)$$

$$8\pi G = \kappa^2 \left(2b_0 \int_0^1 e^{-2A_0} dy \right)^{-1} \cong \kappa^2 k (1-\Omega^2)^{-1}. \quad (24)$$

The approximations in eqs. (21) and (24) hold when the back reaction of the scalar field on the metric can be neglected.

In the Friedmann equations (22-23), we note that ρ is the bare value of the energy density on the TeV brane, naturally of order M_p^4, while $\Omega^4 \rho$ is the physically observable value, of order $(\text{TeV})^4$. Since ρ_* has no such suppression, it seems highly unlikely that ρ_* should be nonzero today; otherwise it would tend to vastly dominate the present expansion of the universe. We also point out that these equations are consistent only if energy is separately conserved on each brane: $\dot{\rho} + 3H(\rho + p) = 0$ and $\dot{\rho}_* + 3H(\rho_* + p_*) = 0$. This can be derived directly by considering the (05) Einstein equation, evaluated at either of the branes. The equations of state on the two branes are completely independent; there is no relation between p/ρ and p_*/ρ_*.

The above solutions are quite general, but they are not complete because we have not yet solved for the scalar field perturbation, $\delta\Phi$. This would generically be intractable, but there is a special case in which things simplify, namely, when the brane potentials $V_i(\Phi)$ become stiff. In this case, the boundary condition for the scalar fluctuation becomes $\delta\Phi = 0$ at either brane. There is no information about the derivative $\delta\Phi'$ in this case; although $\delta\Phi \to 0$, at the same $V''(\Phi) \to \infty$ in such a way that the product $\delta\Phi V''(\Phi)$ remains finite, and eq. (16) is automatically satisfied.

Notice that the shift in $\delta\Phi$, eq. (17), respects the boundary conditions on $\delta\Phi$. Moreover, Φ_0' is always nonzero for our solution. It is therefore always possible, given some solution $\delta\Phi$ which vanishes at the branes, to choose an f such that $\delta\Phi$ becomes zero. This is a convenient choice of gauge because it simplifies the equations of motion, and we will make it for the remainder of this letter.[1] Thus far we have satisfied the (00), (ii) and (05) Einstein equations. As noted above, eqs. (12) and (13) are equivalent, so either one just determines the shift in the radius. Using the former, and defining

$$G(y) = \left[\tfrac{1}{2}e^{2A_0(y)} + A_0' e^{4A_0(y)} \int_0^y e^{-2A_0} dy\right] \Big/ \int_0^1 e^{-2A_0} dy \cong \frac{kb_0 e^{4kb_0 y}}{1-\Omega^2}, \quad (25)$$

we find that

$$\begin{aligned}\frac{\delta b}{b_0} &= \frac{b_0}{2\Phi_0'^2}\left[\Omega^4(\rho-3p)G + (\rho_* - 3p_*)(G - A_0' e^{4A_0})\right] \\ &\cong \frac{kb_0^2 e^{4kb_0 y}}{2\Phi_0'^2(1-\Omega^2)}\left[\Omega^4(\rho-3p) + \Omega^2(\rho_* - 3p_*)\right];\end{aligned} \quad (26)$$

the last expression is found by approximating $A_0 = kb_0 y$ everywhere, which means neglecting the back reaction. Using the zeroth order solution (7) for Φ_0, and integrating over y, we can obtain the shift in the size of the extra dimension,

$$\int_0^1 \delta b\, dy \cong \frac{[\Omega^4(\rho-3p) + \Omega^2(\rho_* - 3p_*)]}{8(\varepsilon k v_0)^2 \Omega^{4+2\varepsilon}} \quad (27)$$

We can compare this to the result of ref. [16] by using their result for the radion mass, $m_r^2 \cong (4/3)\kappa^2(\varepsilon v_0 k)^2 \Omega^{2+2\varepsilon}$, and the relation $k\kappa^2 \cong 1/M_p^2$. Then

$$\frac{\int_0^1 \delta b\, dy}{b_0} \cong \frac{[\Omega^4(\rho-3p) + \Omega^2(\rho_* - 3p_*)]}{6kb_0 m_r^2 M_p^2 \Omega^2} \quad (28)$$

which agrees with ref. [16], except for small corrections of order $(1+\Omega^2)$. As is well known, the shift in the radion vanishes when the universe is radiation dominated, because

---

[1] The above argument is strictly true only for diffeomorphisms which are constant in time, while for our problem we need $f(t,y) \sim \rho(t), \rho_*(t)$. However, the time variation of such an f is of higher order in ρ and ρ_*, so we can neglect it to leading order in the perturbations.

the radion couples to the trace of the stress energy tensor, which vanishes if the matter is conformally invariant.

Above we focused on the shift in the size of the extra dimension due to cosmological expansion, but the more experimentally relevant quantity is the shift in the lapse function, $n(t,1)$, evaluated on the TeV brane. As emphasized in ref. [11], the change in $n(t,1)$ between the present and the past determines how much physical energy scales on our brane, like the weak scale, M_W, have evolved. The time dependence of M_W is given by $M_W(t)/M_W(t_0) = e^{-\delta N(t,1) + \delta N(t_0,1) + \delta N(t,0) - \delta N(t_0,0)}$. In terms of the variables of the previous section, $\delta N' = \Psi + \Upsilon + A_0' \delta b/b_0$. We find that

$$M_W(t) \cong M_W(t_0) \left(1 - \frac{\rho_p(t)}{8\Omega^4 M_p^2 k^2}\right), \qquad (29)$$

assuming that $\rho_* = p_* = 0$. With $k \sim M_p/30$ and assuming $g_* \sim 100$ relativistic degrees of freedom and $\Omega M_p = 1$ TeV, the correction to M_W becomes of order unity at a temperature of 130 GeV. Thus the temporal variation in fundamental mass scales might have some relevance for the electroweak phase transition and baryogenesis.

We thank C. Csaki, M. Graesser and G. Kribs for helpful discussions. JC thanks Nordita for its hospitality while this work was being finished.

REFERENCES

1. L. Randall and R. Sundrum, Phys. Rev. Lett. **83**, 3370 (1999) [hep-ph/9905221]; Phys. Rev. Lett. **83**, 4690 (1999) [hep-th/9906064].
2. W. D. Goldberger and M. B. Wise, Phys. Rev. Lett. **83**, 4922 (1999) [hep-ph/9907447].
3. C. Csaki, M. Graesser, L. Randall and J. Terning, Phys. Rev. **D62**, 045015 (2000) [hep-ph/9911406].
4. J. M. Cline and H. Firouzjahi, hep-ph/0005235, to be published in Phys. Rev. D.
5. C. Csaki, M. Graesser, C. Kolda and J. Terning, Phys. Lett. **B462**, 34 (1999) [hep-ph/9906513].
6. J. M. Cline, C. Grojean and G. Servant, Phys. Rev. Lett. **83**, 4245 (1999) [hep-ph/9906523].
7. P. Kanti, I. I. Kogan, K. A. Olive and M. Pospelov, Phys. Lett. **B468**, 31 (1999) [hep-ph/9909481]; Phys. Rev. **D61**, 106004 (2000) [hep-ph/9912266]; P. Kanti, K. A. Olive and M. Pospelov, Phys. Lett. **B481**, 386 (2000) [hep-ph/0002229].
8. R. N. Mohapatra, A. Perez-Lorenzana and C. A. de Sousa Pires, hep-ph/0003328.
9. J. Lesgourgues, S. Pastor, M. Peloso and L. Sorbo, hep-ph/0004086.
10. J. E. Kim and B. Kyae, Phys. Lett. **B486**, 165 (2000) [hep-th/0005139].
11. P. Kanti, K. A. Olive and M. Pospelov, hep-ph/0005146.
12. C. Kennedy and E. M. Prodanov, Phys. Lett. **B488**, 11 (2000) [hep-th/0003299].
13. K. Enqvist, E. Keski-Vakkuri and S. Rasanen, hep-th/0007254.
14. A. Mennim and R. A. Battye, hep-th/0008192.
15. O. DeWolfe, D. Z. Freedman, S. S. Gubser and A. Karch, Phys. Rev. **D62**, 046008 (2000) [hep-th/9909134].
16. C. Csaki, M. L. Graesser and G. D. Kribs, hep-th/0008151.

B-PHYSICS

Determining $|V_{ub}|$ from the $\bar{B} \to X_u \ell \bar{\nu}$ dilepton invariant mass spectrum[1]

Christian W. Bauer*, Zoltan Ligeti[†] and Michael Luke**

*Department of Physics, University of California, San Diego, 9500 Gilman Drive, La Jolla CA USA 92093

[†]Theoretical Physics Group, Ernest Orlando Lawrence Berkeley National Laboratory, University of California, Berkeley, CA USA 94720

**Department of Physics, University of Toronto, 60 St. George Street, Toronto, Ontario, Canada M5S 1A7

Abstract. The invariant mass spectrum of the lepton pair in inclusive semileptonic $\bar{B} \to X_u \ell \bar{\nu}$ decay yields a model independent determination of $|V_{ub}|$ [1]. Unlike the lepton energy and hadronic invariant mass spectra, nonperturbative effects are only important in the resonance region, and play a parametrically suppressed role when $d\Gamma/dq^2$ is integrated over $q^2 > (m_B - m_D)^2$, which is required to eliminate the $\bar{B} \to X_c \ell \bar{\nu}$ background. We discuss these backgrounds for q^2 slightly below $(m_B - m_D)^2$, and point out that instead of $q^2 > (m_B - m_D)^2 = 11.6\,\text{GeV}^2$, the cut can be lowered to $q^2 \gtrsim 10.5\,\text{GeV}^2$. This is important experimentally, particularly when effects of a finite neutrino reconstruction resolution are included.

INTRODUCTION

A precise and model independent determination of the Cabibbo-Kobayashi-Maskawa (CKM) matrix element V_{ub} is important for testing the Standard Model at B factories via the comparison of the angles and the sides of the unitarity triangle.

If it were not for the huge background from decays to charm, it would be straightforward to determine $|V_{ub}|$ from inclusive semileptonic decays. Inclusive B decay rates can be computed model independently in a series in Λ_{QCD}/m_b and $\alpha_s(m_b)$ using an operator product expansion (OPE) [2, 3, 4, 5], and the result may schematically be written as

$$d\Gamma = \binom{b \text{ quark}}{\text{decay}} \times \left\{ 1 + \frac{0}{m_b} + \frac{f(\lambda_1, \lambda_2)}{m_b^2} + \ldots + \frac{\alpha_s}{\pi}(\ldots) + \frac{\alpha_s^2}{\pi^2}(\ldots) + \ldots \right\}. \quad (1)$$

At leading order, the B meson decay rate is equal to the b quark decay rate. The leading nonperturbative corrections of order $\Lambda_{\text{QCD}}^2/m_b^2$ are characterized by two heavy quark effective theory (HQET) matrix elements, usually called λ_1 and λ_2. These matrix

---

[1] Talk presented by M.L.

elements also occur in the expansion of the B and B^* masses in powers of Λ_{QCD}/m_b,

$$m_B = m_b + \bar{\Lambda} - \frac{\lambda_1 + 3\lambda_2}{2m_b} + \ldots, \qquad m_{B^*} = m_b + \bar{\Lambda} - \frac{\lambda_1 - \lambda_2}{2m_b} + \ldots. \tag{2}$$

Similar formulae hold for the D and D^* masses. The parameters $\bar{\Lambda}$ and λ_1 are independent of the heavy b quark mass, while there is a weak logarithmic scale dependence in λ_2. The measured $B^* - B$ mass splitting fixes $\lambda_2(m_b) = 0.12\,\text{GeV}^2$, while $\bar{\Lambda}$ and λ_1 (or, equivalently, a short distance b quark mass and λ_1) may be determined from other physical quantities [6, 7, 8]. Thus, a measurement of the total $B \to X_u \ell \bar{\nu}$ rate would provide a $\sim 5\%$ determination of $|V_{ub}|$ [9, 10].

Unfortunately, the $\bar{B} \to X_u \ell \bar{\nu}$ rate can only be measured imposing cuts on the phase space to eliminate the ~ 100 times larger $\bar{B} \to X_c \ell \bar{\nu}$ background. Since the predictions of the OPE are only model independent for *sufficiently inclusive* observables, these cuts can destroy the convergence of the expansion. This is the case for two kinematic regions for which the charm background is absent and which have received much attention: the large lepton energy region, $E_\ell > (m_B^2 - m_D^2)/2m_B$, and the small hadronic invariant mass region, $m_X < m_D$ [11, 12, 13].

The poor behaviour of the OPE for these quantities is slightly subtle, because in both cases there is sufficient phase space for many different resonances to be produced in the final state, so an inclusive description of the decays is still appropriate. However, in both of these regions of phase space the $\bar{B} \to X_u \ell \bar{\nu}$ decay products are dominated by high energy, low invariant mass hadronic states,

$$E_X \sim m_b, \; m_X^2 \sim \Lambda_{\text{QCD}} m_b \gg \Lambda_{\text{QCD}}^2 \tag{3}$$

(where E_X and m_X are the energy and invariant mass of the final hadronic state). In this region the differential rate is very sensitive to the details of the wave function of the b quark in the B meson. Since the OPE is just sensitive to local matrix elements corresponding to expectation values of operators in the meson, the first few orders in the OPE do not contain enough information to describe the decay, and as a result the OPE does not converge.

This is simple to see by considering the kinematics. A b quark in a B meson has momentum

$$p_b^\mu = m_b v^\mu + k^\mu \tag{4}$$

where v^μ is the four-velocity of the quark, and k^μ is a small residual momentum of order Λ_{QCD}. If the hadron decays to leptons with momentum q and light hadrons with total momentum p_X, the invariant mass of the light hadrons may be written

$$m_X^2 = (m_b v + k - q)^2 = (m_b v - q)^2 + 2k \cdot (m_b v - q) + O(\Lambda_{\text{QCD}}^2). \tag{5}$$

The first term in the expansion is $O(m_b^2)$ over most of phase space, while the second is $O(\Lambda_{\text{QCD}} m_b)$, and so is suppressed over most of phase space. The OPE presumes that this power counting holds, so that the second term may be treated as a small perturbation. However, if E_X is large and m_X is small, $m_b v - q$ is almost light-like,

$$m_b v^\mu - q^\mu = (E_X, 0, 0, E_X) + O(\Lambda_{\text{QCD}}) \tag{6}$$

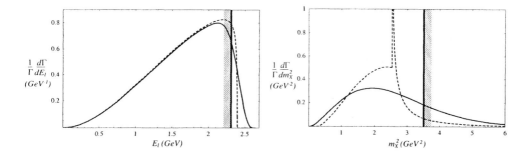

FIGURE 1. The shapes of the lepton energy and hadronic invariant mass spectra. The dashed curves are the b quark decay results to $O(\alpha_s)$, while the solid curves are obtained by smearing with the model distribution function $f(k_+)$ in Eq. (8). The unshaded side of the vertical lines indicate the region free from charm background.

in the b rest frame where $v^\mu = (1,0,0,0)$. Since $E_X \sim O(m_b)$, $(m_b v - q)^2 = O(\Lambda_{QCD} m_b)$. Thus, in this region the first two terms in (5) are of the same order (but still parametrically larger than the remaining terms), and the invariant mass of the final hadronic state reflects the distribution of the light-cone component of the residual momentum of the heavy quark in the hadron,

$$m_X^2 = (m_b v - q)^2 + 2 E_X k_+ + \ldots, \quad k_+ \equiv k_0 + k_3. \tag{7}$$

Since the differential rate in this region depends on the invariant mass of the final state, it is therefore sensitive at leading order to the light-cone wave function of the heavy quark in the meson, $f(k_+)$.

In terms of the OPE, this light-cone wave function arises because of subleading terms in the OPE proportional to $E_X \Lambda_{QCD}/m_X^2$, which are suppressed over most of phase space but are $O(1)$ in the region (3). It has been shown that the most singular terms in the OPE may be resummed into a nonlocal operator whose matrix element in a B meson is the light-cone structure function of the meson. Since $f(k_+)$ is a nonperturbative function, it cannot be calculated analytically, so the rate in the region (3) is model-dependent even at leading order in Λ_{QCD}/m_b.

The situation is illustrated in Fig. 1, where we have plotted the lepton energy and hadronic invariant mass spectra in the parton model (dashed curves) and incorporating a simple one-parameter model for the distribution function (solid curves) [17]

$$f(k_+) = \frac{32}{\pi^2 \Lambda}(1-x)^2 \exp\left[-\frac{4}{\pi}(1-x)^2\right] \Theta(1-x), \quad x \equiv \frac{k_+}{\Lambda}, \quad \Lambda = 0.48\,\text{GeV}. \tag{8}$$

The differences between the curves in the regions of interest indicate the sensitivity of the spectrum to the precise form of $f(k_+)$. Currently, there are measurements of $|V_{ub}|$ from both methods. From the lepton energy cut, the PDG reports $|V_{ub}/V_{cb}| = 0.08 \pm 0.02$, while a recent DELPHI measurement using the hadronic invariant mass cut

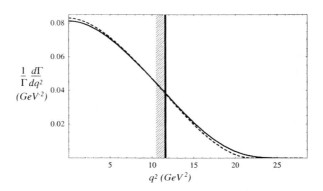

FIGURE 2. The dilepton invariant mass spectrum. The notation is the same as in Fig. 1.

gives $|V_{ub}/V_{cb}| = 0.103^{+0.011}_{-0.012}$ (syst.) ± 0.016 (stat.) ± 0.010 (theory) [16]. In both cases, the theoretical error is an estimate based on varying different models of $f(k_+)$, and so these measurements are no more model-independent than the exclusive measurement from $B \to \rho\ell\bar\nu$. While it may be possible in the future to extract $f(k_+)$ from the $B \to X_s\gamma$ photon spectrum [14, 18], unknown order $\Lambda_{\rm QCD}/m_b$ corrections arise when relating this to semileptonic $b \to u$ decay, limiting the accuracy with which $|V_{ub}|$ may be obtained.

Clearly, one would like to be able to find a cut which eliminates the charm background but does not destroy the convergence of the OPE, so that the distribution function $f(k_+)$ is not required. In Ref. [1] we pointed out that this is the situation for a cut on the dilepton invariant mass. Decays with $q^2 \equiv (p_\ell + p_{\bar\nu})^2 > (m_B - m_D)^2$ must arise from $b \to u$ transition. Such a cut forbids the hadronic final state from moving fast in the B rest frame, and simultaneously imposes $m_X < m_D$ and $E_X < m_D$. Thus, the light-cone expansion which gives rise to the shape function is not relevant in this region of phase space [13, 19]. The effect of smearing the q^2 spectrum with the model distribution function in Eq. (8) is illustrated in Fig. 2. It is clearly a subleading effect. The Dalitz plots relevant for the charged lepton energy and hadronic invariant mass cuts are shown in Fig. 3. Note that the region selected by a q^2 cut is entirely contained within the m_X^2 cut, but because the dangerous region of high energy, low invariant mass final states is not included with the q^2 cut, the OPE does not break down. It is also important to note, however, that the q^2 cut does make the OPE worse than for the full rate; as we will show, the relative size of the unknown $\Lambda_{\rm QCD}^3/m_b^3$ terms grows as the q^2 cut is raised. Equivalently, as was stressed in [20], the effective expansion parameter for this region is $\Lambda_{\rm QCD}/m_c$, not $\Lambda_{\rm QCD}/m_b$.

The $\bar B \to X_u \ell\bar\nu$ decay rate with lepton invariant mass above a given cutoff can therefore be reliably computed working to a fixed order in the OPE (i.e., ignoring the light-cone

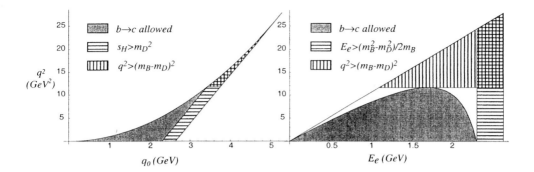

FIGURE 3. Dalitz plots relevant for $\bar{B} \to X_u \ell \bar{\nu}_\ell$. The shaded regions indicate the part of phase space where $\bar{B} \to X_c \ell \bar{\nu}_\ell$ background is present, and the vertical dashed regions corresponds to the cut $q^2 > (m_B - m_D)^2$. In the $q^2 - q_0$ plane, the horizontal dashed region corresponds to an invariant mass cut $m_X^2 > m_D^2$, whereas in the $q^2 - E_\ell$ plane the horizontal dashed region corresponds to the charged lepton energy cut $E_\ell > (m_B^2 - m_D^2)/2m_B$. Note that at tree level, $b \to u$ semileptonic decay populates the entire triangle on the right-hand plot, but only the right boundary of the left-hand plot.

distribution function),

$$\frac{1}{\Gamma_0} \frac{d\Gamma}{d\hat{q}^2} = \left(1 + \frac{\lambda_1}{2m_b^2}\right) 2(1 - \hat{q}^2)^2 (1 + 2\hat{q}^2) + \frac{\lambda_2}{m_b^2}(3 - 45\hat{q}^4 + 30\hat{q}^6) + \frac{\alpha_s(m_b)}{\pi} X(\hat{q}^2) + \left(\frac{\alpha_s(m_b)}{\pi}\right)^2 \beta_0 Y(\hat{q}^2) + \ldots, \quad (9)$$

where $\hat{q}^2 = q^2/m_b^2$, $\beta_0 = 11 - 2n_f/3$, and $\Gamma_0 = G_F^2 |V_{ub}|^2 m_b^5/(192\pi^3)$ is the tree level $b \to u$ decay rate. The ellipses in Eq. (9) denote terms of order $(\Lambda_{QCD}/m_b)^3$ and order α_s^2 terms not enhanced by β_0. The function $X(\hat{q}^2)$ is known analytically [21], whereas $Y(\hat{q}^2)$ was computed numerically [22]. The order $1/m_b^3$ nonperturbative corrections are also known [23], as are the leading logarithmic perturbative corrections proportional to $\alpha_s^n \log^n(m_c/m_b)$ [20]. The matrix element of the kinetic energy operator, λ_1, only enters the $\hat{q}^2$ spectrum in a very simple form, because the unit operator and the kinetic energy operator are related by reparameterization invariance [24].

The relation between the total $\bar{B} \to X_u \ell \bar{\nu}$ decay rate and $|V_{ub}|$ is known at the $\sim 5\%$ level [9, 10],

$$|V_{ub}| = (3.04 \pm 0.06 \pm 0.08) \times 10^{-3} \left(\frac{\mathcal{B}(\bar{B} \to X_u \ell \bar{\nu})|_{q^2 > q_0^2}}{0.001 \times F(q_0^2)} \frac{1.6 \, \text{ps}}{\tau_B}\right)^{1/2}, \quad (10)$$

where $F(q_0^2)$ is the fraction of $\bar{B} \to X_u \ell \bar{\nu}$ events with $q^2 > q_0^2$, satisfying $F(0) = 1$. The errors explicitly shown in Eq. (10) are the estimates of the perturbative and nonperturbative uncertainties in the upsilon expansion [9] respectively. At the present time the

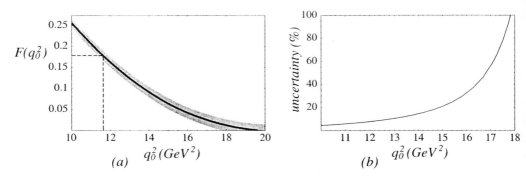

FIGURE 4. (a) The fraction of $\bar B \to X_u \ell \bar\nu$ events with $q^2 > q_0^2$, $F(q_0^2)$, in the upsilon expansion. The dashed line indicates the lower cut $q_0^2 = (m_B - m_D)^2 \simeq 11.6\,\text{GeV}^2$, which corresponds to $F = 0.178 \pm 0.012$. The shaded region is the estimated uncertainty due to $\Lambda_{\text{QCD}}^3/m_b^3$ terms; which is shown in (b) as a percentage of $F(q_0^2)$.

biggest uncertainty is due to the error of a short distance b quark mass, whichever way it is defined [20]. (This can be cast into an uncertainty in an appropriately defined $\bar\Lambda$, or the nonperturbative contribution to the $\Upsilon(1S)$ mass, etc.) By the time the q^2 spectrum in $\bar B \to X_u \ell \bar\nu$ is measured, this uncertainty should be reduced from extracting m_b from the hadron mass [6] or lepton energy [7] spectra in $\bar B \to X_c \ell \bar\nu$, or from the photon energy spectrum [8] in $B \to X_s \gamma$. The uncertainty in the perturbation theory calculation will be largely reduced by computing the full order α_s^2 correction in Eq. (10). The largest "irreducible" uncertainty is from order $\Lambda_{\text{QCD}}^3/m_b^3$ terms in the OPE, the estimated size of which is shown in Fig. 4, together with our central value for $F(q_0^2)$, as functions of q_0^2.

There is another advantage of the q^2 spectrum over the m_X spectrum to measure $|V_{ub}|$. In the variable m_X, about 20% of the charm background is located right next to the $b \to u$ "signal region", $m_X < m_D$, namely $\bar B \to D \ell \bar\nu$ at $m_X = m_D$. In the variable q^2, the charm background just below $q^2 = (m_B - m_D)^2$ comes from the lowest mass X_c states. Their q^2 distributions are well understood based on heavy quark symmetry [25], since this region corresponds to near zero recoil. Fig. 5 shows the $\bar B \to D \ell \bar\nu$ and $\bar B \to D^* \ell \bar\nu$ decay rates using the measured form factors [26] (and $|V_{ub}| = 0.0035$). The $\bar B \to X_u \ell \bar\nu$ rate is the flat curve. Integrated over the region $q^2 > (m_B - m_{D^*})^2 \simeq 10.7\,\text{GeV}^2$, the uncertainty of the $B \to D$ background is small due to its $(w^2 - 1)^{3/2}$ suppression compared to the $\bar B \to X_u \ell \bar\nu$ signal. This uncertainty will be further reduced in the near future. This increases the $b \to u$ region relevant for measuring $|V_{ub}|$ by $\sim 1\,\text{GeV}^2$. The $B \to D^*$ rate is only suppressed by $(w^2 - 1)^{1/2}$ near zero recoil, and therefore it is more difficult to subtract it reliably from the $b \to u$ signal. The nonresonant $D\pi$ final state contributes in the same region as $\bar B \to D^*$, and it is reliably predicted to be small near maximal q^2 (zero recoil) based on chiral perturbation theory [27]. The D^{**} states only contribute for $q^2 < 9\,\text{GeV}^2$, and some aspects of their q^2 spectra are also known model independently [28].

Concerning experimental considerations, measuring the q^2 spectrum requires reconstruction of the neutrino four-momentum, just like measuring the hadronic invariant

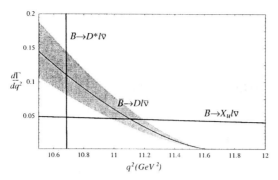

FIGURE 5. Charm backgrounds near $q^2 = (m_B - m_D)^2$ (arbitrary units). The shaded region denotes the uncertainty on the $\bar{B} \to D\ell\bar{\nu}$ rate.

mass spectrum. A lepton energy cut may be required for this technique, however, the constraint $q^2 > (m_B - m_D)^2$ automatically implies $E_\ell > (m_B - m_D)^2/2m_B \simeq 1.1\,\text{GeV}$ in the B rest frame. Even if the E_ℓ cut has to be slightly larger than this, the utility of our method will not be affected, but a calculation including the effects of arbitrary E_ℓ and q^2 cuts would be required. If experimental resolution on the reconstruction of the neutrino momentum necessitates a significantly larger cut than $q_0^2 = (m_B - m_D)^2$, then the uncertainties in the OPE calculation of $F(q_0^2)$ increase. In this case, it may be possible to obtain useful model independent information on the q^2 spectrum in the region $q^2 > m_{\psi(2S)}^2 \simeq 13.6\,\text{GeV}^2$ from the q^2 spectrum in the rare decay $\bar{B} \to X_s \ell^+ \ell^-$, which may be measured in the upcoming Tevatron Run-II.

In conclusion, we have shown that the q^2 spectrum in inclusive semileptonic $\bar{B} \to X_u \ell \bar{\nu}$ decay gives a model independent determination of $|V_{ub}|$ with small theoretical uncertainty. Nonperturbative effects are only important in the resonance region, and play a parametrically suppressed role when $d\Gamma/dq^2$ is integrated over $q^2 > (m_B - m_D)^2$, which is required to eliminate the charm background. This is a qualitatively better situation than other extractions of $|V_{ub}|$ from inclusive charmless semileptonic B decay.

ACKNOWLEDGEMENTS

This work was supported in part by the Natural Sciences and Engineering Research Council of Canada and by the Director, Office of Science, Office of High Energy and Nuclear Physics, Division of High Energy Physics, of the U.S. Department of Energy under Contract DE-AC03-76SF00098.

REFERENCES

1. C.W. Bauer, Z. Ligeti, and M. Luke, Phys. Lett. B479 (2000) 395.

2. J. Chay, H. Georgi, and B. Grinstein, Phys. Lett. B247 (1990) 399; M. Voloshin and M. Shifman, Sov. J. Nucl. Phys. 41 (1985) 120.
3. I.I. Bigi *et al.*, Phys. Lett. B293 (1992) 430; Phys. Lett B297 (1993) 477 (E); I.I. Bigi *et al.*, Phys. Rev. Lett. 71 (1993) 496.
4. A.V. Manohar and M.B. Wise, Phys. Rev. D49 (1994) 1310.
5. B. Blok *et al.*, Phys. Rev. D49 (1994) 3356.
6. A.F. Falk, M. Luke, and M.J. Savage, Phys. Rev. D53 (1996) 2491; Phys. Rev. D53 (1996) 6316; A.F. Falk and M. Luke, Phys. Rev. D57 (1998) 424.
7. M. Gremm *et al.*, Phys. Rev. Lett. 77 (1996) 20; M.B. Voloshin, Phys. Rev. D51 (1995) 4934.
8. A. Kapustin and Z. Ligeti, Phys. Lett. B355 (1995) 318; C. Bauer, Phys. Rev. D57 (1998) 5611; Erratum ibid. D60 (1999) 099907; Z. Ligeti, M. Luke, A.V. Manohar, and M.B. Wise, Phys. Rev. D60 (1999) 034019; A.L. Kagan and M. Neubert, Eur. Phys. J. C7 (1999) 5.
9. A.H. Hoang, Z. Ligeti, and A.V. Manohar, Phys. Rev. Lett. 82 (1999) 277; Phys. Rev. D59 (1999) 074017.
10. N. Uraltsev, Int. J. Mod. Phys. A14 (1999) 4641.
11. V. Barger, C. S. Kim and R. J. Phillips, Phys. Lett. B251, (1990) 629.
12. A.F. Falk, Z. Ligeti, and M.B. Wise, Phys. Lett. B406 (1997) 225; I. Bigi, R.D. Dikeman, and N. Uraltsev, Eur. Phys. J. C4 (1998) 453.
13. R. D. Dikeman and N. Uraltsev, Nucl. Phys. B509 (1998) 378.
14. M. Neubert, Phys. Rev. D49 (1994) 3392; D49 (1994) 4623; I.I. Bigi *et al.*, Int. J. Mod. Phys. A9 (1994) 2467.
15. F. De Fazio and M. Neubert, JHEP06 (1999) 017.
16. P. Abreu *et al.* [DELPHI Collaboration], Phys. Lett. B478 (2000) 14.
17. T. Mannel and M. Neubert, Phys. Rev. D50 (1994) 2037.
18. A.K. Leibovich, I. Low, and I.Z. Rothstein, Phys. Rev. D61 (2000) 053006; Phys. Lett. B486 (2000) 86.
19. G. Buchalla and G. Isidori, Nucl. Phys. B525 (1998) 333.
20. M. Neubert, JHEP 0007 (2000) 022.
21. M. Jezabek and J.H. Kuhn, Nucl. Phys. B314 (1989) 1.
22. M. Luke, M. Savage, and M.B. Wise, Phys. Lett. B343 (1995) 329.
23. C.W. Bauer and C.N. Burrell, Phys. Lett. B469 (1999) 248; hep-ph/9911404.
24. M. Luke and A.V. Manohar, Phys. Lett. B286 (1992) 348.
25. N. Isgur and M.B. Wise, Phys. Lett. B232 (1989) 113; Phys. Lett. B237 (1990) 527.
26. J. Bartelt *et al.*, CLEO Collaboration, hep-ex/9811042; the LEP average for $B \to D^*$ is taken from http://lepvcb.web.cern.ch/LEPVCB/Tampere.html
27. C. Lee, M. Lu, and M.B. Wise, Phys. Rev. D46 (1992) 5040.
28. A.K. Leibovich *et al.*, Phys. Rev. Lett. 78 (1997) 3995; Phys. Rev. D57 (1998) 308.

$B \to X_s \ell^+ \ell^-$ in the vectorlike quark model

[a]M. R. Ahmady[1], [b]M. Nagashima[2], and [b]A. Sugamoto[3]

[a]*Department of Physics, Mount Allison University, Sackville, NB E4L 1E6 Canada*
[b]*Department of Physics, Ochanomizu University, 1-1 Otsuka 2, Bunkyo-ku, Tokyo 112, Japan*

Abstract. We extend the standard model by adding an extra generation of isosinglet up- and down-type quark pair which engage in weak interactions only via mixing with the three ordinary quark families. It is shown that the generalized 4 × 4 quark mixing matrix, which is necessarily nonunitary, leads to nonvanishing flavor changing neutral currents. We then proceed to investigate various distributions and total branching ratio of the inclusive $B \to X_s \ell^+ \ell^-$ ($\ell = e, \mu$) rare B decays in the context of this model. It is shown that the shapes of the differential branching ratio and forward-backward asymmetry distribution are very sensitive to the value of the model parameters which are constrained by the experimental upper bound on $BR(B \to X_s \mu^+ \mu^-)$. We also indicate that, for certain values of the dileptonic invariant mass, CP asymmetries up to 10% can be obtained.

B factories, CLEO III and other dedicated B experiments are expected to observe new rare B decay channels and to improve the precision of those which have already been measured. Radiative B decays, which proceed via loop effects, are quite crucial for observing the signals of new physics beyond the standard model (SM) in the near future. One example of such models is the extension of the SM to contain an extra generation of isosinglet quarks [1]:

$$\psi^i \equiv \begin{pmatrix} u^\alpha \\ d^\alpha \end{pmatrix}, \ \alpha = 1...4 , \tag{1}$$

where

$$\begin{pmatrix} u^4 \\ d^4 \end{pmatrix} \equiv \begin{pmatrix} U \\ D \end{pmatrix} . \tag{2}$$

In other words, both left- and right-handed components of the additional pair of quarks, which are denoted U and D with charges $+2/3$ and $-1/3$, respectively, are $SU(2)_L$ singlets. As a result, the Dirac mass terms of vectorlike quarks, i.e.,

---

[1]) Email: mahmady@mta.ca
[2]) Email: g0070508@edu.cc.ocha.ac.jp
[3]) Email: sugamoto@phys.ocha.ac.jp

$$m_U(\bar{U}_L U_R + \bar{U}_R U_L) + m_D(\bar{D}_L D_R + \bar{D}_R D_L) \;, \tag{3}$$

are allowed by electroweak gauge symmetry. However, for ordinary quarks, gauge invariant Yukawa couplings to an isodoublet scalar Higgs field ϕ,i.e.,

$$-f_d^{ij}\bar{\psi}_L^i d_R^j \phi - f_u^{ij}\bar{\psi}_L^i u_R^j \tilde{\phi} + H.C. \;, \; i=1,2,3 \;, \tag{4}$$

are responsible for the mass generation via spontaneous symmetry breaking. The link between the vectorlike and the ordinary quarks are provided by the extra gauge invariant Yukawa couplings which can be written as follows:

$$-f_d^{i4}\bar{\psi}_L^i D_R \phi - f_u^{i4}\bar{\psi}_L^i U_R \tilde{\phi} + H.C. \;. \tag{5}$$

After spontaneous symmetry breaking, Eqs. (4) and (5) along with Eq. (3) lead to 4×4 mass matrices for the up- and down-type quarks:

$$\bar{d}_L^\alpha M_d^{\alpha\beta} d_R^\beta + \bar{u}_L^\alpha M_u^{\alpha\beta} u_R^\beta + H.C. \;, \tag{6}$$

In general, M_d and M_u are not diagonal, and to achieve diagonalization, unitary transformations from weak to mass eigenstates are necessary,i.e.,

$$u_{L,R}^\alpha = A_{L,R}^{u\;\alpha\beta} u'^\beta_{L,R} \;, \; d_{L,R}^\alpha = A_{L,R}^{d\;\alpha\beta} d'^\beta_{L,R} \;, \tag{7}$$

where we use prime to denote the mass eigenstates. The interesting property of the vectorlike quark model (VQM) is that the above transformations result in the intergenerational mixing among quarks not only in the charged current sector but also in the neutral current interactions. For example, the charged current interaction term which, when written in terms of weak eigenstates, involves only the three generations of ordinary quarks,i.e.,

$$J_{CC}^{W\;\mu} = \sum_{i=1}^{3} I\frac{g}{\sqrt{2}} \bar{u}_L^i \gamma^\mu d_L^i W_\mu^+ + H.C. \;, \tag{8}$$

transforms to

$$J_{CC}^{W\;\mu} = \sum_{\alpha,\beta=1}^{4} I\frac{g}{\sqrt{2}} \bar{u}'^\alpha_L V^{\alpha\beta} \gamma^\mu d'^\beta_L W_\mu^+ + H.C. \;, \tag{9}$$

where

$$V^{\alpha\beta} = \sum_{i=1}^{3} (A_L^{u\dagger})^{\alpha i} (A_L^d)^{i\beta} \;, \tag{10}$$

when expressed in terms of the mass eigenstates. It is straightforward to show that the 4×4 quark mixing matrix V is nonunitary

$$(V^\dagger V)^{\alpha\beta} = \delta^{\alpha\beta} - \left(A_L^{d\,4\alpha}\right)^* A_L^{d\,4\beta} ,$$
$$(VV^\dagger)^{\alpha\beta} = \delta^{\alpha\beta} - \left(A_L^{u\,4\alpha}\right)^* A_L^{u\,4\beta} . \tag{11}$$

Therefore, a close examination of the neutral current sector reveals that flavor changing neutral currents (FCNC) like

$$J_{NC}^{Z\,\mu} = I \frac{g}{\cos\theta_w} \sum_{\alpha,\beta=1}^{4} \left(I_w^q U^{\alpha\beta} \bar{q}_L'^\alpha \gamma^\mu q_L'^\beta - Q_q \sin^2\theta_w \delta^{\alpha\beta} \bar{q}'^\alpha \gamma^\mu q'^\beta \right) , \tag{12}$$

where

$$U^{\alpha\beta} = \sum_{i=1}^{3} \left(A_L^{q\,i\alpha}\right)^* A_L^{q\,i\beta} = \delta^{\alpha\beta} - \left(A_L^{q\,4\alpha}\right)^* A_L^{q\,4\beta} = \begin{cases} (V^\dagger V)^{\alpha\beta}, & q \equiv \text{down} - \text{type} \\ (VV^\dagger)^{\alpha\beta}, & q \equiv \text{up} - \text{type} \end{cases}, \tag{13}$$

are consequently developed in the VQM. It is exactly for this reason that rare B decays can serve as excellent venues to see the effects of the vectorlike quarks [2].
The effective Lagrangian for $B \to X_s \ell^+ \ell^-$ is the following

$$L_{\text{eff}} = \frac{G_F}{\sqrt{2}} \left(A \bar{s} L_\mu b \bar{\ell} L^\mu \ell + B \bar{s} L_\mu b \bar{\ell} R^\mu \ell + 2 m_b C \bar{s} T_\mu b \bar{\ell} \gamma^\mu \ell \right) , \tag{14}$$

where

$$L_\mu = \gamma_\mu (1 - \gamma_5) , \quad R_\mu = \gamma_\mu (1 - \gamma_5) , \tag{15}$$

$$T_\mu = i\sigma_{\mu\nu}(1 + \gamma_5) q^\nu / q^2 . \tag{16}$$

The coefficients A and B receive contributions from long-distance charm-quark loop ($c\bar{c}$ continuum) and the intermediate ψ and ψ' resonances as well as short-distance tree and one-loop diagrams (Fig.1). However, C, the coefficient of the magnetic moment operator, is purely short-disctance. The details of the contributing terms to the effective Lagrangian can be found in Reference [3]. The total branching ratio of the dileptonic B decay is dominated by the resonance contributions. However, by using cuts in the differential branching ratio around ψ and ψ' invariant mass, one can get information on the contributing short-distance physics [4].

The VQM parameters which appear in our calculations are: the U-quark mass m_U, the nonunitarity parameter $U^{sb} = |U^{sb}| e^{i\theta}$, where θ is a weak phase, and $V_{4s}^* V_{4b} = U^{sb} - (V_{\text{CKM}}^\dagger V_{\text{CKM}})^{sb}$. V_{CKM}, which is the 3 × 3 submatrix of the matrix V, consists of the elements representing mixing among three ordinary generations of quarks. $(V_{\text{CKM}}^\dagger V_{\text{CKM}})^{sb}$ presents the deviation from unitarity of the 3-generation CKM mixing matrix in the VQM context. The experimental upper bound $BR(B \to X_s \mu^+ \mu^-) \le 5.8 \times 10^{-5}$ [5] is used, to put a rough constraint on the magnitude

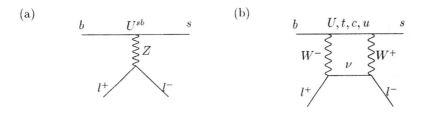
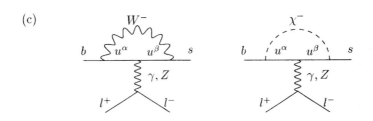
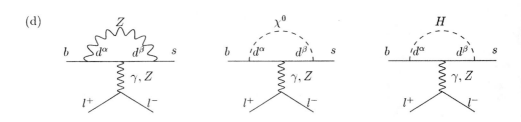

Figure 1: The Feynmann diagrams which contribute to $B \to X_s \ell^+ \ell^-$ in the VQM. (a) The tree level contribution (b) The box diagram (c) The penguin diagrams mediated by charged particles (d) The penguin diagrams mediated by neutral particles. Z and γ coupling to the internal quarks in (c) and (d) are proportional to $U^{\beta\alpha}$ and $\delta^{\beta\alpha}$, respectively.

of U^{sb}, i.e., $|U^{sb}| \lesssim 10^{-3}$. Since the tree level diagram proportional to U^{sb} (Fig. (1a)) contributes to the rare dileptonic decay channels, the one-loop terms with coefficient $V_{4s}^* V_{4b}$ are significant only if $V_{4s}^* V_{4b} \approx -(V_{\text{CKM}}^\dagger V_{\text{CKM}})^{sb} \gg U^{sb}$. Thus, we parametrize our results in terms of $(V_{\text{CKM}}^\dagger V_{\text{CKM}})^{sb}/|V_{cb}| = \epsilon e^{i\phi}$ instead of $V_{4s}^* V_{4b}$, where $\epsilon = |(V_{\text{CKM}}^\dagger V_{\text{CKM}})^{sb}|/|V_{cb}|$ and ϕ is another weak phase of the model.

We use $|V_{cs}| \approx 0.97$, $|V_{cb}| \approx 0.04$, $|V_{ts}|/|V_{cb}| \approx 1.1$, and $V_{us}^* V_{ub} \approx 0$ [6], which are extracted from various experimental measurements and are not affected by the presence of the new physics. $V_{cs}^* V_{cb}$ is taken to be real, as is the case, to a good accuracy, in the "standard" parametrization of the CKM matrix. Consequently, $V_{ts}^* V_{tb}$, which is not known experimentally, can be expressed in terms of the VQM parameters as:

$$\frac{V_{ts}^* V_{tb}}{|V_{cb}|} \approx \epsilon e^{i\phi} - |V_{cs}| \ . \tag{17}$$

Figure 2 illustrates the differential branching ratio for some values of the VQM parameters as campared to the SM prediction. We use $\epsilon = 0.3$ and all constructive/destructive contribution possibilities of the extra, beyond the SM, terms are considered. We observe that away from the resonances, where SD operators are dominant, the shift from the SM expectation, depending on the parameter values, can be quite significant. To constrain the model parameters by using the experimental results on $BR(B \to X_s \mu^+ \mu^-)$ reported in Ref. [5], we calculate the total branching but excluding the resonances ψ and ψ' with a $\delta = \pm 0.1$GeV cut. Our results are depicted in Fig. 3 in the form of acceptable regions in the $|U^{sb}|$ versus m_U plane for various choices of the relative sign of the extra contributions. The most stringent constraint is obtained if the relative phases θ and ϕ both vanish (Fig. 3(a)).

Figure 4 illustrates our results for the forward-backward asymmetry distribution of the decay $B \to X_s \mu^+ \mu^-$ in the VQM as compared to the SM. From Figs. 4(a) and 4(b) we observe that, when the tree level FCNC contributes constructively, even though the sign of the asymmetry remains the same as in the SM, its shape, away from the resonances, can be significantly different. On the other hand, as is shown in Figs. 4(c) and 4(d), for the destructive contribution of the nonunitarity induced tree level term and large enough values of $|U^{sb}|$ or m_U, the sign of A_{FB} can be opposite to what is predicted by the SM. One important observable in this decay channel is the point of zero asymmetry in the forward-backward asymmetry distribution, which occurs somewhere below the resonance ψ. Our investigation reveals that, as far as the VQM is concerned, the position of this point is quite stable and is not shifted very much from its SM value for various choices of the model parameters.

Since in the VQM, the effective Lagrangian contains terms with different CP-odd weak phases, as well as, LD continuum and resonance contributions which are sources of perturbative and nonperturbative CP-even strong phases, the direct CP asymmetry, unlike the SM, is expected to be non-zero. The size of this asymmetry,

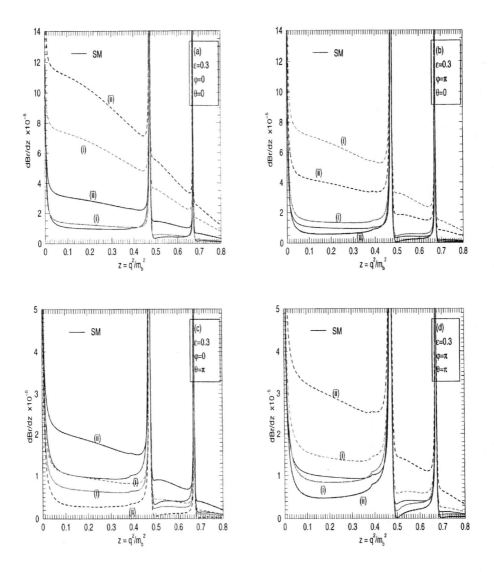

Figure 2: The differential branching ratio versus dilepton invariant mass for various values of the VQM parameters is compared to the SM prediction. Results with $|U^{sb}| = 10^{-4}$ and $|U^{sb}| = 10^{-3}$, for two different values of the U-quark mass, (i) m_U=200GeV, and (ii) m_U=400GeV, are shown by solid lines and dashed lines, respectively.

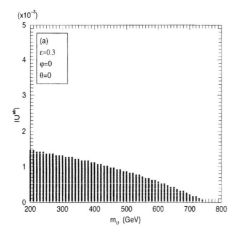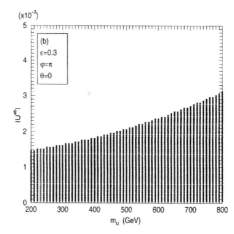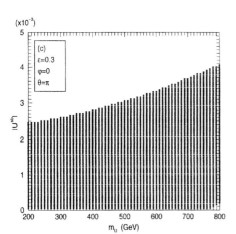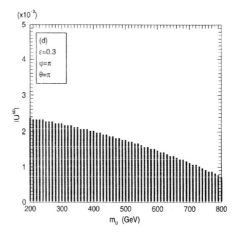

Figure 3: The acceptable region (shaded area) of the $|U^{sb}|$ versus m_U plane for various values of the VQM parameters obtained by using the experimental upper bound on $BR(B \to X_s \mu^+ \mu^-)$.

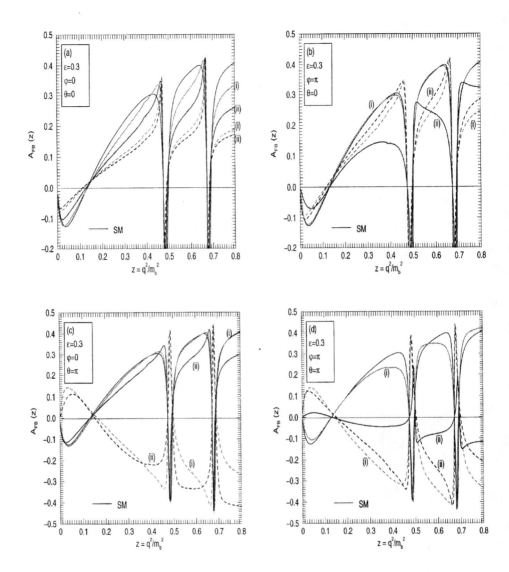

Figure 4: The forward-backward asymmetry distribution of $B \to X_s \mu^+ \mu^-$ for various values of the VQM parameters is compared to the SM prediction. Results with $|U^{sb}| = 10^{-4}$ and $|U^{sb}| = 10^{-3}$, for two different values of the U-quark mass, (i) m_U=200GeV, and (ii) m_U=400GeV, are shown by solid lines and dashed lines, respectively.

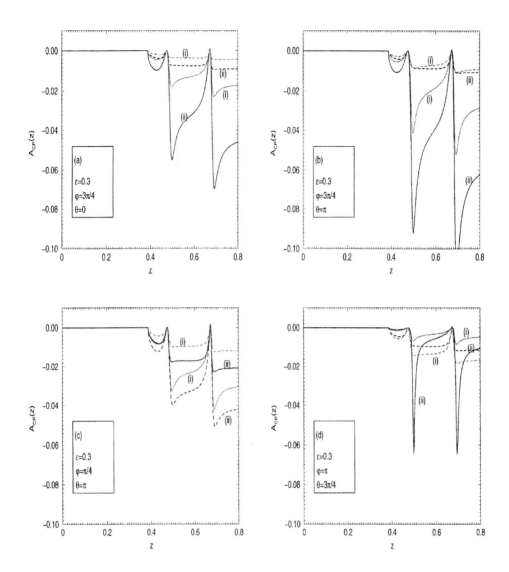

Figure 5: The CP asymmetry distribution of $B \to X_s \mu^+ \mu^-$ for various values of the VQM parameters. Results with $|U^{sb}| = 10^{-4}$ and $|U^{sb}| = 10^{-3}$, for two different values of the U-quark mass, (i) $m_U=200$GeV, and (ii) $m_U=400$GeV, are shown by solid lines and dashed lines, respectively.

which is zero for invariant dilepton masses below $2m_c$ threshold due to the vanishing strong phase, depends on the interplay of the various contributing terms. Our results for certain choices of the model parameters are depicted in Fig. 5. From Figs. 5(a) and 5(b), we observe that when the weak phase ϕ is large, smaller values of the nonunitarity parameter lead to significantly larger CP asymmetries sensitive to the U quark mass, m_U. In fact, for a negative relative sign of the tree level contribution ($\theta = \pi$) and $m_U = 400$GeV, as is shown in Fig. 5(b), asymmetries of the order of 10% can be achieved.

In conclusion, we have investigated various observables of the dileptonic rare B decay $B \to X_s \ell^+ \ell^-$ in the presence of an extra generation of vectorlike quarks. The measurement of this decay channel in the near future should provide more stringent constraints on the model parameters.

Acknowledgement

Special thanks to the Natural Sciences and Engineering research Council of Canada for support under the International Opportunity Fund grant number 237905-00.

REFERENCES

1. L. Bento and G. C. Branco, Phys. Lett. **B245**, 599 (1990);
   L. Bento, G. C. Branco and P. A. Parada, Phys. Lett. **B267**, 95 (1991).
2. Y. Nir and D. Silverman, Phys. Rev. **D42**, 1477 (1990); G. C. Branco, T. Morozumi, P. A. Parada and M. N. Rebelo, Phys. Rev. **D48**, 1167 (1993); L. T. Handoko and T. Morozumi, Mod. Phys. Lett. **A10**, 309 (1995); M. Aoki, E. Asakawa, M. Nagashima, N. Oshimo and A. Sugamoto, Phys. Lett. **B487**, 321 (2000); M. Aoki, G-C. Cho, M. Nagashima and N. Oshimo, OCHA-PP-170, hep-ph/0102165; C.-H. V. Chang, D. Chang and W. Y. Keung, Phys. Rev. **D61**, 053007 (2000); G. Barenboim, F.J. Botella, O. Vives, Phys. Rev. **D64**, 015007 (2001).
3. M. R. Ahmady, M. Nagashima and A. Sugamoto, OCHA-PP-172,hep-ph/0105049, to appear in Phys. Rev. D.
4. M. R. Ahmady, Phys. Rev. **D53**, 2843 (1996); M. R. Ahmady, OCHA-PP-69, hep-ph/9511212.
5. CLEO Collaboration, S. Glenn *et al.*, Phys. Rev. Lett. **80**, 2289 (1998).
6. Particle Data Group, D. E. Groom *et al.*, Eur. Phys. J. **C15**, 1 (2000).

Renormalon analysis of Heavy-Light Exclusive B Decays

Alexander R. Williamson

Department of Physics, University of Toronto, 60 St. George Street, Toronto, ON, Canada M5S 1A7

Abstract. We study two-body exclusive decays of the form $\bar{B} \to D^{(*)}L$ ($L = \pi, \rho, \ldots$) in the heavy-quark limit. We perform a renormalon analysis of such processes to determine the order at which nonperturbative factorization-breaking power corrections enter the amplitude. We find that a class of leading power corrections to the color octet matrix element, of $O(\Lambda_{\text{QCD}}/m_b)$, vanish in the limit of a symmetric light meson parton distribution function. We discuss the phenomenological significance of this result.

INTRODUCTION

The weak decays of B mesons into hadronic final states are important for an understanding of the CKM sector of the standard model, and particularly for the study of CP violation. These decays involve a mixture of calculable weak physics, perturbative QCD, and nonperturbative QCD. It is the latter component which contains the bulk of the theoretical difficulty, and it is primarily contained in the evaluation of low energy matrix elements of quark operators.

A common assumption used to simplify the nonperturbative component of hadronic decays is factorization, which specifies that the hadronic matrix element of a four-quark operator be factored into two matrix elements of simple currents. For example,

$$\langle \pi^- D^+ | (\bar{c}b)_{V-A} (\bar{d}u)_{V-A} | \bar{B} \rangle \to \langle \pi^- | (\bar{d}u)_{V-A} | 0 \rangle \langle D^+ | (\bar{c}b)_{V-A} | \bar{B} \rangle. \quad (1)$$

This prescription (which, following the authors of [1, 2], we refer to as 'naive factorization') considerably simplifies matters because the factored structures on the right hand side may be parameterized in terms of decay constants and form factors. It amounts, however, to ignoring corrections which connect the (π^-) to the $(\bar{B}D^+)$ system. Since these corrections are responsible for final-state rescattering and strong interaction phase shifts, leaving them out ignores important physics. Also, the left hand side of (1) is renormalization scale dependent, while the right-hand side is not — a clear indication that relevant physics is being lost.

Recently it was argued that, for certain B decays to heavy-light final states $\bar{B} \to HL$, the strong interactions which break factorization are hard in the heavy quark limit [1, 2], and can therefore be calculated perturbatively. This proposal has been explicitly verified to two-loop order [2]. This idea allows one to include perturbative corrections missing in (1) without introducing any new nonperturbative parameters. It thus provides a remarkably attractive means to study rescattering and strong phases in two-body hadronic decays. A generalization of this idea has also been proposed for decays to two light mesons [1, 2]. In this talk we will restrict our attention to final states with one heavy meson (D, D^*) and one light meson $(\pi, \rho, \ldots)$, and present work that was done in collaboration with Craig Burrell [3], and subsequently confirmed by other authors [4].

This proposal is valid only in the strict heavy quark limit [1]. It receives power corrections of the form $(\Lambda_{QCD}/m_b)^n$ from a variety of sources including hard spectator interactions, non-factorizable soft and collinear gluon exchange, and transverse momenta of quarks in the light meson. Unlike power corrections in inclusive B decays, there is as yet no systematic way to compute these corrections for exclusive decays. By naive power counting one expects such corrections at $O(\Lambda_{QCD}/m_b)$, but situations are known where the naively expected corrections vanish. For instance, in the zero recoil $B \to D$ transition matrix element in Heavy Quark Effective Theory the leading $1/m_b$ corrections vanish [5]. Therefore rather than relying on the naive expectation, one would like to calculate the power corrections directly.

In the absence of a general theory of power corrections, we aim to determine at least the order at which power corrections enter. This may be done by carrying out a renormalon analysis, which involves calculating a subset of Feynman graphs at each order in perturbation theory. It exploits the fact that the perturbative series in quantum field theory is asymptotic, and permits one to extract from its large-order perturbative behavior information about the scaling of nonperturbative power corrections. In this paper, we use renormalons to assess the parametric size of a subset of power corrections to the separation (1) for the class of $\bar{B} \to HL$ decays discussed in [2].

The remainder of this paper is organized as follows: in Section we discuss the phenomenological context and motivation of our study. In Section we review the theoretical framework of renormalon analyses. In Section we describe the renormalon calculation and give the result. Section contains our conclusions.

FACTORIZATION FORMALISM

We work in an effective theory where the weak bosons and top quark have been integrated out. The relevant part of the effective Hamiltonian, valid below M_W, is

$$\mathcal{H}_{\text{eff}} = \frac{G_F}{\sqrt{2}} V_{ud}^* V_{cb} (C_1 O_1 + C_8 O_8) \quad (2)$$

where the singlet and octet operators are, respectively,

$$O_1 = \bar{c}\gamma_\mu(1-\gamma_5)b\,\bar{d}\gamma^\mu(1-\gamma_5)u, \quad (3)$$
$$O_8 = \bar{c}\gamma_\mu(1-\gamma_5)T^a b\,\bar{d}\gamma^\mu(1-\gamma_5)T^a u. \quad (4)$$

The running of the Wilson coefficients C_i has been calculated at next-to-leading order [14, 15].

---

[1] Following the authors of [1, 2], we assume that the physical b quark mass is not so large that Sudakov form factors modify the power counting.

The computation of any Feynman amplitude in this theory involves evaluating matrix elements of the operators $O_{1,8}$. In general such matrix elements contain both hard and soft physics. One would like to disentangle these energy scales; the hard physics could be calculated directly and the soft physics parameterized by form factors and decay constants. Though it is not obvious that the physics can be separated in this way, it has been argued [1, 2] that in certain situations it is possible to do so.

More specifically, consider a decay of the form $\bar{B} \to HL$, where H is a heavy meson ($H = D, D^*$) and L is a light meson ($L = \pi, \rho$). We require that the topology of the decay be such that the light quark in the initial state is transferred to the heavy final state meson. In this case, and in the heavy quark limit, it is argued that 'non-factorizable' corrections are perturbative and may be calculated. These statements are summarized in the factorization equation [1, 2]

$$\langle HL|O_i|\bar{B}\rangle = F^{\bar{B}\to H}(m_L^2) f_L \int_0^1 dx\, T_i^I(x) \Phi_L(x) + \cdots \qquad (5)$$

where the ellipsis denotes contributions suppressed by powers of Λ_{QCD}/m_b. In this expression, the B decay form factor $F^{\bar{B}\to H}$ and the light meson decay constant f_L are the nonperturbative parameters present in the case of naive factorization (1). The 'non-factorizable' physics is contained in the convolution of the perturbatively calculable hard-scattering kernel $T_i^I(x)$ with the light-cone momentum distribution of the leading Fock state (quark-antiquark) of the light meson $\Phi_L(x)$. The parameter x is the momentum fraction of one of the quarks inside the light meson.

To leading order in α_s, one finds [2, 16]

$$T_1^I(x) = 1 + O(\alpha_s^2), \qquad T_8^I(x) = 0 + O(\alpha_s). \qquad (6)$$

Given that the light meson distribution function $\Phi_L(x)$ is normalized to unity, naive factorization (1) is restored as the leading term in a perturbative expansion.

There are other decay topologies (penguin, annihilation) for which certain assumptions leading to (5) are invalid. However, detailed arguments show that these topologies are suppressed by powers of Λ_{QCD}/m_b, and are therefore irrelevant in the heavy quark limit [2].

In the next section, we present the renormalon analysis of the 'non-factorizable' corrections. This will involve studying the large-order perturbative properties of (6) as a means of determining the order at which power corrections enter the factorization equation (5).

RENORMALON ANALYSIS

The purpose of the renormalon analysis is to extract non-perturbative information from the large-order behaviour of a perturbative series. To see how this is possible consider the gluon self-energy depicted in Figure 1. Inserting this dressed 'gluon' into a loop diagram is equivalent to evaluating the one-loop diagram with a running coupling. When the loop momentum is large, the coupling constant is small and perturbative. However, when the loop momentum is small, the coupling constant is large and thus non-perturbative.

A typical amplitude in QCD perturbation theory may be expressed in the general form

$$R(\alpha_s) = \sum_{n=0}^{\infty} R_n \alpha_s^{n+1}. \tag{7}$$

Normally one only calculates a few terms in this series. However, one may ask about the general behavior of this series at large orders in perturbation theory. It has been argued [6] that quantum field theories of phenomenological interest have large order coefficients of the form

$$R_n \stackrel{n \to \infty}{\sim} a^n n! n^b \tag{8}$$

for some constants a and b. Clearly, such a series is factorially divergent. It might appear that, as a result, a sum for the series cannot be defined. However, one may define the Borel sum $\tilde{R}$ in the following way. Perform a Borel transformation on the series:

$$R(\alpha_s) = \sum_{n=0}^{\infty} R_n \alpha_s^{n+1} \implies B[R](t) = \sum_{n=0}^{\infty} \frac{R_n}{n!} t^n. \tag{9}$$

This series in terms of the Borel parameter t is convergent and may be explicitly summed. One may then perform an inverse Borel transformation to obtain the Borel sum

$$\tilde{R} = \int_0^{\infty} dt \, e^{-t/\alpha_s} B[R](t). \tag{10}$$

The original series R and the Borel sum $\tilde{R}$ have the same series expansion.

In some cases, the transformed series $B[R](t)$ has poles along the positive real axis [6]. When these poles are encountered in the inverse Borel transformation (10), one is forced to deform the integration contour either above or below the real axis. Nothing specifies which choice to make, yet the result of the integration depends on the choice. As a result, the Borel sum acquires an ambiguity. For a simple pole located at $t_0 > 0$, the ambiguity is

$$\delta \tilde{R} \sim e^{-t_0/\alpha_s(\mu)}$$
$$\sim \left(\frac{\Lambda_{QCD}}{\mu}\right)^{-2\beta_0 t_0} = \left(\frac{\Lambda_{QCD}}{\mu}\right)^{2u_0} \tag{11}$$

where we have defined $u_0 = -\beta_0 t_0 > 0$. The ambiguity has the form of a nonperturbative power correction. For physical quantities, which cannot be ambiguous, there must be present power corrections to remove this ambiguity. Note that for $\mu > \Lambda_{QCD}$ it is the pole nearest the origin which gives the leading power correction. This simple sketch illustrates how the large-order perturbative behavior of the theory reveals something about the nonperturbative sector of the theory [6].

In general, this approach only permits one to determine the order of the power corrections and not their coefficients or analytic form. Furthermore, while a pole at u_0 in the Borel plane definitely indicates the presence of power corrections $\sim (\Lambda_{QCD}/m_b)^{2u_0}$, the absence of a pole does not necessarily imply the absence of power corrections of that order. The absence of a pole is, rather, suggestive that power corrections of the corresponding order are absent [6]. The renormalon technique has been applied in a variety of contexts where a general theory of power corrections has not been available [7, 8, 9, 10, 11, 12]. In Section we study power corrections to factorization in this way.

In practice one cannot sum the entire perturbative series (7) to obtain an exact expression for the Borel transformed amplitude (9). Instead, one sums a subset of the Feynman diagrams at each order, implicitly taking the resulting analytic structure to be characteristic of the full result. Typically, all-orders contributions are obtained by inserting into graphs 'bubble chain' propagators of the kind shown in Figure 1. One may take the formal limit $N_f \to -\infty$ with $\alpha_s N_f$ fixed, in which case the set of graphs with a single 'bubble chain' insertion are dominant [6]. This 'bubble chain' propagator, which we denote by a dashed gluon line, has the form [9, 13]

$$D_{\mu\nu}(k) = \frac{i}{k^2}\left(-g_{\mu\nu} + \frac{k_\mu k_\nu}{k^2}\right) \sum_{n=0}^{\infty} (\beta_0 N_f \alpha_s)^n (\ln(-k^2/\mu^2) + C)^n. \quad (12)$$

The factors in the sum arise from the fermion loops depicted in Figure 1. These loops have been renormalized in an MS-like scheme, and C is a scheme dependent constant. In the $\overline{\text{MS}}$ scheme, $C = -5/3$. We discuss renormalization of amplitudes containing the renormalon propagator in more detail in Section . The Borel transform of this propagator with respect to $\alpha_s N_f$ is

$$\begin{aligned}B[D_{\mu\nu}(k)](u) &= \frac{1}{\alpha_s N_f}\frac{i}{k^2}\left(-g_{\mu\nu} + \frac{k_\mu k_\nu}{k^2}\right) \sum_{n=0}^{\infty} \frac{(-u)^n}{n!}(\ln(-k^2/\mu^2) + C)^n \\ &= \frac{1}{\alpha_s N_f}\left(\frac{\mu^2}{e^C}\right)^u \frac{i}{(-k^2)^{2+u}}(k_\mu k_\nu - k^2 g_{\mu\nu}). \quad (13)\end{aligned}$$

The limit $u \to 0$ of this expression, equivalent to retaining only the first term in the expansion depicted in Figure 1, reduces to the usual gluon propagator as expected. For Feynman graphs in which the α_s dependence arises only from gluon exchange, the Borel transformed graph is obtained by replacing the gluon propagator by this renormalon propagator.

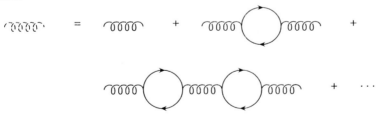

FIGURE 1. The renormalon 'bubble chain' propagator.

THE CALCULATION

For a particular heavy final state H (D, D^*) and light final state L $(\pi, \rho, \ldots)$ we must calculate the matrix elements

$$\langle O_{1,8} \rangle \equiv \langle H(p')L(q)|O_{1,8}|\bar{B}(p)\rangle. \tag{14}$$

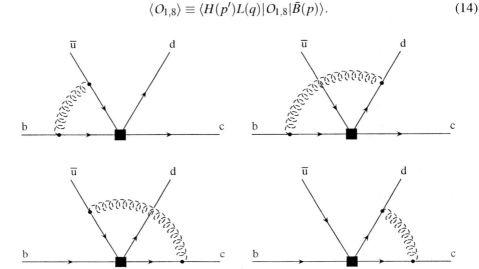

FIGURE 2. The factorization-breaking corrections with renormalon propagators.

The matrix element for the singlet operator is the simplest, so we consider it first. At leading order in α_s the operator factorizes cleanly into a product of currents. The leading factorization-breaking QCD corrections are shown in Figure 2. In order to create a color singlet structure from the resulting graph, one would have to consider the light meson to be in a Fock state higher than $(q'\bar{q})$; this situation, however, is power suppressed by Λ_{QCD}/m_b [2]. Alternatively, one could retain the leading Fock state and consider the exchange of two gluons rather than just one; this situation is suppressed by α_s relative to the graphs in Figure 2. In our calculation, then, the singlet matrix element is simply

$$\begin{aligned}\langle O_1 \rangle &= \langle L(q)|\bar{d}\gamma^\mu(1-\gamma_5)u|0\rangle \langle H(p')|\bar{c}\gamma_\mu(1-\gamma_5)b|\bar{B}(p)\rangle \\ &= if_L\Big(\langle J_V\rangle - \langle J_A\rangle\Big)\end{aligned} \tag{15}$$

where we define two matrix elements $\langle J_V \rangle = \langle H(p')|\bar{c}\slashed{q}b|\bar{B}(p)\rangle$ and $\langle J_A \rangle = \langle H(p')|\bar{c}\slashed{q}\gamma_5 b|\bar{B}(p)\rangle$.

The octet matrix element $\langle O_8 \rangle$ has a richer structure. It vanishes at leading order, but the color structures of the graphs shown in Figure 2 are such that $\langle O_8 \rangle$ receives perturbative corrections at all higher orders. Let us denote the individual amplitudes of the diagrams in Figure 2 as $\mathcal{A}_i$, and the corresponding Borel transformed amplitudes as $B[\mathcal{A}_i]$.

We have evaluated each of these graphs in $d = 4$ dimensions, as the divergences which would normally be present are regulated by the Borel parameter u in the renormalon propagator (13). In the limit $u \to 0$, however, each of the $B[\mathcal{A}_i](u)$ contain both infrared and ultraviolet divergences. It was a central result of [1, 2] to show that in the heavy quark limit the sum

$$B_0[\mathcal{A}](u) = \sum_{i=1}^{4} B[\mathcal{A}_i](u) \tag{16}$$

is infrared finite for amplitudes of the type we are considering. This cancellation may be seen explicitly from our amplitudes in the vicinity of $u \sim 0$.

$$B[\mathcal{A}_1](u) = \frac{if_L C_F}{2N_c} \langle J_{V-A} \rangle \left[\frac{1}{u^2} + \frac{1}{u} \left(-2\log x \left(1 - \frac{m_c^2}{m_b^2}\right) - C + \log \frac{\mu^2}{m_b^2} - 2 \right) \right] + \cdots$$

$$B[\mathcal{A}_2](u) = \frac{if_L C_F}{2N_c} \langle J_{V-A} \rangle \left[-\frac{1}{u^2} + \frac{1}{u} \left(2\log \bar{x} \left(1 - \frac{m_c^2}{m_b^2}\right) + C - \log \frac{\mu^2}{m_b^2} - 1 \right) \right] + \cdots$$

$$B[\mathcal{A}_3](u) = \frac{if_L C_F}{2N_c} \langle J_{V-A} \rangle \left[-\frac{1}{u^2} + \frac{1}{u} \left(2\log x \left(1 - \frac{m_b^2}{m_c^2}\right) + C - \log \frac{\mu^2}{m_c^2} - 1 \right) \right] + \cdots$$

$$B[\mathcal{A}_4](u) = \frac{if_L C_F}{2N_c} \langle J_{V-A} \rangle \left[\frac{1}{u^2} + \frac{1}{u} \left(-2\log \bar{x} \left(1 - \frac{m_b^2}{m_c^2}\right) - C + \log \frac{\mu^2}{m_c^2} - 2 \right) \right] + (17)$$

where the ellipses denote terms finite as $u \to 0$, and we introduce the shorthand $\langle J_{V-A} \rangle = \langle J_V \rangle - \langle J_A \rangle$. The collinear divergences cancel in pairs among the even and odd numbered amplitudes, while the remaining soft divergences cancel in the manner prescribed by Bjorken's color transparency argument [2, 17].

For the sum of the four amplitudes we find

$$B_0[\mathcal{A}](u \sim 0) = -\frac{if_L C_F}{2N_c} \langle J_{V-A} \rangle \frac{6}{u} + \text{finite}. \tag{18}$$

This remaining divergence is ultraviolet in origin, and may be renormalized in a manner consistent with MS-like subtraction schemes. To this end, we follow the prescription of [9, 18] by defining a renormalized amplitude

$$B[\mathcal{A}](u) = B_0[\mathcal{A}](u) + S_\mathcal{A}(u) \tag{19}$$

where $S_\mathcal{A}(u)$ contains a divergence which cancels that in $B_0[\mathcal{A}](u)$ at the origin but is finite elsewhere. This regulating function may be written as

$$S_\mathcal{A}(u) = \frac{1}{u} \sum_{n=0}^{\infty} \frac{g_n}{n!} u^n \tag{20}$$

where the coefficients g_n are the expansion coefficients of another function $G(\varepsilon) = \sum_{n=0} g_n \varepsilon^n$ related to the amplitudes computed in $d = 4 - 2\varepsilon$ dimensions. We refer the reader to the relevant literature for an explanation of this method [9, 18]. For the sum of amplitudes $\mathcal{A}$ we find

$$G(\varepsilon) = \frac{if_L C_F}{2N_c} \frac{2(1+\varepsilon)(1+2\varepsilon)(3+2\varepsilon)\Gamma(4+2\varepsilon)}{3\Gamma(1-\varepsilon)\Gamma^2(2+\varepsilon)\Gamma(3+\varepsilon)} \langle J_{V-A} \rangle$$

$$= \frac{if_L C_F}{2N_c} \langle J_{V-A} \rangle \left(6 + 23\varepsilon + \frac{127}{6}\varepsilon^2 + \cdots \right). \tag{21}$$

The first term in this expansion cancels the ultraviolet divergence remaining in (18); $B[\mathcal{A}](u)$ is infrared and ultraviolet finite.

In this notation, the Borel transform of the factorization equation (5) is

$$B[\langle O_8\rangle](u) = \int_0^1 dx\, B[\mathcal{A}](u)\Phi_L(x). \tag{22}$$

Comparing to (5), we see that $B[\mathcal{A}](u)$ is proportional to the Borel transform of the hard scattering kernel $T_8^I(x)$.

Borel poles and power corrections

Recall from (11) that it is the pole nearest the origin on the positive real axis in the Borel plane which indicates the leading power correction. A pole at the origin would indicate an $O(1)$ correction, but the renormalization procedure outlined in the previous section ensures that no such pole is present. We find that the first pole is located at $u = 1/2$, corresponding to a power correction of $O(\Lambda_{\rm QCD}/m_b)$. This is not a very surprising result, as power corrections of this order are known to be present from a variety of sources [2]. However, in the vicinity of this pole the Borel transformed amplitude has the form

$$B[\langle O_8\rangle](u \sim 1/2) \propto \frac{1}{u - 1/2} \int_0^1 dx\, \Phi_L(x) \frac{(x - \bar{x})}{x\bar{x}} \langle J_{V-A}\rangle + \cdots \tag{23}$$

where $\bar{x} = 1 - x$, and the ellipsis denotes nonsingular terms.

Before interpreting this result, we must specify the form of the light meson momentum distribution $\Phi_L(x)$. It is customary to write it as [19]

$$\Phi_L(x) = 6x(1-x)\left[1 + \sum_{n=1}^{\infty} \alpha_n^L(\mu) C_n^{3/2}(2x-1)\right] \tag{24}$$

where the Gegenbauer polynomials $C_n^{3/2}(y)$ are given by

$$C_n^{3/2}(y) = \frac{1}{n!} \frac{d^n}{dh^n}(1 - 2hy + h^2)^{-3/2}\bigg|_{h=0}. \tag{25}$$

The distribution function $\Phi_L(x)$ is a nonperturbative object for which the Gegenbauer moments $\alpha_n^L(\mu)$ are unspecified. It is known, however, that $\alpha_n^L(\mu \to \infty) = 0$ [19]. For $\mu \sim m_b \gg \Lambda_{\rm QCD}$, one may take the distribution function to have the asymptotic form $\Phi_L^0(x) = 6x(1-x)$ up to power corrections of $O(\Lambda_{\rm QCD}/m_b)$. In this case the wavefunction is symmetric under $x \to 1-x$, and the integration over x in (23) vanishes, removing the pole.

More generally, note that the Gegenbauer polynomials $C_n^{3/2}(2x-1)$ in (24) with n even are, like $\Phi_L^0(x)$, even under $x \to 1-x$, while those with n odd are odd under the same replacement. The result of the integration (23) may then be written as

$$B[\langle O_8\rangle](u \sim 1/2) \propto \frac{6}{u-1/2}\langle J_{V-A}\rangle \left(\alpha_1^L(\mu)+\alpha_3^L(\mu)+\alpha_5^L(\mu)+\cdots\right). \qquad (26)$$

Only the asymmetric terms of $\Phi_L(x)$ contribute to the residue. Thus the renormalon analysis indicates the presence of $O(\Lambda_{QCD}/m_b)$ factorization-breaking power corrections only for asymmetric light meson wavefunctions.

For certain final states of interest ($L = \pi, \rho, \ldots$), SU(2) symmetry ensures that the wavefunction is symmetric [20, 21]. In these cases the pole at $u = 1/2$ disappears, and the renormalon analysis gives no indication of factorization-breaking $O(\Lambda_{QCD}/m_b)$ corrections, suggesting that such power corrections are likely absent. This result implies that power corrections due to the 'non-factorizable' vertex diagrams in Figure 2 should be suppressed in $\bar{B} \to D^{(*)+}\pi^-$ and $\bar{B} \to D^{(*)+}\rho^-$ decays relative to decays into a light meson with an asymmetric distribution function, such as $\bar{B} \to D^{(*)+}K^-$.

Experimental data for these decays has recently become available and is presented in Table 1. The theoretical values are those calculated in [2], and are in good agreement with the experimental central values. It is clear that the data will need to be more precise before it will begin to give us theoretical information about the relative importance of power corrections in different channels.

TABLE 1. Branching Ratios for $\bar{B}_d \to D^{(*)+}L^-$ decays in units of 10^{-3}. The last two columns show the experimental data from the Belle collaboration [22] and the Particle Data Group [23].

Decay mode	Theory	Belle data	PDG2000
$\bar{B}_d \to D^+\pi^-$	3.27	-	3.0 ± 0.4
$\bar{B}_d \to D^+K^-$	0.25	0.20 ± 0.06	-
$\bar{B}_d \to D^+\rho^-$	7.64	-	7.9 ± 1.4
$\bar{B}_d \to D^{*+}\pi^-$	3.05	-	2.8 ± 0.2
$\bar{B}_d \to D^{*+}K^-$	0.22	0.21 ± 0.05	-
$\bar{B}_d \to D^{*+}\rho^-$	7.59	-	6.7 ± 3.3

In assessing this result, one should keep in mind that there are other sources of power corrections to these decays which our analysis does not address. For example, we have not considered soft interactions with the spectator quark, annihilation diagrams, or interactions with sub-leading Fock states involving additional soft partons [1, 2]. Our conclusions about the vanishing $O(\Lambda_{QCD}/m_b)$ corrections in the symmetric limit apply only to the class of diagrams we have considered.

The next renormalon pole is located at $u = 1$, corresponding to an $O(\Lambda_{QCD}^2/m_b^2)$ correction, and here there is a nonzero residue even in the case of a symmetric wavefunction $\Phi_L(x)$. Therefore one should expect power corrections of this order for all decay modes.

CONCLUSIONS

In this paper we have carried out a renormalon analysis of factorization-breaking effects in $\bar{B}$ meson decays to certain hadronic heavy-light final states. The renormalon approach, which probes the theory at high orders in perturbation theory, allows us to learn about nonperturbative power corrections. In the low energy effective theory governing the $\bar{B}$ decays there are two operators, a color singlet and a color octet. The factorization-breaking corrections to the singlet operator are, however, suppressed by powers of α_s or Λ_{QCD}/m_b relative to the octet operator; for this reason we focus our analysis on the octet.

We find that the renormalon analysis of 'non-factorizable' corrections to the octet matrix element indicates the presence of power corrections of $O(\Lambda_{QCD}/m_b)$. We also find that these leading power corrections are sensitive only to asymmetries in the light meson light-cone parton distribution function. Thus for a symmetric distribution function the leading renormalon pole vanishes, suggesting the absence of $O(\Lambda_{QCD}/m_b)$ 'non-factorizable' corrections in such cases. The next power corrections are present at $O(\Lambda_{QCD}^2/m_b^2)$.

As it is natural to expect the light meson wave function to be symmetric for certain final states ($L = \pi, \rho, \ldots$), the potential vanishing of the $O(\Lambda_{QCD}/m_b)$ power corrections in the symmetric limit is an interesting result that warrants a few additional comments. First, it should be noted that a particular decay $\bar{B} \to HL$ of the type we have considered in this paper always involves a combination of the singlet and octet operator matrix elements. Our result, however, applies only to the octet matrix element — we are unable to conclude from our analysis whether the leading power corrections to the singlet matrix element vanish in an analogous way.

Second, the power corrections we can probe through the renormalon analysis are physically due to soft and collinear 'non-factorizable' gluon exchange, and our comments about suppression of the leading corrections should be understood as referring to these effects only. There are, however, other $O(\Lambda_{QCD}/m_b)$ corrections originating from diagrams which were not considered in our analysis. A more detailed examination of these effects would have to be undertaken to accurately assess the size of $O(\Lambda_{QCD}/m_b)$ corrections to a physical decay amplitude.

REFERENCES

1. M. Beneke et al., Phys. Rev. Lett. **83**, 1914 (1999).
2. M. Beneke et al., Nucl. Phys. **B591**, 313 (2000).
3. C. Burrell and A. Williamson, Phys. Rev. D **64**, 034009 (2001).
4. T. Becher, M. Neubert and B. Pecjak, [hep-ph/0102219].
5. M. Luke, Phys. Lett. **B252** (1990) 447.
6. M. Beneke, *Phys.Rept.*, 317 (1999) 1.
7. A.H. Mueller, Nucl. Phys. **B250**, 327 (1985).
8. A.V. Manohar and M.B. Wise, Phys. Lett. B **344**, 407 (1995).
9. M. Beneke and V.M. Braun, Nucl. Phys. **B426**, 301 (1994).
10. I.I. Bigi et al., Phys. Rev. D **50**, 2234 (1994).
11. B.R. Webber, Phys. Lett. B **339**, 148 (1994).
12. M. Luke, A.V. Manohar, and M.J. Savage, Phys. Rev. D **51**, 4924 (1995).

13. M. Beneke, Nucl. Phys. **B405**, 424 (1993).
14. G. Altarelli *et al.*, Nucl. Phys. **B187**, 461 (1981).
15. A. Buras and P. Weisz, Nucl. Phys. **B333**, 66 (1990).
16. H.D. Politzer and M.B. Wise, Phys. Lett. B **257**, 399 (1991).
17. J.D. Bjorken, Nucl.Phys. (Proc.Suppl.) **B11** (1989) 325.
18. P. Ball, M. Beneke, and V.M. Braun, Nucl. Phys. **B452**, 563 (1995).
19. S.J. Brodsky and G.P. Lepage, Phys. Rev. D **22**, 2157 (1980).
20. P. Ball and V.M. Braun, Phys. Rev. D **54**, 2182 (1996).
21. E. Bagan, P. Ball, and V.M. Braun, Phys. Lett. B **417**, 154 (1998).
22. T. Iijima, Belle Collaboration, [hep-ph/0105005].
23. Particle Data Group, Eur. Phys. J. **C15** (2000) 1-878.

MIGNERON TRIBUTE SESSION II

Composites in color superconducting phase of dense QCD with two quark flavors

V.A. Miransky

Department of Applied Mathematics, University of Western Ontario, London, Ontario N6A 5B7, Canada

Abstract

We study the Bethe-Salpeter equations for spin zero diquark composites in the color superconducting phase of $N_f = 2$ cold dense QCD. The explicit form of the spectrum of the diquarks, containing an infinite tower of narrow (at high density) resonances, is derived. It is argued that there are five pseudo-Nambu-Goldstone bosons (pseudoscalars) that remain almost massless at large chemical potential. These five pseudoscalars should play an important role in the infrared dynamics of $N_f = 2$ dense QCD.

Only a few years ago, not much was known about the properties of different phases in dense quark matter (see, however, Refs. [1, 2]). The situation drastically changed after the ground breaking estimates of the color superconducting order parameter were obtained in Refs. [3, 4]. Within the framework of a phenomenological model, it was shown that the order parameter could be as large as 100 MeV. Afterwards, the same estimates were also obtained within the microscopic theory, quantum chromodynamics [5, 6, 7, 8, 9, 10, 11, 12]. The further progress in the field was mostly motivated by the hope that the color superconducting phase could be produced either in heavy ion experiment, or in the interior of neutron (or rather quark) stars.

Despite many advances [13, 14, 15, 16] in study of the color superconducting phase of dense quark matter, the detailed spectrum of the diquark bound states (mesons) is still poorly known. In fact, most of the studies deal with the NG bosons of the three flavor QCD. At best, the indirect methods of Refs. [13, 14, 15, 16] could probe the properties of the pseudo-NG bosons. It was argued in Ref. [17], however, that, because of long-range interactions mediated by the gluons of the magnetic type [5, 6], the presence of an infinite tower of massive diquark states could be the key signature of the color superconducting phase of dense quark matter.

Here we will consider the problem of spin zero bound states in the two flavor color superconductor using the Bethe-Salpeter (BS) equations. We find that the spectrum contains five (nearly) massless states and an infinite tower of massive singlets with respect to the unbroken $SU(2)_c$ subgroup. Furthermore, in the hard dense loop improved ladder approximation, the following mass formula is derived for the singlets:

$$M_n^2 \simeq 4|\Delta|^2 \left(1 - \frac{\alpha_s^2 \kappa}{(2n+1)^4}\right), \quad n = 1, 2, \ldots, \tag{1}$$

where κ is a constant of order 1 (we find that $\kappa \simeq 0.27$), $|\Delta|$ is the dynamical Majorana mass of quarks in the color superconducting phase, and α_s is the value of the running coupling constant related to the scale of the chemical potential μ.

At large chemical potential, we also notice an approximate degeneracy between scalar and pseudoscalar channels. As a result of this parity doubling, the massive diquark states come in pairs. In addition, there also exist five massless scalars and five (nearly) massless pseudoscalars [a doublet, an antidoublet and a singlet under $SU(2)_c$]. While the scalars are removed from the spectrum of physical particles by the Higgs mechanism, the pseudoscalars remain in the spectrum, and they are the relevant degrees of freedom of the infrared dynamics. At high density, the massive and (nearly) massless states are narrow resonances.

In the case of two flavor dense QCD, the original gauge symmetry $SU(3)_c$ breaks down to the $SU(2)_c$ by Higgs mechanism. The flavor $SU(2)_L \times SU(2)_R$ group remains intact at the vacuum. The appropriate order parameter is an antitriplet in color and a singlet in flavor. Without loss of generality, we assume that the order parameter points in the third direction of the color space. In order to have a convenient description of the bound states at the true vacuum, we introduce the following Majorana spinors,

$$\Psi_a^i = \psi_a^i + \varepsilon_{3ab}\varepsilon^{ij}(\psi^C)_j^b, \quad a = 1, 2, \tag{2}$$

$$\Phi_a^i = \phi_a^i - \varepsilon_{3ab}\varepsilon^{ij}(\phi^C)_j^b, \quad a = 1, 2, \tag{3}$$

made of the Weyl spinors of the first two colors,

$$\psi_a^i = \mathcal{P}_+(\Psi_D)_a^i, \qquad (\psi^C)_j^b = \mathcal{P}_-(\Psi_D^C)_j^b, \tag{4}$$

$$\phi_a^i = \mathcal{P}_-(\Psi_D)_a^i, \qquad (\phi^C)_j^b = \mathcal{P}_+(\Psi_D^C)_j^b. \tag{5}$$

Here $i, j = 1, 2$ are flavor indices, $\mathcal{P}_\pm = (1 \pm \gamma^5)/2$ are the left- and right-handed projectors, Ψ_D is the Dirac spinor, and $\Psi_D^C = C\bar{\Psi}_D^T$ is its charge conjugate. Regarding the quark of the third color, we use the Weyl spinors, ψ^i and ϕ^i, for left and right components, respectively (notice that the color index is omitted).

The BS wave functions of the bound diquark states in the channels of interest are given by

$$(2\pi)^4 \delta^4(p_+ - p_- - P)\chi_a^{(\tilde{b})}(p,P) =$$
$$= \langle 0|T\Psi_a^i(p_+)\bar{\psi}_i(-p_-)|P;\tilde{b}\rangle_L, \tag{6}$$
$$(2\pi)^4 \delta^4(p_+ - p_- - P)\lambda_{(\tilde{a})}^b(p,P) =$$
$$= \langle 0|T\psi^i(p_+)\bar{\Psi}_i^b(-p_-)|P;\tilde{a}\rangle_L, \tag{7}$$
$$(2\pi)^4 \delta^4(p_+ - p_- - P)\boldsymbol{\eta}(p,P) =$$
$$= \langle 0|T\Psi_a^i(p_+)\bar{\Psi}_i^a(-p_-)|P\rangle_L, \tag{8}$$
$$(2\pi)^4 \delta^4(p_+ - p_- - P)\boldsymbol{\sigma}(p,P) =$$
$$= \langle 0|T\psi^i(p_+)\bar{\psi}_i(-p_-)|P\rangle_L, \tag{9}$$

where $p = (p_+ + p_-)/2$ and the quantities on the right hand side of these equations are defined as the Fourier transforms of the corresponding BS wave functions in the coordinate space. There are also the BS wave functions constructed out of the right handed fields Φ_a^i and ϕ^i. One might notice that there is another diquark channel, a triplet under $SU(2)_c$, that we do not consider here. The reason is that the repulsion dominates in such a channel, and no bound states are expected [the triplet comes from the $SU(3)_c$ sextet].

In order to derive the BS equations, we use the method developed in Ref. [18] for the case of zero chemical potential. To this end, we need to know the quark propagators and the quark-gluon interactions.

By introducing the multicomponent spinor that combines the Majorana spinors of the first two colors and the Weyl spinors of the third color, $\left(\Psi_b^j, \psi^j, \psi_j^C\right)^T$, we find that the inverse propagator takes the following block-diagonal form:

$$G_p^{-1} = \text{diag}\left(S_p^{-1}\delta_a{}^b\delta^i{}_j, \; s_p^{-1}\delta^i{}_j, \; \bar{s}_p^{-1}\delta_i{}^j\right), \tag{10}$$

where, upon neglecting the wave functions renormalization of quarks [6, 7, 8, 9, 10, 11, 12],

$$S_p^{-1} = -i\left(\slashed{p} + \mu\gamma^0\gamma^5 + \Delta_p\mathcal{P}_- + \tilde{\Delta}_p\mathcal{P}_+\right), \tag{11}$$

$$s_p^{-1} = -i\left(\slashed{p} + \mu\gamma^0\right)\mathcal{P}_+, \tag{12}$$

$$\bar{s}_p^{-1} = -i\left(\slashed{p} - \mu\gamma^0\right)\mathcal{P}_-. \tag{13}$$

Here the notation, $\Delta_p = \Delta_p^+ \Lambda_p^+ + \Delta_p^- \Lambda_p^-$, $\tilde{\Delta}_p = \gamma^0 \Delta_p^\dagger \gamma^0$, and $\Lambda_p^\pm = (1 \pm \vec{\alpha} \cdot \vec{p}/|p|)/2$ are the same as in Ref. [7].

The bare vertex, $\gamma^{A\mu}$, is also a 3×3 matrix,

$$\gamma^{A\mu} = \gamma^\mu \begin{pmatrix} \bar{\gamma}_{11}^A \delta^i{}_j & \bar{\gamma}_{12}^A \delta^i{}_j & \bar{\gamma}_{13}^A \varepsilon^{ij} \\ \bar{\gamma}_{21}^A \delta^i{}_j & \bar{\gamma}_{22}^A \delta^i{}_j & \bar{\gamma}_{23}^A \varepsilon^{ij} \\ \bar{\gamma}_{31}^A \varepsilon_{ij} & \bar{\gamma}_{32}^A \varepsilon_{ij} & \bar{\gamma}_{33}^A \delta_i{}^j \end{pmatrix}, \tag{14}$$

with

$$\left(\bar{\gamma}_{11}^A\right)_a^b = T_a^{Ab} - 2\delta_8^A T_a^{8b} \mathcal{P}_-, \tag{15}$$

$$\bar{\gamma}_{22}^A = T_3^{A3} \mathcal{P}_+, \tag{16}$$

$$\bar{\gamma}_{33}^A = -T_3^{A3} \mathcal{P}_-, \tag{17}$$

$$\left(\bar{\gamma}_{12}^A\right)_a = T_a^{A3} \mathcal{P}_+, \tag{18}$$

$$\left(\bar{\gamma}_{13}^A\right)_a = -\varepsilon_{3ac} T_3^{Ac} \mathcal{P}_-, \tag{19}$$

$$\left(\bar{\gamma}_{21}^A\right)^b = T_3^{Ab} \mathcal{P}_+, \tag{20}$$

$$\left(\bar{\gamma}_{31}^A\right)^b = -T_c^{A3} \varepsilon^{3cb} \mathcal{P}_-, \tag{21}$$

$$\left(\bar{\gamma}_{23}^A\right)_a = 0, \tag{22}$$

$$\left(\bar{\gamma}_{32}^A\right)^b = 0, \tag{23}$$

where T^A are the $SU(3)_c$ generators in the fundamental representation. By making use of this vertex and the propagator in Eq. (10), it is straightforward to derive the BS equations in the (hard dense loop improved) ladder approximation. The details of the derivation, as well as the explicit form of equations are given elsewhere [19]. Here we just note that the most transparent form of the equations appears for the amputated BS wave functions, defined by

$$\chi(p,P) = S^{-1}(p + \frac{P}{2})\boldsymbol{\chi}(p,P)s^{-1}(p - \frac{P}{2}), \tag{24}$$

$$\lambda(p,P) = s^{-1}(p + \frac{P}{2})\boldsymbol{\lambda}(p,P)S^{-1}(p - \frac{P}{2}), \tag{25}$$

$$\eta(p, P) = S^{-1}(p + \frac{P}{2})\boldsymbol{\eta}(p, P)S^{-1}(p - \frac{P}{2}), \qquad (26)$$

$$\sigma(p, P) = s^{-1}(p + \frac{P}{2})\boldsymbol{\sigma}(p, P)s^{-1}(p - \frac{P}{2}). \qquad (27)$$

In order to get a feeling of the problem at hand, let us briefly discuss the analysis of the BS equation for the χ-doublet. In general, the BS wave function contains eight different Dirac structures [20]. It is of great advantage to notice that only four of them survive in the center of mass frame, $P = (M_b, \vec{0})$,

$$\chi_a^{(\tilde{b})}(p, 0) = \delta_a^{\tilde{b}} \hat{\chi}(p), \qquad (28)$$

where

$$\hat{\chi}(p) = \left[\chi_1^- \Lambda_p^+ + (p_0 - \epsilon_p^- + \frac{M_b}{2})\chi_2^- \gamma^0 \Lambda_p^+ \right.$$
$$\left. + \chi_1^+ \Lambda_p^- + (p_0 + \epsilon_p^+ + \frac{M_b}{2})\chi_2^+ \gamma^0 \Lambda_p^- \right] \mathcal{P}_+, \qquad (29)$$

with $\epsilon_p^\pm = |\vec{p}| \pm \mu$ [the factors $(p_0 \pm \epsilon_p^\pm + M_b/2)$ are introduced here for convenience]. This is the most general structure that is allowed by the space-time symmetries of the model.

Now, in the particular case of the NG bosons, $M_b = 0$, we will show that the BS wave function is fixed by the Ward identities. Indeed, let us consider the following non-amputated vertex:

$$\boldsymbol{\Gamma}_{aj,\mu}^{A,i}(x, y) = \langle 0 | T j_\mu^A(0) \Psi_a^i(x) \bar{\psi}_j(y) | 0 \rangle, \qquad (30)$$

where, for our purposes, it is sufficient to consider $A = 4, \ldots, 8$ (that correspond to the five broken generators). In the (hard dense loop improved) ladder approximation, the vertex satisfies the following Ward identity [19]:

$$P^\mu \boldsymbol{\Gamma}_{aj,\mu}^{A,i}(k+P, k) = i T_a^{A3} \delta_j^i [s_k - S_{k+P}] \mathcal{P}_-. \qquad (31)$$

As in the case of the BS wave functions, it is more convenient to deal with the corresponding amputated quantity,

$$\Gamma_{aj,\mu}^{A,i}(k+P, k) = S_{k+P}^{-1} \boldsymbol{\Gamma}_{aj,\mu}^{A,i}(k+P, k) s_k^{-1}. \qquad (32)$$

This latter satisfies the following identity:

$$P^\mu \Gamma_{aj,\mu}^{A,i}(k+P, k) = i T_a^{A3} \delta_j^i \left[S_{k+P}^{-1} - s_k^{-1} \right] \mathcal{P}_+. \qquad (33)$$

By making use of the explicit form of the quark propagators in Eqs. (11) and (12), we could check that the right hand side of Eq. (33) is non-zero in the limit $P \to 0$. This is possible only if the vertex on the left hand side develops a pole as $P \to 0$. After a simple calculation, we obtain

$$\Gamma^{A,i}_{aj,\mu}(k+P,k)\Big|_{P\to 0} \simeq \frac{\tilde{P}^\mu}{P_\nu \tilde{P}^\nu} T_a^{A3} \delta_j^i \tilde{\Delta}_k \mathcal{P}_+, \qquad (34)$$

where, we introduced $\tilde{P}^\mu = (P_0, c_\chi^2 \vec{P})$ with c_χ being the velocity of the NG boson in the χ-doublet channel.

By making use of the definition in Eqs. (6) and (24), it is also not difficult to show that the pole contribution to the vertex function (34) is directly related to the BS wave function. By omitting the details,

$$\chi_a^{(\tilde{a})}(p,0) \equiv \delta_a^{(\tilde{a})} \chi(p,0) = \delta_a^{(\tilde{a})} \frac{\tilde{\Delta}_p}{F^{(\chi)}} \mathcal{P}_+, \qquad (35)$$

where $F^{(\chi)}$ is the decay constant of the corresponding doublet whose formal definition is given by

$$\langle 0| \sum_{A=4}^{7} T_a^{A3} j_\mu^A(0) |P,\tilde{b}\rangle_L = i\delta_a^{\tilde{b}} \tilde{P}_\mu F^{(\chi)}. \qquad (36)$$

By comparing the Dirac structures in Eqs. (29) and (35), we see that no components of the $\chi_2^\pm$ type appear in Eq. (35) which follows from the Ward identities. It was rewarding to establish that, in this approximation, the structure of the BS wave function required by the Ward identity is indeed a solution to the BS equation for the χ-doublet. A similar situation takes place for the η-singlet [21].

Now let us discuss the fate of the massless states that we obtain. Altogether, there are five scalars and five pseudoscalars (a doublet, an antidoublet and a singlet). Because of the Higgs mechanism, the scalars are removed from the spectrum. Nevertheless, these scalar bound states exist in the theory as "ghosts" [22], and one cannot get rid of them completely, unless a unitary gauge is found. In fact, these ghosts play a very important role in getting rid of unphysical poles from the on-shell scattering amplitudes [22].

As for the pseudoscalars, they remain in the spectrum as pseudo-NG bosons. In the (hard dense loop improved) ladder approximation, they look like NG bosons because the left and right sectors of quarks decouple. One could think of this as an effective enlargement of the original color symmetry from $SU(3)_c$ to an approximate $SU(3)_{c,L} \times SU(3)_{c,R}$. Then, since the approximate symmetry of the ground state is $SU(2)_{c,L} \times SU(2)_{c,R}$, five scalar NG bosons (which are removed by the Higgs mechanism) and five pseudoscalar NG bosons (which remain in the spectrum) should appear. Of course, in the full theory, the pseudoscalars are only pseudo-NG bosons. Indeed, they should get non-zero masses due to higher orders corrections that are beyond the improved ladder approximation [23]. Since the theory is weakly coupled at large chemical potential, it is natural to expect that the masses of the pseudo-NG bosons are small compared to the value of the dynamical quark mass.

We conclude our discussion of the massless diquarks by emphasizing that the low-energy dynamics of the two flavor QCD is dominated by massless quarks of the third color (which might eventually get a small mass too if another (non-scalar) condensate is generated [3, 24]) and by the five pseudoscalars that remain almost massless in the dense quark matter. Of course, the gluons (glueballs) of the unbroken $SU(2)_c$ may also be of some relevance but we do not study this question here.

Now, let us consider massive diquarks. The structure of the BS equations becomes even more complicated in this case. In addition, one does not have a rigorous argument to neglect the component functions like $\chi_2^{\pm}$ in Eq. (29). In spite of this, we argue that all the approximations made before might still be reliable. Indeed, from the experience of solving the gap equation (which coincides with the BS equation for the massless states), we know that the most important region of momenta in the integral equation is $|\Delta| \ll p \ll \mu$. In this region, the kernel of the BS equations for massive states, $M_b \sim |\Delta|$, is almost the same. The deviations appear only in the infrared region where $p < |\Delta|$.

Therefore, in our analysis of the BS equations for massive states, we closely follow the approximation used for the massless diquarks. By assuming that the component functions depend only on the time component of the momentum (compare with the analysis of the gap equation in Refs. [6, 7, 8, 9, 10, 11, 12]), we arrive at the following equation for the BS wave function of the massive singlets:

$$\eta_1^-(p) = \frac{\alpha_s}{4\pi} \int_0^\Lambda dq K^{(\eta)}(q) \eta_1^-(q) \ln \frac{\Lambda}{|q-p|}, \qquad (37)$$

where $\Lambda = (4\pi)^{3/2}\mu/\alpha_s^{5/2}$, and the kernel reads

$$K^{(\eta)}(q) = \frac{\sqrt{q^2 + |\Delta|^2}}{q^2 + |\Delta|^2 - (M_\eta/2)^2}, \qquad (38)$$

$[\eta_1^-(p)$ is a scalar function that appears in the decomposition of the BS wave function η over the Dirac matrices (compare with Eq. (29)]. At this point it is appropriate to emphasize that the Meissner effect plays an important role in the analysis of the massive bound states. Indeed, our analysis shows that these massive states are quasiclassical in nature, i.e., their binding energy is small compared to the value of the gap [see Eq. (1)]. As a result, only the long range interaction mediated by the unscreened gluons of the unbroken $SU(2)_c$ is strong enough to produce these diquark states. We took this into account in Eq. (37). For completeness, we mention that the (nearly) massless (pseudo-) NG bosons are tightly bound, and the Meissner effect is not so important for their binding dynamics.

By approximating the kernel (38) in each of the following three regions: $0 < q < \sqrt{|\Delta|^2 - (M_\eta/2)^2}$, $\sqrt{|\Delta|^2 - (M_\eta/2)^2} < q < |\Delta|$ and $|\Delta| < q < \Lambda$, we could solve the BS equation (37) analytically. Then, by matching the logarithmic derivatives of the separate solutions, we obtain the spectrum of the massive diquarks. By omitting the details, it is presented in Eq. (1).

We would like to emphasize that for large μ the hard dense loop improved ladder approximation is reliable for the description of those bound states. The point is that a) the region of momenta primarily responsible for the formation of these composites is $E_{bind} < q \ll \mu$, where the binding energy $E_{bind} \sim \alpha_s^2 \Delta$, and b) $E_{bind} \to \infty$ as $\mu \to \infty$ [6, 7, 8, 9]. Therefore the vacuum effects are higher order ones in α_s in that region. Because of that, the hard dense loop improved ladder approximation, in which the contribution of the vacuum effects to the running of the coupling constant is neglected and only the running due to the polarization effects provided by the quark matter (non-zero μ) is taken into account, is justifiable for large μ.

Now, let us consider the case of massive diquarks in the doublet channel. As is easy to check, the binding interaction in this channel is exclusively due to the five gluons affected by the Meissner effect. The approximate BS equation looks similar to Eq. (37), but with a different kernel and $|q - p|$ in the logarithm replaced by $|q - p| + c\Delta$ where $c = O(1)$ is a constant [7]. At high density when the coupling constant is weak, this equation does not allow a non-trivial solution for $M \neq 0$. From the physical viewpoint, this indicates that the heavy gluons, with $M_{gl} \sim (\alpha_s \mu^2 \Delta)^{1/3} \gg \Delta$, cannot provide a sufficiently strong attraction to form massive radial excitations of the NG and pseudo-NG bosons.

At the end, let us note that the massive diquark states may truly be just resonances in the full theory, since they could decay into the pseudo-NG bosons and/or gluons (glueballs) of the unbroken $SU(2)_c$. At high density, however, both the running coupling $\alpha_s(\mu)$ and the effective Yukawa coupling $g_Y = |\Delta|/F \sim |\Delta|/\mu$ [14, 15, 16, 25]) are small, and, therefore, these massive resonances are narrow.

In conclusion, we considered the problem of diquark bound states in the color superconducting phase of $N_f = 2$ dense QCD. While the scalar NG bosons are ghosts in the theory, the pseudoscalar pseudo-NG bosons are physical particles that should play an important role in the infrared. We also obtained the spectrum of the massive narrow diquark resonances, whose existence would be a clear signature of the unscreened long range forces in dense QCD.

Acknowledgments. This talk is based on the paper [19] done in collaboration with I.A. Shovkovy and L.C.R. Wijewardhana.

References

[1] B.C. Barrois, Nucl. Phys. **B129** (1977) 390; S.C. Frautschi, in "Hadronic matter at extreme energy density", edited by N. Cabibbo and L. Sertorio (Plenum Press, 1980).

[2] D. Bailin and A. Love, Nucl. Phys. **B190** (1981) 175; *ibid.* **B205** (1982) 119; Phys. Rep. **107** (1984) 325.

[3] M. Alford, K. Rajagopal and F. Wilczek, Phys. Lett. B **422** (1998) 247.

[4] R. Rapp, T. Schaefer, E.V. Shuryak and M. Velkovsky, Phys. Rev. Lett. **81** (1998) 53.

[5] R.D. Pisarski and D.H. Rischke, Phys. Rev. Lett. **83** (1999) 37.

[6] D.T. Son, Phys. Rev. D **59** (1999) 094019.

[7] D.K. Hong, V.A. Miransky, I.A. Shovkovy and L.C.R. Wijewardhana, Phys. Rev. D **61** (2000) 056001.

[8] T. Schafer and F. Wilczek, Phys. Rev. D **60** (1999) 114033.

[9] R.D. Pisarski and D.H. Rischke, Phys. Rev. D **61** (2000) 05150.

[10] S.D.H. Hsu and M. Schwetz, Nucl. Phys. **B572** (2000) 211.

[11] W.E. Brown, J.T. Liu and H.-C. Ren, Phys. Rev. D **61** (2000) 114012; *ibid.* D **62** (2000) 054016.

[12] I.A. Shovkovy and L.C.R. Wijewardhana, Phys. Lett. B **470** (1999) 189; T. Schafer, Nucl. Phys. **B575** (2000) 269.

[13] R. Casalbuoni and R. Gatto, Phys. Lett. B **464** (1999) 111; hep-ph/9911223; D.K. Hong, M. Rho, and I. Zahed, Phys. Lett. B **468** (1999) 261.

[14] D.T. Son and M.A. Stephanov, Phys. Rev. D **61** (2000) 074012.

[15] M. Rho, A. Wirzba and I. Zahed, Phys. Lett. B **473** (2000) 126; M. Rho, E. Shuryak, A. Wirzba and I. Zahed, Nucl. Phys. **A676** (2000) 273.

[16] D.K. Hong, T. Lee and D.-P. Min, Phys. Lett. B **477** (2000) 137; C. Manuel and M.G.H. Tytgat, *ibid.* B **479** (2000) 190; K. Zarembo, Phys. Rev. D **62** (2000) 054003; S.R. Beane, P.F. Bedaque and M.J. Savage, Phys. Lett. B **483** (2000) 131.

[17] V.A. Miransky, I.A. Shovkovy and L.C.R. Wijewardhana, Phys. Lett. B **468** (1999) 270.

[18] P.I. Fomin, V.P. Gusynin, V.A. Miransky and Yu.A. Sitenko, Riv. Nuovo Cimento, **6** No. 5 (1983) 1.

[19] V.A. Miransky, I.A. Shovkovy and L.C.R. Wijewardhana, Phys. Rev. D **62** (2000) 085025.

[20] Note that in the case of zero chemical potential there are only four different Dirac structures, see Ref. [18].

[21] In connection with the Ward identities, it is appropriate to mention here the complementary analysis in W.E. Brown, J.T. Liu and H.-C. Ren, Phys. Rev. D **62** (2000) 054013. The authors of this paper consider the contribution to the Ward identity that is directly related to the quark wave function renormalization.

[22] R. Jackiw and K. Johnson, Phys. Rev. D **8** (1973) 2386; J.M. Cornwall and R.E. Norton, *ibid.* D **8** (1973) 3338.

[23] An example of such higher order corrections is the box diagram in the BS kernel with two intermediate gluons. We would like to point out that the phenomenon of the pseudo-NG bosons was first considered in S. Weinberg, Phys. Rev. Lett. **29** (1972) 1698; Phys. Rev. D **7** (1973) 2887. As in those papers, the existence of the pseudo-NG bosons in two flavor dense QCD is connected with the presence of an extended symmetry in the leading order, which is *not* a symmetry in the full theory.

[24] T. Schaefer, Phys. Rev. D **62** (2000) 094007.

[25] D.H. Rischke, Phys. Rev. D **62** (2000) 034007; *ibid.* D **62** (2000) 054017.

Infrared Dynamics in Vector-Like Gauge Theories: QCD and Beyond

Victor Elias

*Department of Applied Mathematics, The University of Western Ontario
London, Ontario N6A 5B7 Canada*

Abstract

Padé-approximant methods are used to extract information about leading positive zeros or poles of QCD and SQCD β-functions from the known terms of their perturbative series. For QCD, such methods are seen to corroborate the flavour-threshold behaviour obtained via lattice approaches for the occurrence of infrared-stable fixed points. All possible Padé-approximant versions of the known (one- to four-loop) terms of the QCD $\overline{MS}$ β-function series are consistent with this threshold occurring at or above $n_f = 6$. This conclusion continues to be true even if higher-degree Padé-approximants are introduced to accommodate an *arbitrary* five-loop contribution to the QCD β-function for a given number of flavours.

The dynamics characterising the infrared region of asymptotically-free gauge theories are becoming an area of active investigation. Historically speaking, the property of asymptotic freedom has been linked with "infrared slavery" at large distances – the non-observability of gauge-group nonsinglet particles as asymptotic free-particle states because of the growth of the gauge coupling constant to arbitrarily large values at some infrared momentum scale. Such methodological simplification, however, rests upon the existence of a Landau singularity in the evolution of the gauge couplant, a singularity that is likely an artefact of truncating the gauge couplant's β-function to any given order. Consider, for example, the QCD β-function to two-loop order for three active flavours:

$$\mu^2 \frac{dx}{d\mu^2} = -\frac{9}{4}x^2 - 4x^3, \quad x \equiv \alpha_s(\mu)/\pi. \tag{1}$$

Given some empirical initial value, e.g. $x_\tau = \alpha_s(m_\tau)/\pi$, eq.(1) has the analytic solution

$$\log\left(\mu^2/m_\tau^2\right) = \frac{4}{9}\left[\frac{1}{x(\mu)} - \frac{1}{x_\tau}\right] + \frac{64}{81}\log\left[\frac{x(\mu)(x_\tau + 9/16)}{x_\tau(x(\mu) + 9/16)}\right]. \tag{2}$$

As μ decreases from m_τ in this expression, the couplant $x(\mu)$ grows from x_τ to become infinite at

$$\mu_L = m_\tau \left(1 + 9/(16 x_\tau)\right)^{32/81} \exp\left[-2/(9 x_\tau)\right] \tag{3}$$

the Landau singularity. The temptation is to identify μ_L with the infrared boundary of QCD as a perturbative gauge theory of quarks and gluons. Note that if $\alpha_s(m_\tau) = \pi x_\tau = 0.35$, the central PDG value [1], then μ_L is found from (3) to be 493 MeV.

However, this Landau singularity is really a consequence of truncating the perturbative β-function series (1) to its renormalization scheme independent two-loop order terms. Subsequent β-function terms are renormalization-scheme-dependent and presumably negotiable (I will have more to say on this later on). For example, we can conjecture two toy-model β-functions

$$\mu^2 \frac{dx}{d\mu^2} = -\frac{9}{4}\frac{x^2}{(1-16x/9)} \tag{4}$$

$$\mu^2 \frac{dx}{d\mu^2} = -\frac{9}{4}x^2\left(1 + \frac{16}{9}x + R_2 x^2\right), \quad R_2 < 0 \tag{5}$$

which both agree to two-loop order with eq. (2), but have infrared behaviour manifestly different from the Landau pole characterising (2). The former equation has a β-function pole at $x = 9/16 = 0.5625$. Consequently, as μ decreases from m_τ, x increases from x_τ to a maximum value of 0.5625, which is achieved at a critical infrared momentum scale

$$\mu_c = m_\tau \left(\frac{9}{16x_\tau}\right)^{32/81} \exp\left[\frac{32}{81} - \frac{2}{9x_\tau}\right] \tag{6}$$

The domain of $x(\mu)$ is restricted to $\mu > \mu_c$, in which case μ_c represents an infrared-boundary of QCD at which the gauge couplant achieves its largest possible *but finite* value. The existence of an infrared boundary is no longer coupled with the arbitrary growth of the interaction coupling.

By contrast, the β-function of eq. (5) has a zero at $x_{IRFP} = -8/(9R_2) + \sqrt{64/(81R_2^2) - 1/R_2}$ (R_2 is negative), meaning that $x(\mu)$ increases to eventually level off at x_{IRFP} as $\mu \to 0$. For this case, which corresponds to the existence of an infrared-stable fixed point (IRFP), *there is no infrared boundary* to gauge-theoretical QCD – the domain of $x(\mu)$ includes all (positive) values of μ.

A point that needs to be made here is that if (1), (4), and (5) were to represent all-orders β-functions arising from legitimate but differing renormalization schemes, [1] then the infrared behaviour of the couplant would itself be scheme-dependent. Indeed, one could easily construct a β-function candidate consistent with (1) to two-loop order whose higher order terms are sufficiently large and negative [*e.g.* an arbitrarily large negative R_2 in (5)] to ensure that an IRFP occurs at an arbitrarily *small* value for x, suggesting that QCD in such a scheme would remain a tractable perturbative gauge theory at arbitrarily large

---
[1] Equation (1) the β-function truncated after scheme-independent two-loop order terms, has already been used to generate an apparently self-consistent renormalization scheme [2].

distances. Since the infrared behaviour of QCD is empirically *known* to be confining at sufficiently large distances, such apparent scheme-dependence must be overstated. We therefore take the point of view here that differing but self-consistently realized renormalization schemes will necessarily lead to the same physical results – in particular, to equivalent infrared dynamics.

A vector-like gauge theory that illustrates all this is that of $N = 1$ supersymmetric SU(3) Yang-Mills theory, supersymmetric gluodynamics characterised only by gluon and gluino fields. Because of its supersymmetry, the all-orders β-function for this theory can be obtained algebraically by requiring that the anomaly multiplet within the theory respect the Adler-Bardeen theorem [3,4], or alternatively via instanton-calculus considerations [5]. Surprisingly, the same β-function arises from either approach, and even more surprisingly, this β-function, like (4), is characterised *by a pole*:

$$\mu^2 \frac{dx}{d\mu^2} = -\frac{9x^2}{1 - 6x}, \quad x \equiv g^2(\mu)/16\pi^2 \tag{7}$$

As noted in [6], such a pole necessarily implies a critical momentum scale μ_c for the asymptotically free phase of the theory that would constitute an infrared boundary to the region for which the theory is perturbative. The pole value of the couplant [$x(\mu_c) = 1/6$] constitutes an infrared-attractive point terminating the evolution of the (real) gauge coupling constant in its asymptotically-free phase. We henceforth denote (7) as the NSVZ β-function, after the authors of ref. 5.

The β-function for supersymmetric gluodynamics has also been calculated to three non-leading orders within the dimensional-reduction (DRED) renormalization scheme [7]:

$$\mu^2 \frac{dx}{d\mu^2} = -9x^2[1 + 6x + 63x^2 + 918x^3 + ...] \tag{8}$$

The higher order terms of the β-function (8) differ from those of the geometric series implicit in (7) once one gets beyond their equivalent leading and next-to-leading terms. Both NSVZ and DRED schemes are presumably valid ones (although only the former upholds the Adler-Bardeen Theorem to all orders). One can even obtain a perturbative road-map between couplants in the two schemes: if $z \equiv x^{NSVZ}$ and $y \equiv x^{DRED}$, then [4]

$$y = z\left[1 + \sigma z + (27 + 6\sigma + \sigma^2)z^2 + (351 + 117\sigma + 15\sigma^2 + \sigma^3)z^3 + ...\right] \tag{9}$$

where the constant σ is *arbitrary* as a consequence of both (asymptotically-free) couplants having identical leading and next-to-leading terms within their respective β-function series.[2] Consequently, unless one specifies initial values

---

[2] An arbitrary constant in the relation between perturbative couplants in two different renormalization schemes of a given theory will occur provided the leading and next-to-leading terms of the β-function series in both schemes are the same. Such is the case, of course, for conventional QCD.

for couplants in each scheme, one cannot say anything at all about the relative size of couplants in the two different schemes at a given momentum scale. If, for example, the all-order extension of the DRED series (8) exhibits a pole, such a pole need not occur at its $x = 1/6$ NSVZ location. Nevertheless, consistent infrared dynamics *does require* that a pole indeed occur in the DRED scheme. In other words, the all-orders extension of (8) should not exhibit the Landau-pole or IRFP behaviour seen to characterise of the β-function examples (1) and (5), respectively. Rather, the DRED scheme should the same infrared dynamics as are evident from (7): a finite infrared-attractive point $x(\mu_c)$ terminating the evolution of the asymptotically-free phase of the couplant, with μ_c serving as an infrared bound on the domain of $x(\mu)$.

To get some insight into the infrared behaviour of DRED supersymmetric gluodynamics, the four known terms of the β-function series (8) have been utilised to construct Padé approximants – ratios of polynomials in x whose power-series expansions reproduce these known terms. Padé approximants are often employed not only to predict next order terms of a series, such as the β-function series for QCD [8], scalar field theory [8,9,10], and supersymmetric QCD [4,11], but also to explore whether infinite series with only a few known terms can be expected to have zeros *or poles* [12]. [3] In reference [4], for example, it is demonstrated that every Padé approximant (except the truncated series itself) that reproduces the first four known terms of a series which differs infinitesimally from the geometric series $1/(1 - |r|x)$ exhibits a pole that

1) differs infinitesimally from the true pole at $x = 1/|r|$, and

2) occurs prior to any positive spurious Padé-approximant zeros.

In other words, the pole driven infrared dynamics of the NSVZ β-function (7) could have been predicted from Padé approximants constructed from that series' first four terms.

Relevant Padé approximants whose power series reproduce the four known terms of the DRED β-function (8) are [4]

$$\mu^2 \frac{dx}{d\mu^2} = -9x^2 \frac{(1 - 14x)}{(1 - 20x + 57x^2)} \tag{10}$$

$$\mu^2 \frac{dx}{d\mu^2} = -9x^2 \frac{(1 - 8.5714x - 24.4286x^2)}{(1 - 14.5714x)} \tag{11}$$

$$\mu^2 \frac{dx}{d\mu^2} = -9x^2 \frac{1}{(1 - 6x - 27x^2 - 810x^3)} \tag{12}$$

The first positive pole of (10) occurs at $x = 0.0604$ and precedes the zero at $x = 1/14$. Similarly, the pole of (11) at $x = 1/14.5714 = 0.0686$ precedes

---

[3] Straightforward examples of this method for recovering the first positive pole or zero of $\sec(x) \pm \tan(x)$ are presented in ref. [13].

that approximant's first positive zero ($x = 0.0924$), and the approximant (12) exhibits a first positive pole at $x = 0.0773$. Although (12) is constructed to have no zeros other than $x = 0$, there is no *a priori* reason that this approximant should exhibit a positive pole of comparable magnitude to those of (10) and (11). These results (and further analysis of higher approximants presented in [4]) are clearly indicative of the same pole-driven infrared dynamics that characterise the NSVZ β-function (7) for the same $N = 1$ supersymmetric SU(3) Yang-Mills field theory.

The techniques illustrated above are applicable to QCD itself. There is considerable controversy as to the infrared dynamics which characterise QCD, as well as the flavour dependence of such dynamics. The idea that the QCD couplant freezes out to some effective $\alpha_s(\mu = 0)$ at sufficiently low momentum scales appears to have both some theoretical justification [14] (even for the $n_f = 0$ case [15]) and as well as phenomenological utility [16]. However, such IRFP dynamics for $n_f \lesssim 6$ are inconsistent with both theoretical studies based upon a β-function series truncated after its first two scheme independent terms [17] as well as with a lattice study indicative of an $n_f = 7$ threshold for IRFP dynamics [18].

A Padé-approximant approach to the infrared dynamics of QCD similar to the example presented above is formulated in detail in refs. [9], [13], and [19]. In this work, the known (and first unknown) terms of the QCD $\overline{MS}$ β-function series [20]

$$\mu^2 \frac{dx}{d\mu^2} = -\beta_0 x^2 S(x), \quad \beta_0 = 11/12 - n_f/6, \quad x \equiv \alpha_s(\mu)/\pi \qquad (13)$$

$$\begin{aligned} S(x) &= 1 + [(51/8 - 19n_f/24)/\beta_0] x \\ &+ [(2857/2 - 5033n_f/18 + 325n_f^2/54)/64\beta_0] x^2 \\ &+ [(114.23 - 27.134 n_f + 1.5824 n_f^2 + 5.8567 \cdot 10^{-3} n_f^3)/\beta_0] x^3 \\ &+ R_4 x^4 + ... \end{aligned} \qquad (14)$$

are used to construct various Padé-approximants whose leading positive zeros/poles are compared to see which occur first (or occur at all). We choose to use the $\overline{MS}$ renormalization shceme simply because the β-function series in this scheme is known to higher order than in any other perturbative scheme – moreover, phenomenological QCD is overwhelmingly based upon $\overline{MS}$ calculations. We denote by $S^{[N|M]}(x)$ the Padé-approximant to $S(x)$ whose numerator and denominator are respectively degree-N and degree-M polynomials of the couplant x. If $N + M = 3$, the power-series expansion of the approximant $S^{[N|M]}(x)$ is of sufficiently high order to reproduce the known series terms in (14); *e.g.* for $n_f = 3$

$$S^{[2|1]}(x) = \frac{1 - 2.9169 x - 3.8750 x^2}{1 - 4.6947 x} \qquad (15)$$

$$S^{[1|2]}(x) = \frac{1 - 8.1734x}{1 - 9.9511x + 13.220x^2} \tag{16}$$

In both of these approximants, a positive pole precedes any positive zeros, indicative of the same pole-driven infrared dynamics known to characterise NSVZ supersymmetric gluodynamics. If one constructs such approximants for any choice of n_f (as in ref. [13]), one finds for $n_f \leq 5$ that $S^{[2|1]}$ always has a positive pole preceding any positive zeros; moreover, the same statement applies to $S^{[1|2]}$ as well, provided $n_f \leq 6$. These results suggest that pole-driven infrared dynamics characterise QCD for up to five or six flavours. The occurrence of positive zeros that are not preceded by positive poles, (*i.e.*, β-function zeros corresponding to IRFP dynamics) does not characterise $S^{[2|1]}$ until $n_f \geq 7$, and does not characterise $S^{[1|2]}$ until $n_f \geq 9$, corroborating the existence of a flavour threshold of IRFP behaviour [17,18] as well as the absence of such behaviour below this threshold [21].

In ref. [13], these same qualitative conclusions are shown to be upheld by $S^{[N|M]}$ for N+M =5, *i.e.*, for Padé-approximants to $S(x)$ whose power series are constructed to reproduce the four known terms of (14) and the *arbitrary* five loop term $R_4 x^4$. In such approximants, each numerator and denominator polynomial coefficient of powers of x is itself linear in the parameter R_4 – *e.g.* for $n_f = 3$

$$S^{[2|2]}(x) = \frac{1 + (7.1946 - 0.10261 R_4)x + (-11.329 + 0.075644 R_4)x^2}{1 + (5.4168 - 0.10261 R_4)x + (-25.430 + 0.25806 R_4)x^2} \tag{17}$$

Regardless of the value R_4 takes, one finds that positive zeros which precede every positive pole do not occur in the approximants $S^{[1|3]}$, $S^{[2|2]}$, or $S^{[3|1]}$ until $n_f \geq 9, 7,$ and 6, respectively, consistent with the non-occurrence of IRFP dynamics below these threshold values of n_f. Moreover, for arbitrary R_4 a positive pole which precedes every positive zero *does* occur in the approximants $S^{[1|3]}$ and $S^{[2|2]}$, provided $n_f \leq 5$. The approximant $S^{[3|1]}$ has a single positive pole only if $R_4 > 0$, a consequence of having a denominator linear in x, but this pole is found to precede any numerator zeros in the approximant for all positive values of R_4 as long as $n_f \leq 7$.

These results are clearly indicative of the occurrence of pole-driven infrared dynamics for QCD once heavy flavours are decoupled. Similar pole-driven infrared dynamics are also shown in [13] to characterise QCD in the $N_c \to \infty$ 't Hooft limit. Moreover, very recent work [18] has demonstrated that different approximants to the $N_c = 3$, $n_f = 3$ $\overline{MS}$ QCD β-function exhibit surprising consistency in their predictions of infrared boundary coordinates $(\mu_c, x(\mu_c))$ associated with pole-driven infrared dynamics. The implications of such dynamics, in particular the possibility of having both a strong and an asymptotically-free phase of QCD with common infrared properties [6], have only begun to be explored.

Acknowledgements

I am grateful to my research collaborators F. A. Chishtie, V. A. Miransky, and T. G. Steele, who coauthored of much of the research described above. I am also indebted to my deceased research collaborator Mark Samuel, who pioneered the application of Padé approximants to perturbative quantum field theory, and to Roger Migneron, whose final research paper [9] is the first published work in which Padé-approximants are applied to the infrared structure of QCD.

References

1. D. E. Groom et al. [Particle Data Group], Eur. Phys. J. C **15**, 1 (2000).

2. G. 't Hooft, in Recent Developments in Gauge Theories, Vol. 59 of NATO Advanced Study Institute Series B: Physics, edited by G. t Hooft et al. (Plenum, N.Y., 1980).

3. D. R. T. Jones, Phys. Lett. B **123**, 45 (1983).

4. V. Elias, J. Phys. G **27**, 217 (2001).

5. V. Novikov, M. Shifman, A. Vainshtein, and V. Zakharov, Nucl. Phys. B **229**, 381 (1983).

6. I. I. Kogan and M. Shifman, Phys. Rev. Lett. **75**, 2085 (1995).

7. L. N. Avdeev, G. A. Chochia, and A. A. Vladimirov, Phys. Lett. B **105**, 272 (1981); I. Jack, D. R. T. Jones, and A. Pickering, Phys. Lett. B **435**, 61 (1998).

8. J. Ellis, M. Karliner, and M. A. Samuel, Phys. Lett. B **400**, 176 (1997).

9. V. Elias, T. G. Steele, F. Chishtie, R. Migneron, and K. Sprague, Phys. Rev. D **58**, 116007 (1998).

10. F. Chishtie, V. Elias, and T. G. Steele, Phys. Lett. B **466**, 266 (1999); F. A. Chishtie and V. Elias, Phys. Lett. B **499**, 270 (2001).

11. I. Jack, D. R. T. Jones, and M. A. Samuel, Phys. Lett. B **407**, 143 (1997); J. Ellis, I. Jack, D. R. T. Jones, M. Karliner, and M. A. Samuel, Phys. Rev. D **57**, 2665 (1998).

12. G. Baker and P. Graves-Morris, Pad Approximants [Vol. 13 of Encyclopedia of Mathematics and its Applications] (Addison-Wesley, Reading, MA, 1981) pp. 48-57.

13. F. A. Chishtie, V. Elias, V. A. Miransky, and T. G. Steele, Prog. Theor. Phys. **104**, 603 (2000).

14. A. C. Mattingly and P. M. Stevenson, Phys. Rev. Lett. **69**, 1320 (1992); P. M. Stevenson, Phys. Lett. B **331**, 187 (1994).

15. A. C. Mattingly and P. M. Stevenson, Phys. Rev. D **49**, 437 (1994).

16. Yu. L. Dokshitzer, in Proceedings of the 29th International Conference in High Energy Physics, A. Astbury, D. Axen, and J. Robinson, eds. (World Scientific, Singapore, 1999) pp. 305-324.

17. T. Banks and A. Zaks, Nucl. Phys. B **196**, 189 (1982); T. Appelquist, J. Terning, and L. C. R. Wijewardhana, Phys. Rev. Lett. **77**, 1214 (1996); V. A. Miransky and K. Yamawaki, Phys. Rev. D55, 5051 and (Err.) D **56**, 3768 (1997).

18. Y. Iwasaki, K. Kanaya, S. Sakai, and T. Yoshi, Phys. Rev. Lett. **69**, 21 (1992).

19. F. A. Chishtie, V. Elias, and T. G. Steele, Phys. Lett. B **514**, 279 (2001).

20. T. van Ritbergen, J. A. M. Vermaseren, and S. A. Larin, Phys. Lett. B **405**, 323 (1997).

21. E. Gardi, G. Grunberg, and M. Karliner, JHEP 9807, 007 (1998).

Topics on Neutrino Physics

G. Karl[*] and V. Novikov[†]

[*]*Department of Physics, University of Guelph, Guelph, ON N1G 2W1, Canada*

[†]*ITEP, Moscow, Russia*

Abstract. The birefringence of a neutrino sea is discussed in the standard model.

We had intended to discuss two topics: one is on the possibility of 50 GeV neutrinos [1], and the other on the Optical Activity of a Neutrino Sea [2]. It turned out that time constraints only allowed the second topic to be discussed.

PREHISTORY [3]

Optical active media were discovered in 1811-12 by Arago and Biot who studied the propagation of plane polarized light through certain media, and showed that the plane of polarization is rotated through passage in these "active" media. In 1817 Fresnel elucidated that the eigenstates of propagation are the right and left handed circularly polarized states which propagate at different speeds. These correspond to two indices of refraction n_R and n_L for the R and L polarized light, and so one speaks of birefringence. The rotatory angle $\Delta\varphi = k(n_R - n_L) L/2$, where L is the length of the sample and k is the wave number of the light wave. The difference between the two indices of refraction n_R and n_L corresponds to a difference between forward scattering amplitudes f(0) for right and left handed circularly polarized light: $n_L - n_R = \pi N \text{Re}[f_{LL}(0) - f_{RR}(0)]/k^2$, where N is the density of the scattering centers. Pasteur demonstrated in 1848 that the medium is responsible for birefringence, and not a parity violating interaction.

MODERN HISTORY

In the last few years, 1998-1999, a number of authors have speculated on the possibility that the vacuum might be birefringent [4]. Such an effect would indicate a violation of CPT and Lorentz invariance. This could come about if there are gauge invariant terms for electromagnetism of the form:

$$\delta L = (1/2)\, b_\mu \varepsilon^{\mu\alpha\beta\gamma} F_{\alpha\beta} A_\gamma$$

Such a Chern-Simons term is renormalizable and gauge invariant, but parity violating and would result in different speeds for R and L polarized light. The vector b_μ could result from an external axial field. Another possibility to realize birefringence is from a neutrino sea.

BIREFRINGENCE OF A NEUTRINO SEA

In the standard cosmological model one expects a neutrino sea with the number of neutrinos approximately the same as the number of photons in the cosmic microwave background: $N_v \sim 100/cm^3$, or $k_F \sim 2 \times 10^{-5}$ eV. There are also claims [5] that the neutrino sea is much more abundant $N_v \sim 10^{15}/cm^3$ or $k_F \sim 1 eV$. If the neutrino sea is dominated by one handedness (say v_L) then photons propagating through such a medium would show birefringence. This was noted by Royer in 1968. The neutrino plays the role of left-handed sugar in creating birefringence. The photons interact with the neutrino through an electron loop. Royer estimated the rotatory power $\varphi \sim G_F \alpha k_F^2 \sim 10^{-80}$ rad/cm, which is rather tiny. However, this estimate was incorrect. Indeed, there is a theorem due to Gell-Mann (1961), which under the assumption of pointlike weak interactions and massless neutrinos leads to a forward scattering amplitude for photons on neutrinos which vanishes. So the Royer estimate vanishes as well.

The theorem is easy to comprehend in the cross channel: $\gamma\gamma$ to $v\,\bar{v}$ where the neutrino antineutrino state should have J=1 and this is forbidden by Landau's theorem for a two photon state. To escape Gell-Mann's restriction one needs nonlocal interactions and this is the case in the Standard Model (of particle physics) where the nonlocality is of order m_W^{-1}. There are a number of diagrams which contribute to γv scattering, and the amplitude does not vanish but contains small factors (pk/m_W^2) which vanish indeed in the limit m_W going to infinity. The P-even amplitude was evaluated by Levine in 1967. It is second order in photon moments k (due to gauge invariance) and has the form:

$$T = C\, G_F\, \alpha\, (\varepsilon_\mu^* \varepsilon_\mu)\, (pk)\, (pk/m_W^2)$$

This expression was evaluated by Levine through an hours computation on an IBM7090 in 1967. This shows the progress in hardware.

However this is a P even amplitude and does not contribute to birefringence. The P odd amplitude in the standard model was evaluated and is third order in the photon momentum k, so that there are two extra factors of (pk/m_W^2). The exact expression is

$$T = (e^4/8\pi^2 s^2)\, \varepsilon_{\mu\nu\alpha\beta}\, \varepsilon^{\mu*} \varepsilon^\nu\, p^\alpha k^\beta (1/m_W^2)(pk/m_W^2)(4/3)[\ln(m_W/m_e) - 11/3]$$

This amplitude has $T_{LL} = -T_{RR}$, so it leads to birefringence. But the answer is very small.

It is amusing that the two-loop amplitude is larger. The two loop amplitude has two electrons and a neutrino in the intermediate state and can be evaluated in the limit of m_W large, but G_F fixed. The two loop amplitude is large because the range of the interaction is of the order of the electron's Compton wavelength. The form of the two loop amplitude is:

$$T^{(2)} = C(G_F^2 e^2/(16\pi^2)^2)\varepsilon_{\mu\nu\alpha\beta}\varepsilon^{\mu*}\varepsilon^{\nu}p^{\alpha}k^{\beta}(1/m_e^2)(pk)^2$$

where C is a constant. The comparison of the one loop and two loop amplitudes shows that the two loop amplitude is larger by a factor of 10^7, coming from the fact that $(e^2/16\pi^2 m_e^2) \gg (1/m_W^2)$. The resulting birefringence is still small even if we take the optimistic estimate of $k_F \sim 1eV$ and visible light with $\omega \sim 1eV$, since the factor pk/m_W^2 is about 10^{-22}.

These estimates show that the effect is insignificant for radio-waves from galaxies. However, there may be hope for X-ray signals, or for experiments with laser beams scattering from neutrino beams. Another possibility for the study of such effects is in the early Universe when the black-body radiation was interacting with a neutrino sea. The parity violation would then be imprinted in the cosmic microwave background. To conclude: if there is a neutrino sea, the Universe is slightly birefringent.

REFERENCES

[1] M. Maltoni, V. Novikov, M. Vysotsky, L. Okun and A. Rozanov: *Physics Letters* B**476**, 107-115 (2000).

[2] G. Karl and V. Novikov: preprint *hep-ph*/0009012.

[3] see e.g. G. Karl, *Can.J.Phys.***54**, 568 (1976).

[4] R. Jackiw and V.A. Kostelecky, *Phys.Rev.Let.* **82**,3572 (1999), and references therein, S. Coleman and S. Glashow, *hep- ph*/9812418.

[5] V.M. Lobashev et al., *Physics Letters* B**460**, 227 (1999).

QCD Sum-Rule Bounds on the Light Quark Masses

T.G. Steele

Department of Physics & Engineering Physics, University of Saskatchewan
Saskatoon, SK S7N 5E2, Canada

Abstract. QCD sum-rules are related to an integral of a hadronic spectral function, and hence must satisfy integral inequalities which follow from positivity of the spectral function. Development of these Hölder inequalities and their application to the Laplace sum-rule for pions lead to a lower bound on the average of the non-strange 2 GeV light-quark masses in the $\overline{MS}$ scheme.

The light quark masses are fundamental parameters of QCD, and determination of their values is of importance for high-precision QCD phenomenology and lattice simulations involving dynamical quarks. In this paper the development of Hölder inequalities for QCD Laplace sum-rules [1] is briefly reviewed. These techniques are then used to obtain bounds on the non-strange (current) quark masses $m_n = (m_u + m_d)/2$ evaluated at 2 GeV in the $\overline{MS}$ scheme, updating and extending the Hölder inequality results of ref. [2].

Although it is possible to obtain quark mass ratios in various contexts [3], the only methods which have been able to determine the *absolute* non-strange quark mass scales are the lattice (see [4] for recent results with two dynamical flavours) and QCD sum-rules [2, 5, 6, 7, 8, 9].[1]

In sum-rule and lattice approaches, the pseudoscalar or scalar channels are used since they have the strongest dependence on the quark masses. This is exemplified by the correlation function $\Pi_5(Q^2)$ of renormalization-group (RG) invariant pseudoscalar currents with quantum numbers of the pion:

$$\Pi_5(Q^2) = i\int d^4x\, e^{iq\cdot x} \langle O|T[J_5(x)J_5(0)]|O\rangle \quad (1)$$

$$J_5(x) = \frac{1}{\sqrt{2}}(m_u + m_d)\left[\bar{u}(x)i\gamma_5 u(x) - \bar{d}(x)i\gamma_5 d(x)\right] \quad . \quad (2)$$

The Laplace sum-rule is obtained by by applying the Borel transform operator [11] $\hat{B}$

$$\hat{B} \equiv \lim_{\substack{N,\, Q^2 \to \infty \\ Q^2/N \equiv M^2}} \frac{(-Q^2)^N}{\Gamma(N)}\left(\frac{d}{dQ^2}\right)^N \quad (3)$$

---

[1] An overview of selected lattice and sum-rule results for both non-strange and strange masses can be found in [10].

to the dispersion relation for $\Pi_5(Q^2)$

$$\Pi_5(Q^2) = a + bQ^2 + \frac{Q^4}{\pi} \int_{4m_\pi^2}^{\infty} \frac{\rho_5(t)}{t^2(t+Q^2)} dt \quad , \tag{4}$$

where $\rho_5(t)$ is the hadronic spectral function appropriate to the pion quantum numbers, and the quantities a and b represent subtraction constants. The resulting Laplace sum-rule relating the theoretically-determined quantity

$$R_5(M^2) = M^2 \hat{B}[\Pi_5(Q^2)] \tag{5}$$

to phenomenology is

$$R_5(M^2) = \frac{1}{\pi} \int_{4m_\pi^2}^{\infty} \rho_5(t) \exp\left(-\frac{t}{M^2}\right) dt \quad . \tag{6}$$

Perturbative contributions to $R_5(M^2)$ are known up to four-loop order in the $\overline{\text{MS}}$ [5, 12]. Infinite correlation-length vacuum effects in $R_5(M^2)$ are represented by the (non-perturbative) QCD condensate contributions [5, 11, 13]. In addition to the QCD condensate contributions the pseudoscalar (and scalar) correlation functions are sensitive to finite correlation-length vacuum effects described by direct instantons [14] in the instanton liquid model [15]. Combining all these results, the total result for $R_5(M^2)$ to leading order in the light-quark masses is [2]

$$\begin{aligned} R_5(M^2) &= \frac{3m_n^2 M^4}{8\pi^2}\left(1 + 4.821098\frac{\alpha}{\pi} + 21.97646\left(\frac{\alpha}{\pi}\right)^2 + 53.14179\left(\frac{\alpha}{\pi}\right)^3\right) \\ &+ m_n^2\left(-\langle m\bar{q}q\rangle + \frac{1}{8\pi}\langle \alpha G^2\rangle + \frac{\pi\langle O_6\rangle}{4M^2}\right) \\ &+ m_n^2 \frac{3\rho_c^2 M^6}{8\pi^2} e^{-\rho_c^2 M^2/2}\left[K_0\left(\rho_c^2 M^2/2\right) + K_1\left(\rho_c^2 M^2/2\right)\right] \quad , \end{aligned} \tag{7}$$

where α and $m_n = (m_u + m_d)/2$ are the $\overline{\text{MS}}$ running coupling and quark masses at the scale M, and $\rho_c = 1/(600\,\text{MeV})$ represents the instanton size in the instanton liquid model [15]. $SU(2)$ symmetry has been used for the dimension-four quark condensates (i.e. $(m_u + m_d)\langle \bar{u}u\rangle + \bar{d}d\rangle) \equiv 4m\langle \bar{q}q\rangle$), and $\langle O_6\rangle$ denotes the dimension six quark condensates

$$\langle O_6\rangle \equiv \alpha_s\left[\left(2\langle \bar{u}\sigma_{\mu\nu}\gamma_5 T^a u \bar{u}\sigma^{\mu\nu}\gamma_5 T^a u\rangle + u \to d\right) - 4\langle \bar{u}\sigma_{\mu\nu}\gamma_5 T^a u \bar{d}\sigma^{\mu\nu}\gamma_5 T^a d\rangle \right. \\ \left. + \frac{2}{3}\langle \left(\bar{u}\gamma_\mu T^a u + \bar{d}\gamma_\mu T^a d\right)\sum_{u,d,s}\bar{q}\gamma^\mu T^a q\rangle\right] \tag{8}$$

The vacuum saturation hypothesis [11] will be used as a reference value for $\langle O_6\rangle$

$$\langle O_6\rangle = f_{vs}\frac{448}{27}\alpha\langle \bar{q}q\bar{q}q\rangle = f_{vs} 3 \times 10^{-3}\,\text{GeV}^6 \tag{9}$$

where $f_{vs} = 1$ for exact vacuum saturation. Larger values of effective dimension-six operators found in [16, 17] imply that f_{vs} could be as large as $f_{vs} = 2$. The quark condensate is determined by the GMOR (PCAC) relation

$$(m_u + m_d)\langle \bar{u}u + \bar{d}d \rangle = 4m\langle \bar{q}q \rangle = -2f_\pi^2 m_\pi^2 \tag{10}$$

where $f_\pi = 93\,\text{MeV}$. A recent determination of the gluon condensate $\langle \alpha G^2 \rangle$ will be used: [18]

$$\langle \alpha G^2 \rangle = (0.07 \pm 0.01)\,\text{GeV}^4 \quad . \tag{11}$$

However, it should be noted that there is some discrepancy between [18] and the smaller value $\langle \alpha G^2 \rangle = (0.047 \pm 0.014)\,\text{GeV}^4$ found in [17].

Note that *all* the theoretical contributions in (7) are proportional to m_n^2, demonstrating that the quark mass sets the scale of the pseudoscalar channel. This dependence on the quark mass can be singled out as follows:

$$R_5(M^2) = [m_n(M)]^2 \, G_5(M^2) \tag{12}$$

where G_5 is independent of m_n and is trivially extractable from (7). Higher-loop perturbative contributions in (7) are thus significant since they can effectively enhance the quark mass with increasing loop order.

Determinations of the non-strange quark mass m_n using the sum-rule (6) require input of a phenomenological model for the spectral function $\rho_5(t)$. The mass m_n can then be determined by fitting to find the best agreement between the phenomenological model and the theoretical prediction respectively appearing on the right- and left-hand sides of (6). For example, the simple resonance(s) plus continuum model

$$\frac{1}{\pi}\rho_5(t) = 2f_\pi^2 m_\pi^4 \left[\delta(t - m_\pi^2) + \frac{F_\Pi^2 M_\Pi^4}{f_\pi^2 m_\pi^4} \delta(t - M_\Pi^2) \right] + \Theta(t - s_0) \frac{1}{\pi}\rho^{QCD}(t) \tag{13}$$

represents the pion pole (m_π), a narrow-width approximation to the pion excitation (M_Π) such as the $\Pi(1300)$, and a QCD continuum above the continuum threshold $t = s_0$. Of course more detailed phenomenological models can be considered which take into account possible width effects for the pion excitation, further resonances, resonance(s) enhancement of the 3π continuum *etc.* This leads to significant model dependence which partially accounts for the spread of theoretical estimates in [8, 9]. Since the common phenomenological portion of all these models is the pion pole, it is valuable to extract quark mass *bounds* which only rely upon the input of the pion pole on the phenomenological side of (6).

The existence of such bounds is easily seen by separating the pion pole out from $\rho_5(t)$, in which case (6) becomes

$$[m_n(M)]^2 \, G_5(M^2) = 2f_\pi^2 m_\pi^4 + \frac{1}{\pi} \int_{9m_\pi^2}^\infty \rho_5(t) \exp\left(-\frac{t}{M^2}\right) dt \quad . \tag{14}$$

Since $\rho_5(t) \geq 0$ in the integral appearing on the right-hand side of (14), a bound on the quark mass is obtained:

$$m_n(M) \geq \sqrt{\frac{2f_\pi^2 m_\pi^4}{G_5(M^2)}} \qquad (15)$$

Analysis of these bounds following from simple positivity of the "residual" portion on the right-hand side of (14) was studied in [5, 6].

Improvements upon the positivity bound of (15) are achieved by developing more stringent inequalities based on the positivity of $\rho_5(t)$. Since $\rho_5(t) \geq 0$, the right-hand (phenomenological) side of (6) must satisfy integral inequalities over a measure $d\mu = \rho_5(t) dt$. In particular, Hölder's inequality over a measure $d\mu$ is

$$\left| \int_{t_1}^{t_2} f(t)g(t) d\mu \right| \leq \left(\int_{t_1}^{t_2} |f(t)|^p d\mu \right)^{\frac{1}{p}} \left(\int_{t_1}^{t_2} |g(t)|^q d\mu \right)^{\frac{1}{q}} , \frac{1}{p} + \frac{1}{q} = 1 \,;\, p, q \geq 1 \quad , (16)$$

which for $p = q = 2$ reduces to the familiar Schwarz inequality, implying that the Hölder inequality is a more general constraint. The Hölder inequality can be applied to Laplace sum-rules by identifying $d\mu = \rho_5(t) dt$, $\tau = 1/M^2$ and defining

$$S_5(\tau) = \frac{1}{\pi} \int_{\mu_{th}}^{\infty} \rho_5(t) e^{-t\tau} dt \qquad (17)$$

where μ_{th} will later be identified with $9m_\pi^2$. Suitable choices of $f(t)$ and $g(t)$ in the Hölder inequality (16) yield the following inequality for $S_5(\tau)$ [1]:

$$S_5(\tau + (1-\omega)\delta\tau) \leq [S_5(\tau)]^\omega [S_5(\tau + \delta\tau)]^{1-\omega} \quad , \forall\, 0 \leq \omega \leq 1 \quad . \qquad (18)$$

In practical applications of this inequality, $\delta\tau \leq 0.1\,\text{GeV}^{-2}$ is used, in which case this inequality analysis becomes local (depending only on the Borel scale M and not on $\delta\tau$) [1, 2].

To employ the Hölder inequality (18) we separate out the pion pole by setting $\mu_{th} = 9m_\pi^2$ in (17).

$$S_5(M^2) = R_5(M^2) - 2f_\pi^2 m_\pi^4 = \int_{9m_\pi^2}^{\infty} \rho_5(t) e^{-t\tau} dt \qquad (19)$$

which has a right-hand side in the standard form (17) for applying the Hölder inequality. Note that simple positivity of $\rho_5(t)$ gives the inequality

$$S_5(M^2) \geq 0 \qquad (20)$$

which simply rephrases (15). Lower bounds on the quark mass m_n can now be obtained by finding the minimum value of m_n for which the Hölder inequality (18) is satisfied. Introducing further phenomenological contributions (*e.g.* three-pion continuum) give a slightly larger mass bound as will be discussed later. However, if only the pion pole is separated out, then the analysis is not subject to uncertainties introduced by the phenomenological model.

Although the details are still a matter of dispute, the overall validity of QCD predictions at the tau mass is evidenced by the analysis of the tau hadronic width, hadronic contributions to $\alpha_{EM}(M_Z)$ and the muon anomalous magnetic moment [19], so we impose the inequality (18) at the tau mass scale $M = M_\tau = 1.77\,\text{GeV}$. This also has the advantage of minimizing perturbative uncertainties in the running of α and m_n, since the the PDG reference scale for the light-quark masses is at $2\,\text{GeV}$ [20], in close proximity to M_τ, and the result $\alpha_s(M_\tau) = 0.33 \pm 0.02$ [21] can thus be used to its maximum advantage. For the remaining small energy range in which the running of α and m_n is needed, the four-loop β-function [22] and four-loop anomalous mass dimension [23] with three active flavours are used, appropriate to the analysis of [21]. This use of the $2\,\text{GeV}$ reference scale for m_n combined with input of $\alpha(M_\tau)$ improves upon the perturbative uncertainties in [2] which employed $\alpha(M_Z)$ and a $1\,\text{GeV}$ m_n reference scale which necessitated matching through the (uncertain) b and c flavour thresholds.

Further theoretical uncertainties devolve from the QCD condensates as given in (11) and (9) with $1 \leq f_{vs} \leq 2$, along with a 15% uncertainty in the instanton liquid parameter ρ_c [15]. The effect of higher-loop perturbative contributions to $R_5(M^2)$ on the resulting m_n bounds is estimated using an asymptotically-improved Padé estimate [24] of the five-loop term, introducing a $138(\alpha/\pi)^4$ correction into (7). Finally, we allow for the possibility that the overall scale of the instanton is 50% uncertain.

The resulting Hölder inequality bound on the $2.0\,\text{GeV}$ $\overline{\text{MS}}$ quark masses, updating the analysis of [2], is

$$m_n(2\,\text{GeV}) = \frac{1}{2}[m_u(2\,\text{GeV}) + m_d(2\,\text{GeV})] \geq 2.1\,\text{MeV} \quad (21)$$

This final result is identical to previous bounds on $m_n(1\,\text{GeV})$ [2] after conversion to $2\,\text{GeV}$ by the PDG [20], indicative of the consistency of perturbative inputs used in the two analyses. The theoretical uncertainties in the quark mass bound (21) from the QCD parameters and (estimated) higher-order perturbative effects are less than 10%, and the result (21) is the absolute lowest bound resulting from the uncertainty analysis. The dominant sources of uncertainty are $\alpha(M_\tau)$ and potential higher-loop corrections. The instanton size ρ_c is the major source of non-perturbative uncertainty, but its effect is smaller than the perturbative sources of uncertainty.

Compared with the positivity inequality (20), as first used to obtain quark mass bounds from QCD sum-rules [5, 6], the Hölder inequality leads to quark mass bounds 50% larger for identical theoretical and phenomenological inputs at $M = M_\tau$, demonstrating that the Hölder inequality provides stringent constraints on the quark mass.

Finally, we discuss the effects of extending the resonance model to include the 3π continuum calculated using lowest-order chiral perturbation theory [9]

$$\frac{1}{\pi}\rho_5(t) = 2f_\pi^2 m_\pi^4 \left[\delta(t - m_\pi^2) + \Theta(t - 9m_\pi^2)\rho_{3\pi}(t)\frac{t}{18(16\pi^2 f_\pi^2)^2}\right] \quad (22)$$

$$\rho_{3\pi}(t) = \int_{4m_\pi^2}^{(\sqrt{t}-m_\pi)^2} \frac{du}{t} \sqrt{\lambda\left(1,\frac{u}{t},\frac{m_\pi^2}{t}\right)} \sqrt{1 - \frac{4m_\pi^2}{u}} \Bigg\{ 5+$$
$$+\frac{1}{2(t-m_\pi^2)^2}\left[\frac{4}{3}\left(t - 3\left(u - m_\pi^2\right)\right)^2 + \frac{8}{3}\lambda\left(t,u,m_\pi^2\right)\left(1 - \frac{4m_\pi^2}{u}\right) + 10m_\pi^4\right]$$
$$+\frac{1}{t-m_\pi^2}\left[3\left(u - m_\pi^2\right) - t + 10m_\pi^2\right] \Bigg\} \tag{23}$$
$$\lambda(x,y,z) = x^2 + y^2 + z^2 - 2xy - 2yz - 2xz \tag{24}$$

which becomes $\rho_{3\pi}(t) \to 3$ in the limit $m_\pi \to 0$. Inclusion of the 3π continuum (23) is still likely to underestimate the total spectral function since more complicated models of the spectral function involve resonance enhancement of this 3π continuum [9]. If this limiting form is used up to a cutoff of 1 GeV, then the resulting Hölder inequality quark mass bounds are *raised* by approximately 10%, and a 14% effect is observed if the cutoff is moved to infinity.[2] Working with the full form (23) complicates the numerical analysis, but the following simple form (with t in GeV units) is easily verified to be a bound on the 3π continuum in the region below 1 GeV.

$$\rho_{3\pi}(t) \geq \frac{4}{3}\left[\left(\sqrt{t} - m_\pi\right)^2 - 4m_\pi^2\right] \tag{25}$$

This approximate form of the 3π continuum again raises the resulting quark mass bounds by approximately 10%.

This paper is dedicated to the memory of Roger Migneron. Many thanks to Vic Elias, Gerry McKeon, and Voldoya Miransky for their efforts in organizing MRST 2001, which resulted in an enjoyable and interesting conference.

REFERENCES

1. M. Benmerrouche, G. Orlandini, T.G. Steele, Phys. Lett. B356 (1995) 573.
2. T.G. Steele, K. Kostuik, J. Kwan, Phys. Lett. B451 (1999) 201.
3. H. Leutwyler, Phys. Lett. B378 (1996) 313.
4. CP-PACS Collaboration: A. Ali Khan *et al*, Nucl. Phys. Proc. Suppl. 94 (2001) 229.
5. C. Becchi, S. Narison, E. de Rafael, F.J. Yndurain, Z. Phys. C8 (1981) 335.
6. S. Narison, E. de Rafael, Phys. Lett. B103 (1981) 57.
7. L. Lellouch, E. de Rafael, J. Taron Phys. Lett. B414 (1997) 195.
8. W. Hubschmid, S. Mallik, Nucl. Phys. B193 (1981) 368;
   A.L. Kataev, N.V. Krasnikov, A.A. Pivovarov, Phys. Lett. B123 (1983) 93;
   M. Kremer, N.A. Papadopoulos, K. Schilcher, Phys. Lett. B143 (1984) 476;
   C.A. Dominguez, Z. Phys. C26 (1984) 269;

---

[2] The exponential suppression of the large-t region in the Laplace sum-rule (6) minimizes any errors in this region from this approximation to the 3π continuum, and also leads to the observed small difference in extending the cutoff to infinity.

      V. Elias, A. H. Fariborz, M. A. Samuel, Fang Shi, T. G. Steele, Phys. Lett. B412 (1997) 131;
      T.G. Steele, J. Breckenridge, M. Benmerrouche, V. Elias, A. Fariborz, Nucl. Phys. A624 (1997) 517;
      K. Maltman, J. Kambor, hep-ph/0107060.
9. J. Bijnens, J. Prades, E. de Rafael, Phys. Lett. B348 (1995) 226.
10. R. Gupta, K. Maltman, hep-ph/0101132.
11. M.A. Shifman, A.I. Vainshtein, V.I. Zakharov, Nucl. Phys. B147 (1979) 385.
12. K.G. Chetyrkin, Phys. Lett. B390 (1997) 309.
13. E. Bagan, J.I. Latorre, P. Pascual, Z. Phys. C32 (1986) 43.
14. A.E. Dorokhov, S.V. Esaibegyan, N.I. Kochelev, N.G. Stefanis, J. Phys. G23 (1997) 643.
15. E.V. Shuryak, Nucl. Phys. B214 (1983) 237.
16. C.A. Dominguez, J. Sola, Z. Phys. C40 (1988) 63.
17. V. Gimenez, J. Bordes, J.A. Penarrocha, Nucl. Phys. B357 (1991) 3.
18. S. Narison, Nucl. Phys. Proc. Suppl. 54A (1997) 238.
19. E. Braaten, S. Narison, A. Pich, Nucl. Phys. B373 (1992) 581; M. Davier, A. Höcker, Phys. Lett. B435 (1998) 427.
20. D.E. Groom *et al.* (Particle Data Group), Eur. Phys. Jour. C15 (2000) 1.
21. ALEPH Collaboration (R. Barate et al.), Eur. Phys. J. C 4 1998 409;
      T.G. Steele and V. Elias, Mod. Phys. Lett. A 13 1998 3151;
      G. Cvetic and T. Lee, hep-ph/0101297;
      C.J. Maxwell and A. Mirjalili, hep-ph/0103164.
22. T. van Ritbergen, J.A.M. Vermaseren, S.A. Larin, Phys. Lett. B400 (1997) 379
23. K.G. Chetyrkin, Phys. Lett. B404 (1997) 161.
24. J. Ellis, I. Jack, D.R.T. Jones, M .Karliner, M.A. Samuel, Phys. Rev. D57 (1998) 2665;
      M.A. Samuel, J. Ellis, M. Karliner, Phys. Lett. B400 (1997) 176.

QUARKS, GLUONS, AND MESONS

String Model Building at Low String Scale: Towards the Standard Model

Robert G. Leigh

Dept. of Physics, University of Illinois, Urbana IL 61801

Abstract. We consider string model building in scenarios in which the fundamental string scale is low compared to the four-dimensional Planck scale. We present a consistent string theory model which gives a conservative extension of the Standard Model, consisting of a $D3$-brane at a simple orbifold singularity. We envision this, for example, as a *local* singularity within a warped compactification. We describe some of the phenomenology of this model. There are natural hierarchies in the fermion spectrum, which is only realistic for a string scale in the multi-TeV range.

String model building has traditionally assumed that the scale of string physics is large, typically near the Planck scale. Four dimensional physics is obtained through a choice of suitable conformal field theory. In cases where the conformal field theory has a geometric interpretation such as in Calabi-Yau compactifications, the low energy particle content is determined by the zero modes of fields on the corresponding compact space. D-branes provide an alternative method to obtain low energy gauge theories which, in this case, propagate on the world-volume of the brane.

In recent years, there have been several new ideas proposed that may be embedded in string theory. These include large extra dimensions[1] and warped compactification geometries.[2] The latter possibility is particularly natural in string theory, as such effects may be obtained in the geometry around branes. In either of the these situations, the physics is characterized by a fundamental scale (the string scale, M_{str}) that is much smaller than the four-dimensional Planck scale.

In terms of string model building, this is a largely unexplored realm. There has been some preliminary work on some global features of such compactifications. For example, the requirements for obtaining four-dimensional $N = 1$ supersymmetric theories have been examined in Ref. [3], while Giddings, Kachru and Polchinski[4] have examined the supergravity equations of motion for certain warped compactifications, noting that there are important global constraints.

In this talk, I will focus not on these global issues, but rather on finding configurations that give rise to physics which closely resembles the Standard Model within the context of a low string scale. There have been many attempts in the past for arriving at the Standard Model through string compactifications. Recently, there have been several constructions[5, 6, 7] presented which make use of the fact that chiral gauge theories may be obtained on intersecting branes.[8] We will not take that approach here; rather we will consider D-branes at orbifold singularities. This scenario should fit easily into the framework of consistent warped compactifications, although detailed work remains to be done along those lines.

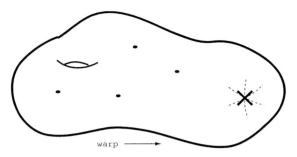

FIGURE 1. The 6-dimensional space. The X marks the local orbifold singularity.

We consider then some 6-dimensional space, X, of finite volume. We will not be concerned here about the details of this space, although we will assume that such a space may be realized with a few important features. First, there will be a *local orbifold singularity* where fractional D-branes reside. The Standard Model physics will take place on these branes. We will consider the case where four-dimensional $N = 1$ supersymmetry is preserved by the orbifold itself. To be precise, we will construct the orbifold C^3/Γ where Γ is a discrete subgroup of $SU(3)$. This is meant to be a local singularity within the space X (*i.e.*, X is not a global orbifold, and may be compact).

Supersymmetry will be assumed to be broken in some way by physics away from the orbifold singularity. The details of how this comes about will not be considered here, and we will assume that supersymmetry breaking is communicated to the brane theory through spurion couplings. The advantage of this scenario is that it relegates questions such as the stability of the configuration to the construction of the six-dimensional space. We have no new solution for stabilizing moduli here, and merely assume that their values are communicated to the brane physics via couplings for open string modes.

Furthermore, in working out the phenomenology of a particular model, we will find it necessary to take a low string scale, in the multi-TeV range. There are many phenomenological successes of the model and several possible experimental signatures. We will see in particular that supersymmetry breaking is intimately tied up in the phenomenology, such as in the structure of lepton masses and mixings.

THE Δ_{27} ORBIFOLD

There are many discrete subgroups of $SU(3)$ that we could consider, but there is one in particular that will lead to a most conservative extension of the Standard Model. There is a series of non-Abelian discrete subgroups, labeled Δ_{3n^2}, which may be understood through the exact sequence

$$0 \to Z_n \times Z_n \to \Delta_{3n^2} \to Z_3 \to 0. \tag{1}$$

This implies that Δ_{3n^2} is a semi-direct product of $Z_n \times Z_n$ and Z_3. We will consider the case $n = 3$ here. The orbifold may be constructed in two steps[9]; first we consider the

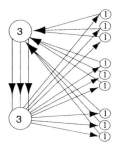

FIGURE 2. The quiver diagram of the Δ_{27} orbifold. Lines represent chiral matter fields.

orbifold by $Z_n \times Z_n$. Denote local complex coordinates by z_j; then we have:

$$e_1 : (z_1, z_2, z_3) \to (\omega z_1, \omega^{-1} z_2, z_3) \tag{2}$$
$$e_2 : (z_1, z_2, z_3) \to (z_1, \omega z_2, \omega^{-1} z_3) \tag{3}$$

where ω is a primitive nth root of unity. The final Z_3 has the action

$$e_3 : (z_1, z_2, z_3) \to (z_2, z_3, z_1). \tag{4}$$

To understand the fractional branes at the orbifold, we construct the *crossed product* algebra, consisting of z_j and e_j, with appropriate relations amongst them. The irreducible representations of this algebra correspond to D-branes– an irreducible representation of dimension r gives a gauge group $U(r)$. If we work out the representation theory of Δ_{27}, we can denote the gauge theory on fractional branes localized at the orbifold singularity by a quiver diagram[10]. This is a theory with gauge group $(U(3)_+ \times U(3)_- \times U(1)^9)/U(1)$,[11, 12] and has the following matter fields: $Q_i = (3_+, \bar{3}_-, 0)$, $L_a = (1_0, 3_+, -a)$, and $\overline{Q}_a = (\bar{3}_-, 1_0, +a)$, where the index a runs over the nine $U(1)$'s and $i = 1, \ldots, 3$. The plus and minus subscripts denote the $U(1)$ charge under the decomposition $U(3) \sim SU(3) \times U(1)$. Each of the fields L_a and $\overline{Q}_a$ are charged under only one of the nine $U(1)$'s. We identify the $SU(3)$ subgroup of $U(3)_+$ with the colour group and $SU(2)_W$ is embedded in the $U(3)_-$ group. The orbifold theory comes with a renormalizable superpotential generated at string tree-level of the form

$$W_0 = \sum_{ia} \lambda_{ia} Q_i L_a \overline{Q}_a, \tag{5}$$

where the λ_{ia} are couplings of order one at the string scale.

We have discussed here the open string fields which propagate on brane world-volumes. There are also a number of closed string fields which are important: there are bulk fields such as the dilaton, graviton and various fluxes, and there are twisted closed string moduli which are restricted to the singularity. The closed string fields show up in the effective action as coefficients of open-string operators. Thus the couplings of the brane theory are set by vacuum expectation values of these fields. In discussing the phenomenology of the model, we will consider these couplings to be as generic as possible, and look for necessary hierarchies elsewhere.

There are moduli which set the values of gauge couplings and θ-terms for each of the gauge groups. There are also moduli (e.g. Fayet-Iliopoulos terms) whose vevs lead to Higgsing of gauge symmetries. We will exploit the latter to break $U(3)_-$ to $SU(2)_W$; the $U(3)_+$ contains the colour group $SU(3)_c$. The breaking can be understood as arising from vevs of some of the $\mathcal{L}$ fields. Upon this breaking, we can decompose the matter fields as

$$\begin{aligned} Q_i &\to Q_i \;,\; q_i \\ \mathcal{L}_{1,i} &\to L_i \;,\; g_i \\ \mathcal{L}_{2,i} &\to H_i \;,\; \bar{e}_i \\ \mathcal{L}_{3,i} &\to \bar{H}_i \;,\; \bar{\nu}_i \end{aligned} \quad (6)$$

and we now make the identification

$$\overline{Q}_{1,i} \to \bar{q}_i, \quad \overline{Q}_{2,i} \to \bar{u}_i, \quad \overline{Q}_{3,i} \to \bar{d}_i. \quad (7)$$

The superpotential contains terms

$$W = \sum_{ij} \left\{ a_{ij} Q_i H_j \bar{u}_j + b_{ij} Q_i \bar{H}_j \bar{d}_j + c_{ij} q_i g_j \bar{q}_j + \ldots \right\}. \quad (8)$$

In the $SU(3)$ breaking refered to above, it is the g_j and $\bar{\nu}_j$ fields that get vevs. Thus, the fields q_j and $\bar{q}_j$ are a massive vector pair, and the remaining matter spectrum is that of the Standard Model, with an extended Higgs sector. This breaking is somewhat subtle: there is one linear combination of charginoés which pairs up with broken gauginoes; three other linear combinations will be identified with the charged leptons. The Higgsing described above, in conjunction with the Green-Schwarz mechanism for anomalous $U(1)$'s leads to a gauge symmetry $SU(3)_c \times SU(2)_W \times U(1)_Y \times U(1)^2$ (above the electroweak scale). In the rest of the talk, we will discuss some rudimentary aspects of the phenomenology of the model.

Supersymmetry Breaking and Lepton Masses:

We should note that the superpotential given above does not contain lepton Yukawa couplings, although generic quark Yukawa couplings are present. The quark Yukawa couplings are of a general matrix form, and we note that there are many Higgs fields which contribute to the masses; for example

$$(m_u)_{ij} = a_{ij} \langle H_j \rangle. \quad (9)$$

One apparently attractive scenario is that the observed hierarchy in quark masses can be induced by a mild hierarchy in Higgs vevs.

To find lepton Yukawa couplings, we must turn to non-renormalizable operators; it is because of this fact that a low string scale will be required (to be definite, we might take $M \sim 10 TeV$). Non-renormalizable operators will appear in the effective Lagrangian and are induced in string perturbation theory. They will have coefficients which are functions of moduli, and are of a size set by the string scale.

We will also need to take into account supersymmetry breaking effects. As we have mentioned, for present purposes, we will assume that supersymmetry breaking is communicated to the brane and has effects which can be parameterized by spurion couplings in the Kähler potential, such as

$$K = \ldots + \frac{1}{M} S \sum \phi_i^\dagger \phi_i + \frac{1}{M^2} \Psi^\dagger \Psi \phi_i^\dagger \phi_i + \ldots, \tag{10}$$

where the ϕ_i are any of the open string modes. The spurions S and Ψ may very well be closed string modes, and are assumed to have F-terms. The couplings to open string fields will induce matter F-terms

$$F_i = (F/M)\langle \phi_i \rangle \equiv m_{susy}\langle \phi_i \rangle. \tag{11}$$

Now let us explore the effects of matter Kähler potential terms. For example, there will be terms of the general form

$$K \supset \frac{1}{M^2} \alpha_{ab}(L_a^\dagger L_b)(L_b^\dagger L_a) + \frac{1}{M^2} \alpha'_{ab}(L_a^\dagger L_a)(L_b^\dagger L_b), \tag{12}$$

which include

$$\alpha_{ij} g_i^\dagger \bar{e}_j H_j^\dagger L_i + \beta_{ij} g_i^\dagger \bar{\nu}_j \bar{H}_j^\dagger L_i. \tag{13}$$

These give rise to charged lepton fermion masses of the form

$$(m_L)_{ij} \sim \alpha_{ij} \frac{F_{g_i}^*}{M^2} \langle H_j^* \rangle = \alpha_{ij} \frac{m_{susy}}{M} \frac{\langle g_i^* \rangle}{M} \langle H_j^* \rangle \tag{14}$$

and neutrino Dirac masses

$$(m_D)_{ij} \sim \beta_{ij} \frac{F_{g_i}^*}{M^2} \langle \bar{H}_j^* \rangle = \beta_{ij} \frac{m_{susy}}{M} \frac{\langle g_i^* \rangle}{M} \langle \bar{H}_j^* \rangle. \tag{15}$$

Thus the charged lepton mass spectrum provides some information on M and $\langle g_j \rangle$. The neutrino Dirac masses are roughly of the same order of magnitude as charged lepton masses, and thus in order to suppress light neutrino masses, we would need to set up a seesaw. There are neutrino Majorana masses, whose dominant contribution comes through mixing with massive gauginoes. We may then estimate

$$(m_M)_{ij} \sim m_{susy}. \tag{16}$$

Although the generic values of the light neutrino masses are in the MeV range, it is quite possible that they can be further suppressed through hierarchies in Higgs vevs, particularly for the first and second generation.

Proton Decay:

If one examines the quiver diagram, one soon realizes that $U(3)_+ \sim SU(3)c \times U(1)_B$, where $U(1)_B$ is baryon number. Thus baryon number is a gauge symmetry in this model, and we conclude that there can be no perturbative violations of the symmetry. This symmetry does however have mixed anomalies, which are cancelled by the Green-Schwarz mechanism. One feature of the Green-Schwarz mechanism is that the global $U(1)_B$ symmetry remains, acting on matter fields. We do expect, because of the anomaly, that there are non-perturbative violations of this symmetry. To see what these are, we need to look more carefully at the anomalies. The relevant non-zero anomaly involving $U(1)_B$ is $SU(3)_- - SU(3)_- - U(1)_B$. Since $SU(3)_-$ is Higgsed to $SU(2)_W$, it is non-perturbative $SU(2)_W$ effects that will lead to violation of $U(1)_B$. Since the $SU(2)_W$ coupling is small, these non-perturbative effects are expected to be negligible, at least at zero temperature.

Flavour Changing Neutral Currents:

A common problem in extensions of the Standard Model is flavour changing neutral currents (FCNC). In the present case, these may be induced through box diagrams via exchange of squark fields. The important thing is the squark mass matrix. For example, there will be contributions like

$$(m_Q^2)_{ij} \sim \frac{\alpha_{ij,a}|F_a|^2}{M^2} \sim \frac{\alpha_{ij,a}|\langle g_a \rangle|^2}{M^2} m_{susy}^2. \qquad (17)$$

Preliminary investigations seem to suggest that these effects can be suppressed to sufficient levels. Given a more detailed analysis, it may turn out to be necessary to seek some other mechanism for FCNC suppression, such as alignment.[13]

Extra $U(1)$'s:

In the model we have presented here, we have not succeeded in obtaining precisely the Standard Model gauge group: there are two extra $U(1)$'s. As there are no fields in the theory that are charged under these $U(1)$'s other than Standard Model fields, they are broken at the weak scale. This is potentially dangerous, as there are fairly stringent bounds on extra Z's. The $U(1)$'s do not couple universally however, and it is possible that they may evade detection.

CONCLUSIONS

We have presented here a possible scenario for string model building at low string scale. The motivation is to provide a candidate brane realization of the Standard Model that may be used in these pursuits. An interesting phenomenological structure emerges in the particular example that we have presented here, based on a Δ_{27} orbifold singularity. It is of interest to further explore the phenomenology of this model; even if it is ruled out by (precision) experiments, it is interesting that we have been able to come so close with a simple construction. Of course, important stringy problems remain, such as the moduli problem, the cosmological constant and the nature of supersymmetry breaking.

We expect that there are other models of this type as well. It would be of interest to understand if there are some general features that may emerge. One can also consider more general constructions, with, say, orientifold planes assembled in such a way as to allow warped compactification.

ACKNOWLEDGMENTS

This talk is based on work[11] with David Berenstein and Vishnu Jejjala. I wish to thank the organizers of the MRST 2001 workshop at the University of Western Ontario for hospitality. This talk has also been presented at the Argonne Theory Institute and the Pacific Institute for Mathematical Sciences String workshop, both in summer 2001. Work supported in part by US Dept. of Energy under contract DE-FG02-91ER40677 and by the Aspen Center for Physics.

REFERENCES

1. Arkani-Hamed, N., Dimopoulos, S., and Dvali, G., *Phys. Lett.*, **B429**, 263–272 (1998), hep-ph/9803315.
2. Randall, L., and Sundrum, R., *Phys. Rev. Lett.*, **83**, 4690 (1999), hep-th/9906064.
3. Gubser, S. S., hep-th/0010010.
4. Giddings, S. B., Kachru, S., and Polchinski, J., hep-th/0105097.
5. Ibanez, L. E., Marchesano, F., and Rabadan, R., hep-th/0105155.
6. Blumenhagen, R., Kors, B., Lust, D., and Ott, T., hep-th/0107138.
7. Cvetic, M., Shiu, G., and Uranga, A. M., hep-th/0107166.
8. Berkooz, M., Douglas, M. R., and Leigh, R. G., *Nucl. Phys.*, **B480**, 265 (1996), hep-th/9606139.
9. Berenstein, D., Jejjala, V., and Leigh, R. G., *Phys. Rev.*, **D64**, 046011 (2001), hep-th/0012050.
10. Douglas, M. R., and Moore, G., hep-th/9603167.
11. Berenstein, D., Jejjala, V., and Leigh, R. G., hep-ph/0105042.
12. Aldazabal, G., Ibanez, L. E., Quevedo, F., and Uranga, A. M., *JHEP*, **08**, 002 (2000), hep-th/0005067.
13. Nir, Y., and Seiberg, N., *Phys. Lett.*, **B309**, 337 (1993), hep-ph/9304307.

Lepton pair production in a charged quark gluon plasma

A. Majumder* and C. Gale*

*Department of Physics, McGill University, Montreal, QC, Canada H3A 2T8

Abstract. We investigate the effects of a charge asymmetry on the spectrum of dileptons radiating from a quark gluon plasma. We demonstrate the existence of a new set of processes in this regime. The dilepton production rate from the corresponding diagrams is shown to be as important as that obtained from the Born-term quark-antiquark annihilation.

INTRODUCTION

The aim of this talk is to show that, when in a medium there is a finite charge density (i.e., a finite chemical potential), a new set of lepton pair-producing processes actually arises [1]. We then calculate a new contribution to the 3-loop photon self-energy. The various cuts of this self-energy contain higher loop contributions to the usual processes of $q\bar{q} \to e^+e^-$, $qg \to qe^+e^-$, $qq \to qqe^+e^-$, and an entirely new channel: $gg \to e^+e^-$. We calculate the contribution of this new reaction to the differential production rate of back-to-back dileptons. It is finally shown that, within reasonable values of parameters, this process may become larger than the differential rate from the bare tree level $q\bar{q} \to e^+e^-$.

At zero temperature, and at finite temperature and zero charge density (note: henceforth, a finite density will imply a finite charge density), diagrams in QED that contain a fermion loop with an odd number of photon vertices (e.g. Fig. 1) are cancelled by an equal and opposite contribution coming from the same diagram with fermion lines running in the opposite direction (Furry's theorem [2, 3, 4]). This statement can also be generalized to QCD for processes with two gluons and an odd number of photon vertices.

A physical perspective is obtained by noting that all these diagrams are are encountered in the perturbative evaluation of Green's functions with an odd number of gauge field operators. At zero (finite) temperature, in the well defined case of QED we observe quantities like $\langle 0|A_{\mu_1}A_{\mu_2}...A_{\mu_{2n+1}}|0\rangle$ ($Tr[\rho(\mu,\beta)A_{\mu_1}A_{\mu_2}...A_{\mu_{2n+1}}]$) under the action of the charge conjugation operator C. In QED we know that $CA_\mu C^{-1} = -A_\mu$. In the case of the vacuum $|0\rangle$, we note that $C|0\rangle = |0\rangle$, as the vacuum is uncharged. As a result

$$\langle 0|A_{\mu_1}A_{\mu_2}...A_{\mu_{2n+1}}|0\rangle = \langle 0|C^{-1}CA_{\mu_1}C^{-1}CA_{\mu_2}...A_{\mu_{2n+1}}C^{-1}C|0\rangle$$

---

[1] Note that a net charge density in a quark gluon plasma does not necessarily imply a net baryon density and vice-versa.

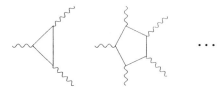

FIGURE 1. Diagrams rendered zero by Furry's theorem. See text for details.

$$= \langle 0|A_{\mu_1}A_{\mu_2}...A_{\mu_{2n+1}}|0\rangle(-1)^{2n+1} = -\langle 0|A_{\mu_1}A_{\mu_2}...A_{\mu_{2n+1}}|0\rangle = 0. \quad (1)$$

At a temperature T, the corresponding quantity to consider is

$$\sum_n \langle n|A_{\mu_1}A_{\mu_2}...A_{\mu_{2n+1}}|n\rangle e^{-\beta(E_n-\mu Q_n)},$$

where $\beta = 1/T$ and μ is a chemical potential. Here, however, $C|n\rangle = e^{i\phi}|-n\rangle$, where $|-n\rangle$ is a state in the ensemble with the same number of antiparticles as there are particles in $|n\rangle$ and vice-versa. If $\mu = 0$ i.e., the ensemble average displays zero density then inserting the operator $C^{-1}C$ as before, we get

$$\langle n|A_{\mu_1}A_{\mu_2}...A_{\mu_{2n+1}}|n\rangle e^{-\beta E_n} = -\langle -n|A_{\mu_1}A_{\mu_2}...A_{\mu_{2n+1}}|-n\rangle e^{-\beta E_n}. \quad (2)$$

The sum over all states will contain the mirror term $\langle -n|A_{\mu_1}A_{\mu_2}...A_{\mu_{2n+1}}|-n\rangle e^{-\beta E_n}$, with the same thermal weight

$$\Rightarrow \sum_n \langle n|A_{\mu_1}A_{\mu_2}...A_{\mu_{2n+1}}|n\rangle e^{-\beta E_n} = 0, \quad (3)$$

and Furry's theorem still holds. However, if $\mu \neq 0$ ($\Rightarrow$ unequal number of particles and antiparticles) then

$$\langle n|A_{\mu_1}A_{\mu_2}...A_{\mu_{2n+1}}|n\rangle e^{-\beta(E_n-\mu Q_n)} = -\langle -n|A_{\mu_1}A_{\mu_2}...A_{\mu_{2n+1}}|-n\rangle e^{-\beta(E_n-\mu Q_n)}, \quad (4)$$

the mirror term this time is $\langle -n|A_{\mu_1}A_{\mu_2}...A_{\mu_{2n+1}}|-n\rangle e^{-\beta(E_n+\mu Q_n)}$, with a different thermal weight, thus

$$\sum_n \langle n|A_{\mu_1}A_{\mu_2}...A_{\mu_{2n+1}}|n\rangle e^{-\beta(E_n-\mu Q_n)} \neq 0, \quad (5)$$

and Furry's theorem will now break down. One may say that the medium, being charged, manifestly breaks charge conjugation invariance and these Green's functions are thus finite, and will lead to the appearance of new processes in a perturbative expansion. The appearance of processes that can be related to symmetry-breaking in a medium has been noted elsewhere [5].

Let us, now, focus our attention on the diagrams of Fig. 2 for the case of two gluons and a photon attached to a quark loop (the analysis is the same even for QED i.e., for three photons connected to an electron loop). Such a process does not exist at zero

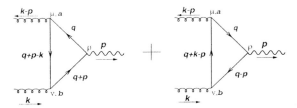

FIGURE 2. The two gluon photon effective vertex as the sum of two diagrams with quark number running in opposite directions.

temperature, or even at finite temperature and zero density. At finite density this leads to a new source of dilepton or photon production ($gg \to l^+l^-$). In order to obtain the full matrix element of a process containing the above as a sub-diagram one must coherently sum contributions from both diagrams which have fermion number running in opposite directions. The amplitude for $\mathcal{T}^{\mu\rho\nu}(=T^{\mu\rho\nu}+T^{\nu\rho\mu})$ is:

$$T^{\mu\rho\nu} = \frac{1}{\beta}\sum_{n=-\infty}^{\infty}\int_{-\infty}^{\infty} eg^2 tr[t^a t^b]\frac{d^3q}{(2\pi)^3} Tr[\gamma^\mu \gamma^\beta \gamma^\rho \gamma^\delta \gamma^\nu \gamma^\alpha]\frac{(q+p-k)_\alpha q_\beta (q+p)_\delta}{(q+p-k)^2 q^2 (q+p)^2},$$

$$T^{\nu\rho\mu} = \frac{1}{\beta}\sum_{n=-\infty}^{\infty}\int_{-\infty}^{\infty} eg^2 tr[t^a t^b]\frac{d^3q}{(2\pi)^3} Tr[\gamma^\nu \gamma^\delta \gamma^\rho \gamma^\beta \gamma^\mu \gamma^\alpha]\frac{(q+k-p)_\alpha q_\beta (q-p)_\delta}{(q+k-p)^2 q^2 (q-p)^2}. \quad (6)$$

Again, the extension of Furry's theorem to finite temperature does not hold at finite density: as, if we set $n \to -n-1$, we note that $q_0 \not\to -q_0$, and, as a result, $T^{\mu\rho\nu}(\mu,T) \neq -T^{\nu\rho\mu}(\mu,T)$. Of course, If we now let the chemical potential go to zero ($\mu \to 0$), we note that for the transformation $n \to -n-1$, we obtain $q_0 \to -q_0$, and, thus, $T^{\mu\rho\nu}(0,T) \to -T^{\nu\rho\mu}(0,T)$. The analysis for fermion loops with larger number of vertices is essentially the same.

A Realistic Calculation

To calculate the contribution made by the diagram of Fig. 2, to the dilepton spectrum emanating from a quark gluon plasma, we calculate the imaginary part of the photon self-energy containing the above diagram as an effective vertex (Fig 3, see reference [1] for details).

$$\Pi_\rho^\rho = \frac{1}{\beta}\sum_{k^0}\int \frac{d^3k}{(2\pi)^3} \mathcal{D}_{\eta\mu}(k)\mathcal{T}^{\mu\rho\nu}(k-p,k;p)\mathcal{D}_{\nu\zeta}(k-p)\mathcal{T}^{\zeta\rho\eta}(k,k-p;-p), \quad (7)$$

where, the $\mathcal{D}$'s represent bare gluon propagators, and the $\mathcal{T}$'s represent the effective vertices. We calculate in the limit of photon three momentum $\vec{p} = 0$. The imaginary part of the considered self-energy contains various cuts. We concentrate, solely, on the cut that represents the process of gluon-gluon to e^+e^- (see Fig.(3)). This is a process, which,

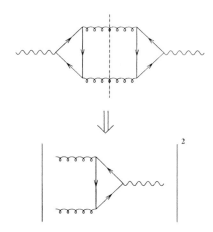

FIGURE 3. The photon self-energy at three loops and the cut that is evaluated

to our knowledge, has not been discussed before. The other possible cuts represent finite density contributions to other known processes of dilepton production.

The differential production rate for pairs of massless leptons, with total energy E and and total momentum $\vec{p} = 0$, is given in terms of the discontinuity in the photon self-energy [6], as

$$\frac{dW}{dEd^3p}(\vec{p}=0) = \frac{\alpha}{12\pi^3} \frac{1}{E^2} \frac{1}{1-e^{E/T}} \frac{1}{2\pi i} \text{Disc}\Pi_\rho^\rho(0), \qquad (8)$$

where α is the electromagnetic coupling constant. The rate of production of a hard lepton pair with total momentum $\vec{p} = 0$, at one-loop order in the photon self-energy (i.e., the Born term), is given for three flavours as

$$\frac{dW}{dEd^3p}(\vec{p}=0) = \frac{5\alpha^2}{6\pi^4}\tilde{n}(E/2-\mu)\tilde{n}(E/2+\mu) + \frac{\alpha^2}{6\pi^4}\tilde{n}^2(E/2) \ . \qquad (9)$$

In the above equation, the first term on the r. h. s. is the contribution from the up and down quark sector; and the second part is the contribution from the strange sector. In a realistic plasma, the net charge and baryon imbalance is caused by the valence quarks brought in by the incoming charged baryon-rich nuclei. The baryon and charge imbalance is, thus, manifested solely in the up and down quark sector; hence, the chemical potential influences only the distribution function of the up and down quarks. The strange and anti-strange quarks are produced in equal numbers in the plasma; resulting in a vanishing strange quark chemical potential. Thus, dilepton production, from the channel indicated by Fig.(3), will only receive contributions from the up and down quark flavours.

The initial temperatures of the plasma, formed at RHIC and LHC, have been predicted to lie in the range from 300-800 MeV [7, 8]. For this calculation we use estimates of $T = 400$ MeV(Fig. 4) and $T = 600$ MeV(Fig. 5). To evaluate the effect of a finite chemical potential, we perform the calculation with two extreme values of chemical potential

$\mu = 0.1T$ (left plots) and $\mu = 0.5T$ (right plots) [9]. The calculation, is performed for three massless flavours of quarks. In this case, the strong coupling constant is (see [10])

$$\alpha_s(T) = \frac{6\pi}{27\ln(T/50\text{MeV})}. \tag{10}$$

The differential rate for the production of dileptons with an invariant mass from 0.5 to 2.5 GeV is presented. On purpose, we avoid regions where the gluons become very soft. In the plots, the dashed line is the rate from tree level $q\bar{q}$ (Eq. (9)); the solid line is that from the process $gg \rightarrow e^+e^-$. We note that in both cases the gluon-gluon process dominates at low energy and dies out at higher energy leaving the $q\bar{q}$ process dominant at higher energy.

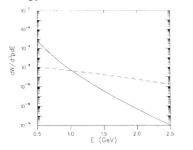
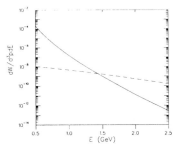

FIGURE 4. The differential production rate of back-to-back dileptons from two processes plotted against dilepton invariant mass. The dashed line represents the contribution from Born term $q\bar{q} \rightarrow e^+e^-$. The solid line corresponds to the process $gg \rightarrow e^+e^-$. Temperature is 400 MeV. Quark chemical potential, in the first figure, is 0.1T. The second figure is the same as the first, but, with $\mu = 0.5T$

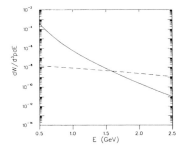
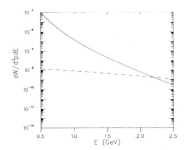

FIGURE 5. Same as above but with T=600MeV.

High Temperature Limit

As the reader may have noted, in the above calculation, $\alpha \simeq 0.3, g \simeq 2$, thus, we are not unequivocally in the perturbative regime. At asymptotically high temperatures ($T \rightarrow \infty$), however, $g \rightarrow 0$, in this limit one may make the Hard Thermal Loop (HTL) approximation [11, 12] (note: only the main results will be quoted here; the details will appear elsewhere [15]). At very high temperature, contributions from loop diagrams are dominated by loop momenta (q) of the order of the temperature T. If the momentum flowing in the external legs is much smaller (of the order of gT, i.e., $q \gg k, p$), then loop corrections are of the same order in g as the bare diagrams, and, thus, must be resummed into the tree amplitudes.

In this limit $(q+p-k)_\alpha \simeq q_\alpha$, and $E_{q+p} \simeq E_q + \vec{p} \cdot \hat{q} + \frac{|\vec{p}|^2}{2E_q} + \frac{|\vec{p} \cdot \hat{q}|^2}{2! E_q}$. On performing the full Matsubara sum and taking the HTL limit for the numerator, we get

$$T^{\mu\rho\nu} = \int \frac{d^3q}{(2\pi)^3} \frac{\delta^{ab}}{2} eg^2 Tr[\gamma^\mu \gamma^\beta \gamma^\rho \gamma^\delta \gamma^\nu \gamma^\alpha] \frac{\hat{q}_{s_1,\alpha} \hat{q}_{-s_2,\beta} \hat{q}_{-s_3,\delta}}{p^0 - s_2 E_2 + s_3 E_3}$$

$$\left[s_2 \frac{s_1(\tilde{n}(E_3 - s_3\mu) - \tilde{n}(E_3 + s_3\mu)) + s_3(\tilde{n}(E_1 + s_1\mu) - \tilde{n}(E_1 - s_1\mu))}{k^0 + s_1 E_1 + s_3 E_3} \right.$$

$$\left. - s_3 \frac{s_1(\tilde{n}(E_2 - s_2\mu) - \tilde{n}(E_2 + s_2\mu)) + s_2(\tilde{n}(E_1 + s_1\mu) - \tilde{n}(E_1 - s_1\mu))}{k^0 - p^0 + s_1 E_1 + s_2 E_2} \right], \quad (11)$$

where, $\hat{q}_+ = (1, \hat{q}_1, \hat{q}_2, \hat{q}_3)$, and $\hat{q}_- = (-1, \hat{q}_1, \hat{q}_2, \hat{q}_3)$. Note that if μ is set to zero this amplitude vanishes identically. The conventional HTL term (i.e., terms proportional to $(gT)^2$) from triangle graphs such as these would come from the terms with $s_1 = +, s_2 = -, s_3 = -$ and $s_1 = -, s_2 = +, s_3 = +$. The term proportional to $(gT)^2$ is

$$T^{\mu\rho\nu}_{(gT)^2} = \int \frac{d^3q}{(2\pi)^3} \frac{\delta^{ab}}{2} eg^2 Tr[\gamma^\mu \gamma^\beta \gamma^\rho \gamma^\delta \gamma^\nu \gamma^\alpha] \left\{ \frac{\hat{q}_{+,\alpha} \hat{q}_{+,\beta} \hat{q}_{+,\delta}}{p^0 - \vec{p} \cdot \hat{q}} \right.$$

$$\left[\frac{\frac{\partial \tilde{n}(E_3+\mu)}{\partial E_3}(-\vec{k} \cdot \hat{q}) - \frac{\partial \tilde{n}(E_3-\mu)}{\partial E_3}(-\vec{k} \cdot \hat{q})}{k^0 - \vec{k} \cdot \hat{q}} + \frac{\frac{\partial \tilde{n}(E_2-\mu)}{\partial E_2}((\vec{p}-\vec{k}) \cdot \hat{q}) - \frac{\partial \tilde{n}(E_2+\mu)}{\partial E_2}((\vec{p}-\vec{k}) \cdot \hat{q})}{k^0 - p^0 + (\vec{p}-\vec{k}) \cdot \hat{q}} \right] +$$

$$\frac{\hat{q}_{-,\alpha} \hat{q}_{-,\beta} \hat{q}_{-,\delta}}{p^0 + \vec{p} \cdot \hat{q}} \left[\frac{\frac{\partial \tilde{n}(E_3-\mu)}{\partial E_3}(-\vec{k} \cdot \hat{q}) - \frac{\partial \tilde{n}(E_3+\mu)}{\partial E_3}(-\vec{k} \cdot \hat{q})}{k^0 + \vec{k} \cdot \hat{q}} + \frac{\frac{\partial \tilde{n}(E_2+\mu)}{\partial E_2}((\vec{p}-\vec{k}) \cdot \hat{q}) - \frac{\partial \tilde{n}(E_2-\mu)}{\partial E_2}((\vec{p}-\vec{k}) \cdot \hat{q})}{k^0 - p^0 - (\vec{p}-\vec{k}) \cdot \hat{q}} \right] \right\}.$$

Note that if we change $\hat{q} \to -\hat{q}$ in the last two terms then $\hat{q}_{+,\alpha} \to -\hat{q}_{-,\alpha}$ and the whole term becomes identically zero. Thus there is no term proportional to $(gT)^2$ in this diagram. The next term is of order $g^2 T$ and is nonzero [15]. However, as is well known, the HTL approximation of the $q\bar{q}\gamma$ vertex is of the order of $(gT)^2$ [14]. Thus, in the high temperature approximation, the $gg\gamma$ HTL vertex is suppressed as compared to the $q\bar{q}\gamma$ HTL-resummed vertex. As a result the dilepton production rate from resummed two-gluon-fusion process will also be suppressed compared to the rate from the resummed Born term, in this limit.

On summing up all the HTL contributions to the self energy of the gluon we get two dispersion relations for the transverse and longitudinal modes of the gluon. A plot of the two modes [15] is as shown in Fig. (6). The upper branch is the dispersion relation for the transverse quasi-particle excitation, the lower one is the longitudinal excitation. The straight line, corresponding to free gluons is shown for reference. Note that there is no minima in either branch (except at $k=0$), unlike in the case of quarks [14]. Hence, we will not see any sharp Van Hove peaks [14] in the resulting dilepton spectra emanating from this process. One may thus expect that the high temperature (and, perhaps, as a result, the very low invariant mass) dilepton production spectra emanating from in-medium gluon-gluon fusion will be suppressed compared to that from in-medium quark

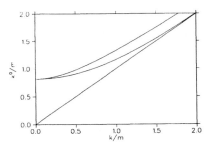

FIGURE 6. HTL resummed dispersion relations of the gluon in a finite temperature, finite density medium.

anti-quark fusion. However, as we have noted in the previous section the intermediate invariant mass rate from bare gluon-gluon fusion is comparable and may be larger than that from bare quark anti-quark fusion.

The entire calculation above is at full chemical and thermal equilibrium. In the early part of an ultrarelativistic heavy ion collision, the gluon number has been predicted to be much higher than at full chemical equilibrium. In such a scenario, the contributions to dilepton spectrum from processes such as those presented here will probably out-shine those from other channels.

The authors wish to thank Y. Aghababaie, A. Bourque, S. Das Gupta , F. Gelis, S. Jeon, D. Kharzeev, C. S. Lam and G. D. Mahlon for helpful discussions. A.M. acknowledges the generous support provided to him by McGill University through the Alexander McFee fellowship, the Hydro-Quebec fellowship and the Neil Croll award. This work was supported in part by the Natural Sciences and Engineering Research Council of Canada and by *le fonds pour la Formation de Chercheurs et l'Aide à la Recherche du Québec*.

REFERENCES

1. A. Majumder and C. Gale, Phys. Rev. D, **63**, 114008 (2001);and erratum, *in press* .
2. W. H. Furry, *Phys. Rev.*, **51**, 125 (1937).
3. C. Itzykson, J. B. Zuber, *Quantum Field Theory*, McGraw Hill, New York, (1980).
4. S. Weinberg, *The Quantum Theory of Fields*, Vol. 1, Cambridge University Press, (1995).
5. See for example, S. A. Chin, Ann. Phys. 108, 301 (1977); H. A. Weldon, Phys. Lett. B, **274**, 133 (1992); G. Wolf, B. Friman and M. Soyeur, Nucl. Phys. **A640**, 129 (1998); O. Teodorescu, A.K. Dutt-Mazumder and C. Gale, Phys. Rev. C **63**, 034903 (2001).
6. C. Gale, and J. I. Kapusta, Nucl. Phys. **B357**, 65 (1991).
7. X. N. Wang, Phys. Rep. **280**, 287 (1997).
8. R. Rapp, hep-ph/0010101.
9. K. Geiger, and J. I. Kapusta, Phys. Rev. D. **47**, 4905 (1993); N. George, for the PHOBOS collaboration, Proceedings of Quark Matter 2001.
10. J. I. Kapusta, and S. M. H. Wong, Phys. Rev. C. **62**, 027901 (2000).
11. E. Braaten, and R. D. Pisarski, Nucl. Phys. **B337** 569 (1990).
12. M. Le Bellac, *Thermal Field Theory*, Cambridge University Press, (1996).
13. U. Heinz, K. Kajantie, and T. Toimela, Ann. Phys. (N.Y.) **176** 218 (1987).
14. E. Braaten, R. D. Pisarski, and T. C. Yuan, Phys. Rev. Lett. **64** 2242 (1990).
15. A. Majumder, and C. Gale, *in preparation*.

$O(\alpha_s^3)$ estimate for the longitudinal cross-section in e^+e^- annihilation to hadrons

F.A. Chishtie

Department of Applied Mathematics
University of Western Ontario

Abstract. Using Renormalization Group (RG) and Pade approximant methods, we estimate the $O(\alpha_s^3)$ contribution to the longitudinal cross-section (σ_L) in the e^+e^- annihilation to hadrons. The $O(\alpha_s^3)$ contribution is expressed as a quadratic polynomial in $L \equiv \ln(\mu^2/Q^2)$. We find that the Pade approximant predictions for the coefficients L and L^2 are within 3.5% of the true values extracted via RG methods. Incorporation of the full $O(\alpha_s^3)$ contribution (including the estimated RG-inaccessible coefficient) to σ_L yields good agreement of the measured cross-section (OPAL[1]) with the current Particle Data Guide value of $\alpha_s(M_Z) = 0.1185 \pm 0.020$, indicative of the importance of higher order perturbative contribution(s). Incorporation of this contribution yields reduced renormalization scale dependence and near agreement of Principal of Minimal Sensitivity (PMS) and Fastest Apparent Convergent (FAC) values of the cross-section.

INTRODUCTION

Experiments carried out in electron positron colliders have given us immense amount of data to test the Standard Model. These experiments, in particular the electron positron annihilation to hadrons, have given us great insight into the validity of QCD as the theory of strong interactions. For the purposes of this paper, we are interested in particular the following semi-inclusive reaction:

$$e^+e^- \xrightarrow{\gamma,Z} H + "X"$$

where H is an outgoing charged hadron or a sum of all charged hadron species, and X denotes any inclusive hadron final state. The unpolarized differential cross section of the above process is given by,

$$\frac{d^2\sigma}{dx_B d\cos\theta} = \frac{3}{8}(1+\cos^2\theta)\frac{d\sigma_T}{dx_B} + \frac{3}{4}\sin^2\theta\frac{d\sigma_L}{dx_B} + \frac{3}{4}\cos\theta\frac{d\sigma_A}{dx_B}$$

where x_B is the Bjorken scaling, σ_T, σ_L, σ_A denote the transverse, longitudinal and the asymmetric cross sections respectively and θ denotes the angle of emission of H with respect to the electron beam direction in the center of mass (CM) frame. In the annihilation scheme (AS) (where the annihilation is purely electromagnetic) the following relationship is shown to hold [1]:

$$\frac{\sigma_T}{\sigma_0} + \frac{\sigma_L}{\sigma_0} = 1 \qquad (1)$$

Measurements by the OPAL [2] collaboration have measured the longitudinal cross-section, σ_L/σ_0 to be 0.057 ± 0.005. There is a slight discrepancy between the extracted value of the strong coupling, $\alpha_s(M_Z)$ by OPAL, (and also the current Particle Data Guide [3] quoted value) and perturbative QCD as stated in [4]. The aim of this paper is to show that higher order QCD corrections are indeed important and do resolve this apparent discrepancy between measurement and theory.

DETERMINATION OF RG-ACCESSIBLE $O(\alpha_s^3)$ COEFFICIENTS

The longitudinal cross-section has been analytically calculated to $O(\alpha_s^2)$ by Neerven et. al [1]. It can be expressed to $O(\alpha_s^3)$ as follows:

$$\frac{\sigma_L}{\sigma_0}(x(\mu), L(\mu)) = -a_o x - (b_0 + b_1 L)x^2 - \underline{(c_0 + c_1 L + c_2 L^2)x^3} \qquad (2)$$

where $x \equiv \dfrac{\alpha_s(\mu)}{\pi}$, $L \equiv \ln\left(\dfrac{\mu^2}{Q^2}\right)$, $a_0 = -1$, $b_0 = -13.583 + 1.028 n_f$, $b_1 = -11/4 + n_f/6$.

Here n_f denotes the number of active quark flavours, Q is the CM energy of the virtual vector boson (γ, Z) and the underlined term indicates the term we seek to estimate.

Using the annihilation scheme, we can then express the transverse cross-section as,

$$\frac{\sigma_T}{\sigma_0}(x(\mu), L(\mu)) \equiv S = 1 + a_o x + (b_0 + b_1 L)x^2 + \underline{(c_0 + c_1 L + c_2 L^2)x^3} \qquad (3)$$

Renormalization scale invariance for a physical observable (like the given cross-section) implies that, the quantity S follows

$$\mu^2 \frac{dS(x(\mu), L(\mu))}{d\mu^2} = 0$$

$$\Rightarrow \left[\frac{\partial}{\partial L} + \beta(x)\frac{\partial}{\partial x}\right] S(x(\mu), L(\mu)) = 0 \qquad (4)$$

where $\beta(x) \equiv -\beta_0 x^2 - \beta_1 x^2 + \ldots$ is the $\overline{MS}$ QCD β-function referenced to n_f active flavours.

Assuming that RG-invariance holds order-by-order, we substitute Eq. (3) into (4) and obtain the following:

$$x^3: \qquad c_1 - 2\beta_0 - \beta_1 a_0 = 0 \Rightarrow c_1 = 2\beta_0 + \beta_1 a_0 \qquad (5)$$
$$x^3 L: \qquad 2c_2 - 2\beta_0 b_1 = 0 \Rightarrow c_2 = \beta_0 b_1 \qquad (6)$$

Hence we have obtained the RG-accessible coefficients c_1 and c_2 and will estimate these along with the RG-inaccessible coefficient c_0 in the following section.

PADE-APPROXIMANT ESTIMATION OF THE $O(\alpha_s^3)$ CONTRIBUTION

Consider the following truncated series,

$$S_{N+1} = 1 + \sum_{n=0}^{N+1} R_n x^n \tag{7}$$

where $\{R_1, R_2, \ldots, R_{N+1}\}$ are known.

An [N|M] Pade approximant to the series S is given as

$$S_{[N|M]} = \frac{1 + \sum_{n=1}^{n=N} a_n x^n}{1 + \sum_{n=1}^{n=N} b_n x^n} \tag{8}$$

where $N + M = n + 1$
The coefficients a_i and b_i are fixed such that the first $n + 1$ terms of the Maclaurin expansion for $S_{[M|M]}$ correspond to the truncated series S_{n+1}. We can generate an [N|1] approximant with knowledge of coefficients $\{R_1, R_2, \ldots R_{N+1}\}$ ($M = 1$)

$$S_{[N|1]} = \frac{1 + \sum_{n=1}^{n=N} a_n x^n}{1 + b_1 x} \tag{9}$$

From (9), the "Pade-predicted" term is:

$$R_{N+2}^{[N|1]} = \frac{R_{N+1}^2}{R_N} \tag{10}$$

To improve this estimate, noting that for series characterized by asymptotic behavior $R_l \sim l! C^l l^\gamma$, the relative error is anticipated to be [5]:

$$\delta_{N+2}^{[N|1]} = \frac{R_{N+2}^{[N|1]} - R_{N+2}}{R_{N+2}} = -\frac{A}{N+1+k} \tag{11}$$

We use the above to generate an "improved" estimate of the unknown coefficient R_3 given our knowledge of R_1 and R_2 in the following series:

$$S = 1 + R_1 x + R_2 x^2 + R_3 x^3 + \ldots \tag{12}$$

where (from (3)),

$$R_1 = a_0$$
$$R_2 = b_0 + b_1 L \tag{13}$$
$$R_3 = c_0 + c_1 L + c_2 L^2$$

Forming a [0|1] and [1|1] Pade approximants of the series above (12), we generate the following relative error(s) via (10) and (11):

$$\delta_2^{[0|1]} = \frac{R_1^2 - R_2}{R_2} = -\frac{A}{1+k} \tag{14}$$

$$\delta_2^{[1|1]} = \frac{R_2^2/R_1 - R_3}{R_3} = -\frac{A}{2+k} \tag{15}$$

Solving for R_3, we finally have,

$$R_3^{Pade} = \frac{(k+2)R_2^3}{R_1\left[(k+1)R_1^2 + R_2\right]} \tag{16}$$

Note that for $k = -1$, we get (10) and for $k = 0$ we get the estimate for R_3 which is used in previous applications of the method [6-14].

Given our knowledge of R_3 via Pade methodology, we estimate c_i by matching the scale dependence of (16) with R_3 in (13) over the entire perturbative region $\mu \geq Q$ through use of the following moment integrals.

$$N_i = (i+1)\int_0^1 dw \, w^i R_3(w) \tag{17}$$

where $w \equiv \frac{\mu^2}{Q^2}$ $[L = -\ln(w)]$

For $n_f = 5$ and $k = -0.965$, the first three moments (and matching) yield:

$$N_0 = -115.27 = c_0 + c_1 + 2c_2 \tag{18}$$

$$N_1 = -92.783 = c_0 + \frac{1}{2}c_1 + \frac{1}{2}c_2 \tag{19}$$

$$N_2 = -86.132 = c_0 + \frac{1}{3}c_1 + \frac{2}{9}c_2 \tag{20}$$

Solving for c_0, c_1 and c_2 yields

$$c_0 = -74.097, \quad c_1 = -33.569, \quad c_2 = -3.802 \tag{21}$$

The exact values of c_1 and c_2 are as follows (from (5) and (6)),

$$c_1 = -34.783, \quad c_2 = -3.674 \tag{22}$$

which indicate an absolute relative error of at most 3.5% on comparison with (21). This gives us confidence in our estimate of the RG-inaccessible coefficient (c_0). We further check the stability of our estimates for c_0 by substituting exact values of RG-accessible coefficients into the moment equations (18), (19) and (20), to give,

$$N_0: \quad c_0 = -73.140$$
$$N_1: \quad c_0 = -73.555 \tag{23}$$
$$N_2: \quad c_0 = -73.724$$

The above results are within 1.3% of the estimated value in (21), which indicates excellent consistency of this method.

Finally, we do a χ^2 fit of R_3, in (13), with our Pade estimate (16). The χ^2 can be expressed as

$$\chi^2[c_0,c_1,c_2,c_3] = \int_0^1 [R_3 - (c_0 - c_1 \ln w + c_2 \ln^2 w)]^2 dw \qquad (24)$$

The requirement $\dfrac{\partial \chi^2}{\partial c_i} = 0$ yields (with the same value of k used earlier)

$$c_0 = -74.098, \quad c_1 = -33.569, \quad c_2 = -3.802 \qquad (25)$$

which is virtually in total agreement with results obtained via moments (21).

Hence, along with the RG results (22) and $c_0 = -73.140$ via (23), we have a full $O(\alpha_s^3)$ estimate for the longitudinal cross-section (2).

RESULTS AND DISCUSSION

As indicated earlier in the Introduction, there was a slight discrepancy of the measured value of the cross-section (at next-to-leading order, obtained in [1]) with the current value of the strong coupling at the Z pole, as seen in Figure 1. Including the $O(\alpha_s^3)$ estimate in (2), we find good agreement with the current value of PDG quoted value of $\alpha_s(M_Z) = 0.1185 \pm 0.020$ (Figure 1). This confirms that inclusion of higher-order contributions are indeed important as stated in [4]. In Figure 2, we show reduced dependence on the renormalization scale, μ. Furthermore, Figure 2 illustrates the near equality of the Principle of Minimal Sensitivity (PMS) and Fastest Apparent Convergence [15] which further increases our confidence in the estimate. The PMS value for the cross section happens at $\mu = 16.5$ GeV and is given by $\sigma_L = 0.055697$. Similarly the Fastest Apparent Convergence value happens at $\mu = 18.5$ GeV and is given by $\sigma_L = 0.055692$. These are in solid agreement with each other and the experimental value, which indicates the usefulness of estimating higher-order perturbative contributions via RG/Pade methodology.

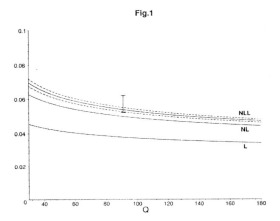

FIGURE 1. Plot of the longitudinal cross-section at $\mu = Q$, at leading (L), next-to-leading (NL) and next-to-next-to-leading (NLL) (estimated) orders versus Q. The solid lines are evaluated at the central value of the strong coupling constant as quoted in PDG [3]. The dashed lines are values of the cross-section (at the NLL order) evaluated at the upper and the lower limit of the strong coupling constant. The curves incorporating our NLL estimate indicate an overall agreement with the experimental value at the Z-pole (shown in figure) [2], indicating the importance of higher-order perturbative effects. Q has the units of GeV.

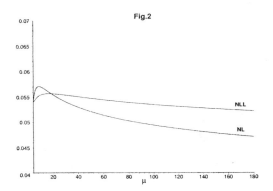

FIGURE 2. Longitudinal cross-section at next-to-leading (NL) and at estimated next-to-next-to-leading order (NLL) versus renormalization scale, μ (GeV).

ACKNOWLEDGMENTS

I would like to thank Victor Elias, Tom Steele, and Mohammad Ahmady for useful discussions and comments. I would also like to thank National Sciences and Research Council of Canada (NSERC) for financial support.

REFERENCES

1. P.J. Rijken, W.L. Neerven, Phys. Lett. **B386** (1996) 422.
2. R. Akers et al. (OPAL), Z.Phys **C68** (1995) 203.
3. D.E. Groom et al. (Particle Data Group) Eur. Phys. J. **C15** (2000) 1.
4. R.K. Ellis, W.J. Sterling, and B.R. Webber, "QCD and Collider Physics", Cambridge University Press 1996, Chapter 6.
5. Ellis J., Gardi E., Karliner M., Samuel M. A., Phys. Lett. **B366** (1996) 268
   Ellis J., Gardi E., Karliner M., Samuel M.A., Phys. Rev. **D54** (1996) 6986.
   Ellis J., Jack I., Jones D.R.T., Karliner M., Samuel M., Phys. Rev. **D57** (1998) 2665.
   Ellis J., Karliner M., Samuel M. A., Phys. Lett. **B400** (1997) 176.
6. Elias V., Steele T.G., Chishtie F., Migneron R., Sprague K., Phys. Rev. **D58** (1998) 116007.
7. Chishtie F., Elias V., Steele T.G., Phys. Rev. **D59** (1999) 105013.
8. F.Chishtie, V. Elias, T.G. Steele, Phys. Lett. B446 (1999) 267.
9. M.R. Ahmady, F.A.Chishtie, V. Elias, T.G. Steele, Phys. Lett. **B479** (2000) 201.
10. F.A.Chishtie, V.Elias, T.G. Steele, J.Phys.**G26** (2000) 93.
11. F.A.Chishtie, V.Elias, T.G. Steele, J.Phys.**G26** (2000) 1239.
12. F.A.Chishtie, V.Elias,, Phys. Rev. **D64** (2001) 016007.
13. F.A.Chishtie, V.Elias,, Phys. Lett. **B499** (2001) 270.
14. F.A.Chishtie, V.Elias,, hep-ph/ 0107052.
15. Stevenson P.M., Phys. Rev. **D23** (1981) 2916
    G. Grunberg, Phys. Lett. **B95** (1980) 70; Phys. Rev. **D29** (1984) 2315.

Exploring pseudoscalar meson scattering in Linear Sigma Models

Deirdre Black (Speaker)*, Amir H. Fariborz[†], Sherif Moussa*, Salah Nasri* and Joseph Schechter*

*Physics Department, Syracuse University, Syracuse, NY 13244
[†]Department of Mathematics/Science, State University of New York Institute of Technology, Utica, New York 13504

Abstract. The three flavor linear sigma model is studied as a toy model for understanding the role of possible light scalar mesons in the $\pi\pi$, πK and $\pi\eta$ elastic scattering channels. We unitarize tree level amplitudes using the K-matrix prescription and, with a sufficiently general model, obtain reasonable fits to the experimental data. The effect of unitarization is very important and leads to the emergence of a nonet of light scalars, with masses below 1 GeV. We compare with a scattering treatment using a more general non-linear sigma model approach and also comment briefly upon how our results fit in with the scalar meson puzzle.

INTRODUCTION

This talk is based on the work of the Syracuse group in [1]. In recent years there has been a renewal of interest in the scalar mesons below and above 1 GeV - for a list of references to the ideas of various authors see [1]. The scalar mesons are an interesting puzzle because their properties do not fit, in an obvious way, those of a conventional $q\bar{q}$ scalar multiplet. In his talk at this meeting Amir Fariborz will address this question further.

Here we look at a basic problem which is that of extracting information about scalar mesons from experiment. This leads to the related problem of describing pseudoscalar meson scattering in a regime which is non-perturbative but at energies which are sufficiently high to make the chiral perturbation scheme difficult to implement. Previously the Syracuse group studied $\pi\pi$, πK and $\pi\eta$ scattering [2] in a $\frac{1}{N_c}$-inspired unitarized non-linear chiral Lagrangian framework and found evidence for nine scalar mesons below 1 GeV which have the quantum numbers of a nonet. These are the isoscalars σ and $f_0(980)$, the isovector $a_0(980)$ and the isospinor $\kappa(900)$ meson. The σ and in particular the κ meson are rather controversial and emerge as extremely broad states requiring a generalization of a Breit-Wigner parameterization. Many authors have found similar or related results (see [1]).

In this context it is interesting to study pseudoscalar meson-meson scattering in a different model in order to investigate the model-dependence of the scalar meson parameters. We will work in the Linear Sigma Model, which is of course the classic chiral symmetric model, in which the scalar mesons are present from the outset.

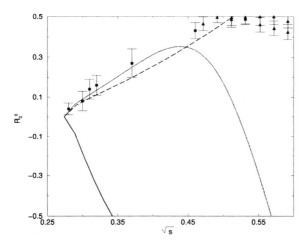

FIGURE 1. Variations of SU(2) Linear Sigma Model predictions for the real part of the $\pi\pi$ scattering amplitude. The dashed line is the current algebra result. The dark solid line shows the effect of adding a width to regularize the σ propagator. This clearly ruins the good behaviour of the tree-level amplitude near threshold. Squares and triangles are experimental data points.

SU(2) LINEAR SIGMA MODEL AND K-MATRIX UNITARIZATION

Let us focus initially on the $\pi\pi$ scattering amplitude. The result in the original two-flavor Linear Sigma Model [3] is well-known. The tree-level invariant $\pi\pi$ amplitude is the sum of a contact diagram and direct and crossed-channel σ-exchange diagrams:

$$A(s,t,u) = \frac{2}{F_\pi^2} \left(m_b^2(\sigma) - m_\pi^2 \right) \left[\frac{m_b^2(\sigma) - m_\pi^2}{m_b^2(\sigma) - s} - 1 \right]. \quad (1)$$

We have put a subscript b, denoting "bare", on the σ mass since this quantity will be shifted by unitarization as we shall see shortly. This amplitude reproduces the current algebra result, which gives good agreement with experiment near threshold, in the limit where $m_b(\sigma)$ is large. We see that even near threshold the σ resonance plays an important role since there is a delicate cancellation between the constant contact term and the σ pole contribution. Above threshold this tree-level amplitude blows up at $s = m_b^2(\sigma)$ and so requires regularization.

One might first try to regularize by adding an imaginary width-type term to the σ propagator. It turns out that this destroys the good threshold behaviour of the amplitude which comes from chiral symmetry. As shown in Fig. 1 including the perturbative width actually reverses the sign of the real part of the low-energy amplitude.

We are therefore led to consider an alternative regularization prescription. We use the well-known K-matrix technique [4] whereby the unitarized partial-wave S-matrix element (we only look at elastic scattering channels) is defined by

$$S \equiv \frac{1+iK}{1-iK}. \qquad (2)$$

Identifying K with the partial wave projection of the tree-level amplitude leads to the following "regularized" partial wave amplitude

$$T_J^I \equiv \frac{T_{J\,tree}^I}{1 - iT_{J\,tree}^I}. \qquad (3)$$

We notice that for $T_{J\,tree}^I \ll 1$ we have $T_J^I \approx T_{J\,tree}^I$ and also that when the tree amplitude diverges, the Real part of the unitarized amplitude is zero. Applying this unitarization to the partial wave projection of Eq. (1) we can get a reasonable fit to the $\pi\pi$ data (shown in Fig. 2) up to about 1 GeV. Similar results were found for the SU(2) case in [5]. In order to try and describe the structure of the $\pi\pi$ scattering amplitude beyond 1 GeV and also to investigate the $I = \frac{1}{2}$ and $I = 1$ scattering channels we next extend our analysis to the SU(3) case.

SU(3) LINEAR SIGMA MODEL TREATMENT

We consider a general non-renormalizable [6] Lagrangian of the form

$$\mathcal{L} = -\frac{1}{2}\text{Tr}\left(\partial_\mu \phi \partial_\mu \phi\right) - \frac{1}{2}\text{Tr}\left(\partial_\mu S \partial_\mu S\right) - V_0 - V_{SB}, \qquad (4)$$

where $M = S + i\phi$ is a 3×3 matrix field ($S = S^\dagger$ represents a scalar nonet and $\phi = \phi^\dagger$ a pseudoscalar nonet) and where V_0 is an arbitrary function of the independent $SU(3)_L \times SU(3)_R \times U(1)_V$ invariants

$$I_1 = \text{Tr}\left(MM^\dagger\right) \quad, \quad I_2 = \text{Tr}\left(MM^\dagger MM^\dagger\right),$$
$$I_3 = \text{Tr}\left((MM^\dagger)^3\right) \quad, \quad I_4 = 6\left(\det M + \det M^\dagger\right). \qquad (5)$$

Of these, only I_4 is not invariant under $U(1)_A$. The symmetry breaker V_{SB} has the minimal form linear in S. Chiral symmetry [6] gives relations among certain parameters of the model, for example amongst many of the masses and trilinear coupling constants. In particular the "bare" mass of the strange scalar state is related to the pseudoscalar meson masses and decay constants by:

$$m_b^2(\kappa) = \frac{F_K m_K^2 - F_\pi m_\pi^2}{F_K - F_\pi}. \qquad (6)$$

If we use a renormalizable potential the model is extremely predictive. Once we fix the pseudoscalar parameters m_π, m_K, m_η or F_K, $m_{\eta'}$ and F_π, it turns out that all but the isoscalar scalar meson masses are fixed. Specifying in addition, for example, the bare

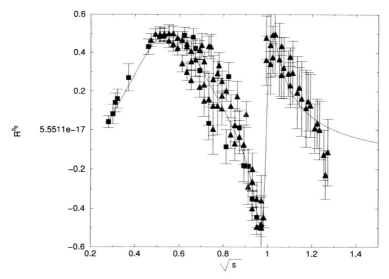

FIGURE 2. Comparison of our best fit for the Real part of the I=J=0 $\pi\pi$ scattering amplitude in the non-renormalizable SU(3) linear sigma model with experiment.

σ mass completely determines all the coupling constants and masses. We calculate the tree-level $\pi\pi$ scattering amplitude and again unitarize the scalar isoscalar partial wave channel according to the K-matrix prescription, which introduces no new parameters. It turns out that this single-parameter (i.e. the bare σ mass) model does not allow a good fit to the experimental data.

If we consider a more general potential then the two isoscalar scalar masses and the isoscalar mixing angle are all independent parameters. We perform a best fit using our unitarized amplitude for this case and show the result in Fig. 3 where it can be seen that we get good agreement with the data up to approximately 1.25 GeV. The best fit values are (as expected from the position of the zeros of the Real part of the amplitude) $m_b(\sigma) = 0.847$ GeV, $m_b(f_0) = 1.3$ GeV and a bare σ-f_0 mixing angle of 48.6^o. We identify the "physical" masses and widths from the poles in the unitarized partial wave amplitude. These are given in Table I.

We also consider the πK channel where the strange scalar meson κ will of course play an important role. Since the bare κ mass is very sensitive to the choice of input parameters, in particular to $\frac{F_\pi}{F_K}$, we consider different values for $m_b(\kappa)$. Once again, we calculate the tree-level amplitude from Eq. (4) and unitarize using the K-matrix method. A plot of the Real part of the resulting amplitude is given in Fig. 3, where the experimental data is also presented. We also looked at $\pi\eta$ scattering where the $a_0(980)$ appears in the direct scalar channel.

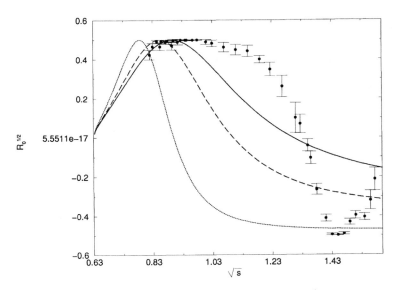

FIGURE 3. Comparison of our prediction for the Real part of the $I=\frac{1}{2}$, $J=0$ πK scattering amplitude in the non-renormalizable SU(3) Linear Sigma Model with experiment. The curves correspond to $m_{\text{BARE}}(\kappa) = 1.3$ GeV (solid), 1.1 GeV (dashed) and 0.9 GeV (dotted).

TABLE 1. Predicted "physical" masses and widths in MeV of the nonet of scalar mesons contrasted with suitable (as discussed in the text) comparison values.

	σ	f_0	κ	a_0
Present Model mass (MeV), width (MeV)	457, 632	993, 51	800, 260-610	890-1010, 110-240
Comparison mass (MeV), width (MeV)	560, 370	980±10, 40-100	900, 275	985, 50-100

DISCUSSION

We studied $\pi\pi$, πK and $\pi\eta$ scattering in a K-matrix unitarized version of the SU(3) Linear Sigma Model. One crucial feature is that the masses of the scalar resonances are shifted by unitarization from their "bare" tree-level values to physical values which are presented in Table I.

For comparison, we have also listed in Table I the $f_0(980)$ and $a_0(980)$ parameters given in the 2001 Review of Particle Properties [7]. For the σ and κ parameters we present for comparison the results of our previous analyses [2]. We see that the masses are consistent to within about 50-100 MeV, while the widths are more model-dependent. In our non-linear chiral Lagrangian approach to pseudoscalar meson meson scattering we included contributions from vector meson exchange which are known to be important and are likely to account for some of the differences between the present model and previous results.

The overall conclusion is that the scalar mesons of the Linear Sigma Model emerge with physical masses below the 1 GeV region. In particular even though we studied a range of bare κ masses between 0.9 and 1.3 GeV, unitarization of the πK scattering amplitude always forced a physical mass of about 800 MeV.

In Fig. 3 we see that although the region just above the πK threshold can be explained in this model, the structure observed around 1.4 GeV is not accounted for. This is associated with the well-established $K_0^*(1430)$ resonance and thus suggests that this state is not described in the simple version of the Linear Sigma Model studied here. Similarly, the heavier isovector $a_0(1450)$ state does not fit in this picture. If there is a nonet of light (masses < 1 GeV) scalar mesons the question remains not only of understanding their quark substructure, but also that of the experimentally observed heavier scalar states such as the $K_0^*(1430)$ and $a_0(1450)$.

ACKNOWLEDGMENTS

The author would like to thank the organizers for their hospitality and a very interesting and enjoyable MRST conference at the University of Western Ontario. This work has been supported in part by the U.S. DOE under contract DE-FG-02-85ER40231. S.M. would like to thank the Egyptian cultural bureau for support and A.H.F. wishes to acknowledge his grant from the State of New York/UUP Professional Development Committee.

REFERENCES

1. D. Black, A. H. Fariborz, S. Moussa, S. Nasri and J. Schechter, Phys. Rev. **D 64**, 014031 (2001).
2. M. Harada, F. Sannino and J. Schechter, Phys. Rev. **D54**, 1991 (1996); D. Black, A.H. Fariborz, F. Sannino and J. Schechter, Phys. Rev. **D58**, 054012 (1998).
3. M. Lévy, Nuovo Cimento **52A**, 23 (1967). See S. Gasiorowicz and D. A. Geffen, Rev. Mod. Phys. **41**, 531 (1969) for a review which contains a large bibliography.
4. See for example, S. U. Chung *et al*, Ann. Physik **4** 404 (1995). See also T.N. Truong, Phys. Rev. Lett. **67**, 2260 (1991).
5. N.N. Achasov and G.N. Shestakov, Phys. Rev. **D49**, 5779 (1994).
6. J. Schechter and Y. Ueda, Phys. Rev. **D3**, 2874, 1971; Erratum D **8** 987 (1973). See also J. Schechter and Y. Ueda, Phys. Rev. **D3**, 168, (1971).
7. Particle Data Group, D. Groom *et al*, Eur. Phys. J. C **15**, 1 (2000).

Probing Scalar Mesons Using a Toy Model in the Linear Sigma Model

Deirdre Black*, Amir H. Fariborz[1][†], Sherif Moussa*, Salah Nasri* and Joseph Schechter*

Department of Physics, Syracuse University, Syracuse, New York 13244-1130
[†] *Department of Mathematics/Science, State University of New York Institute of Technology, Utica, New York 13504-3050*

Abstract.
The toy model of ref. [1] for the lowest and the next-to-lowest lying scalar/pseudoscalar mesons is briefly reviewed. This model, which is in the context of the linear sigma model and is formulated in terms of two meson nonets, provides a possible description of the mass and quark substructure of these mesons.

INTRODUCTION

Scalar mesons play important roles in low-energy QCD, and are at the focus of many theoretical and experimental investigations. Scalars are important from the theoretical point of view because they are Higgs bosons of QCD and induce chiral symmetry breaking, and therefore, are probes of the QCD vacuum. Scalars are also important from a phenomenological point of view, as they are very important intermediate states in Goldstone boson interactions away from threshold, where chiral perturbation theory is not applicable. There are 9 candidates for the lowest-lying scalar mesons ($m < 1$ GeV): $f_0(980)$ [I=0] and $a_0(980)$ [I=1] which are well established experimentally [2]; $\sigma(560)$ or $f_0(400 - 1200)$ [I=0] with uncertain mass and decay width [2]; and $\kappa(900)$ [I=1/2] which is not listed but mentioned in PDG 2000 [2]. The $\kappa(900)$ is only observed in some theoretical models. It is known that a simple $q\bar{q}$ picture does not explain the properties of these mesons. Different theoretical models that go beyond a simple $q\bar{q}$ picture have been developed, including: MIT bag model [3], $K\bar{K}$ molecule [4], unitarized quark model [5], QCD sum-rules [6], and chiral Lagrangians [1, 7, 8, 9, 10, 11].

The next-to-lowest scalars (1 GeV $< m <$ 2 GeV) are: $K_0^*(1430)$ [I=1/2], $a_0(1450)$ [I=1], $f_0(1370)$, $f_0(1500)$, $f_0(1710)$ [I=0], and are all listed in [2]. The $f_0(1500)$ is believed to contain a large glue component and therefore a good candidate for the lowest scalar glueball state. Other states, are generally believed to be closer to $q\bar{q}$ objects; however, some of their properties cannot be explained based on a pure $q\bar{q}$ structure.

Chiral Lagrangians, provide a powerful framework for studying the lowest and the next-to-lowest scalar states probed in different Goldstone boson interactions ($\pi\pi$, πK,

---

[1] Speaker

πη,...) away from threshold [1, 7, 8, 9, 11]. In this approach, a description of scattering amplitudes which are, to a good approximation, both crossing symmetric and unitary is possible. To construct scattering amplitudes, all contributing intermediate resonances up to the energy of interest are considered, and only tree diagrams (motivated by large N_c approximation) are taken into account. In this way, crossing symmetry is satisfied, but the constructed amplitudes should be regularized. Regularization procedure in turn unitarizes the scattering amplitude. By fitting the resulting scattering amplitude to experimental data, the unknown physical properties (mass, decay width, ...) of the light scalar mesons can be extracted. Within a non-linear chiral Lagrangian framework it was shown that:

I. A $\sigma(560)$ and a $\kappa(900)$ are required in order to describe experimental data on $\pi\pi$ and πK scattering amplitudes, respectively, and that these states [as well as $f_0(980)$ and $a_0(980)$] are closer to four-quark objects [7, 8, 9, 10],

II. Some unexpected properties of the next-to-lowest-lying scalars can be explained based on a mixing mechanism between underlying $q\bar{q}$ and $qq\bar{q}\bar{q}$ nonets [11].

Similar results were obtained in the context of a linear sigma model in [1], which are reviewed in the present article, and in [12]. In particular, here we focus on the effects of mixing between the lowest and the next to lowest scalar/pseudoscalar mesons and briefly show how such mixing could probe the quark sub-structure of these states. The complete details can be found in [1].

TWO NONET FORMULATION OF THE LINEAR SIGMA MODEL

In order to study the effect of mixing between the lowest and the next to lowest scalar states, we need to formulate the Lagrangian in terms of two different meson matrices. We start with two nonets M and M' which represent a $q\bar{q}$ and a $qq\bar{q}\bar{q}$ constituent-quark structure, respectively. The linear sigma model Lagrangian for a single nonet M is reviewed in [12] and will not be repeated here. The extension of this Lagrangian to include the nonet M' is written as

$$\mathcal{L} = -\frac{1}{2}\text{Tr}\left(\partial_\mu M \partial_\mu M^\dagger\right) - \frac{1}{2}\text{Tr}\left(\partial_\mu M' \partial_\mu M'^\dagger\right) - V_0(M,M') - V_{SB}, \quad (1)$$

where $V_0(M,M')$ stands for a general polynomial made from $SU(3)_L \times SU(3)_R$ [but not $U(1)_A$] invariants formed out of M and M'. The symmetry breaker V_{SB} has the minimal form

$$V_{SB} = -2\left(A_1 S_1^1 + A_2 S_2^2 + A_3 S_3^3\right), \quad (2)$$

where the A_a are real numbers which turn out to be proportional to the three light ("current" type) quark masses. We consider a very simplified approximation in which the quark mass effective term, V_{SB} is absent and where V_0 is simply given by:

$$V_0 = -c_2 \text{Tr}\left(MM^\dagger\right) + c_4 \text{Tr}\left(MM^\dagger MM^\dagger\right) + d_2 \text{Tr}\left(M'M'^\dagger\right) + e\text{Tr}\left(MM'^\dagger + M'M^\dagger\right). \quad (3)$$

Here c_2, c_4 and d_2 are positive real constants. The M matrix field is chosen to have a wrong sign mass term so that there will be spontaneous breakdown of chiral symmetry.

A pseudoscalar octet will thus be massless. On the other hand, the matrix field M' is being set up to have trivial dynamics except for its mixing term with M. The mixing is controlled by the parameter e and the e-term is the only one which violates $U(1)_A$ symmetry. Its origin is presumably due to instanton effects at the fundamental QCD level. [Other $U(1)_A$-violating terms which contribute to η' mass etc. are not being included for simplicity]. Using the notations $M = S + i\phi$ and $M' = S' + i\phi'$ we may expect vacuum values

$$\left\langle S_a^b \right\rangle = \alpha \delta_a^b, \qquad \left\langle S_a'^b \right\rangle = \beta \delta_a^b. \tag{4}$$

The minimization condition $\left\langle \frac{\partial V_0}{\partial S_a'^b} \right\rangle = 0$ leads to

$$\beta = -\frac{e}{d_2}\alpha \tag{5}$$

while $\left\langle \frac{\partial V_0}{\partial S_a^b} \right\rangle = 0$ yields

$$\alpha^2 = \frac{1}{2c_4}\left(c_2 + \frac{e^2}{d_2}\right). \tag{6}$$

In the absence of mixing the "four-quark" condensate β vanishes while the usual two quark condensate α remains.

The mass spectrum resulting from Eq. (3) has two scalar octets and two pseudoscalar octets, each with an associated SU(3) singlet. Each octet has eight degenerate members since the quark mass terms have been turned off. Let us focus on the I=1, positively charged particles for definiteness and define:

$$\pi^+ = \phi_1^2, \quad \pi'^+ = \phi_1'^2, \quad a^+ = S_1^2, \quad a'^+ = S_1'^2. \tag{7}$$

Then the 2×2 squared mass matrix of π and π' is:

$$2\begin{bmatrix} \frac{e^2}{d_2} & e \\ e & d_2 \end{bmatrix}. \tag{8}$$

This has eigenstates

$$\begin{aligned}\pi_p &= \left(1 + \frac{e^2}{d_2^2}\right)^{-\frac{1}{2}} \left(\pi - \frac{e}{d_2}\pi'\right), \\ \pi_p' &= \left(1 + \frac{e^2}{d_2^2}\right)^{-\frac{1}{2}} \left(\frac{e}{d_2}\pi + \pi'\right),\end{aligned} \tag{9}$$

with masses

$$m^2(\pi_p) = 0, \qquad m_{\text{BARE}}^2(\pi_p') = \frac{2e^2}{d_2} + 2d_2. \tag{10}$$

We put the subscript "BARE" on $m^2(\pi_p')$ to indicate that it may receive non-negligible corrections from K-matrix unitarization as in our detailed treatment of the M only Lagrangian, reviewed in [12]. A possible experimental candidate for such a particle is the $\pi(1300)$.

Computing the axial vector current by Noether's theorem yields

$$\left(J_\mu^{\text{axial}}\right)_1^2 = F_\pi \partial_\mu \pi_p^+ + \ldots,$$

$$F_\pi = 2\alpha \sqrt{1 + \left(\frac{e}{d_2}\right)^2}, \tag{11}$$

where α is given in Eq. (6).

Notice that a term like $\partial_\mu \pi_p'^+$ does not appear in our semi-classical approximation. The 2×2 squared mass matrix of the scalars a and a' is

$$\begin{bmatrix} 4c_2 + \frac{6e^2}{d_2} & 2e \\ 2e & 2d_2 \end{bmatrix}. \tag{12}$$

The eigenstates are defined by

$$\begin{pmatrix} a_p \\ a_p' \end{pmatrix} = \begin{bmatrix} \cos\omega & -\sin\omega \\ \sin\omega & \cos\omega \end{bmatrix} \begin{pmatrix} a \\ a' \end{pmatrix}, \tag{13}$$

with

$$\tan 2\omega = \frac{4e}{2d_2 - 4c_2 - \frac{6e^2}{d_2}}. \tag{14}$$

The corresponding masses are

$$m_{\text{BARE}}^2\left(a_p, a_p'\right) = 2c_2 + d_2 + \frac{3e^2}{d_2} \mp 2e \, \csc 2\omega, \tag{15}$$

where the upper (lower) sign stands for a_p, (a_p').

It is interesting to examine the masses of the degenerate octets in a little more detail. For orientation, first consider the case when the mixing parameter e vanishes. The usual "$q\bar{q}$" pseudoscalars π_p are zero mass Goldstone bosons in this approximation. If $4c_2 > 2d_2$, a_p, the original scalar partner of π_p lies higher than the degenerate "$qq\bar{q}\bar{q}$" scalar and pseudoscalar a_p' and π_p'. When the mixing is turned on, a four quark condensate develops and the mass ordering is

$$m_{\text{BARE}}(a_p) > m_{\text{BARE}}(\pi_p') > m_{\text{BARE}}(a_p') > m_{\text{BARE}}(\pi_p) = 0. \tag{16}$$

This is graphed, as a function of e, in Fig. 1 (with parameter choices $c_2 = 0.25$ GeV^2, $d_2 = 0.32$ GeV^2. In such a scenario, the $qq\bar{q}\bar{q}$ scalar would be the next lightest after the $q\bar{q}$ Goldstone boson. Each particle would be a mixture of $q\bar{q}$ and $qq\bar{q}\bar{q}$ to some extent. For the given parameters the mixing angle remains small however because the denominator of Eq. (14) is always negative and increases in magnitude as e^2 increases. Note especially, that due to the spontaneous breakdown of chiral symmetry, there is no guarantee that the lowest lying scalar is of $q\bar{q}$ type. Also note that π_p' is expected to be more massive than a_p'.

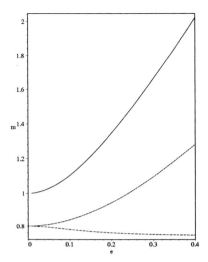

FIGURE 1. Plots of $m_{\text{BARE}}(a_p)$ (solid), $m_{\text{BARE}}(a'_p)$ (dashed) and $m_{\text{BARE}}(\pi'_p)$ versus the mixing parameter e for the choice $c_2 = 0.25$ GeV^2 and $d_2 = 0.32$ GeV^2. The highest lying curve is mainly a "$q\bar{q}$" scalar, while the lowest lying curve is mainly a "$qq\bar{q}\bar{q}$" scalar. The excited pseudoscalar curve is in the middle.

On the other hand, if the QCD dynamics underlying the effective Lagrangian is such that $2d_2 > 4c_2$ we will get a mass ordering $m_{\text{BARE}}(a'_p) > m_{\text{BARE}}(\pi') > m_{\text{BARE}}(a_p)$ in which the four quark scalar appears heaviest. However, in this case we will definitely get a large mixing as e increases since the denominator of Eq. (14) starts out positive when $e = 0$ and will go to zero as e is increased. Thus the next-to-lowest lying a_p can be expected to have a large $qq\bar{q}\bar{q}$ admixture.

All of these remarks pertain to the meson current-quark type operators in the toy model. The important effects of unitarization (i.e. $m_{\text{BARE}} \to m$) are likely, as in our earlier treatment, to favor an interpretation of the low lying physical scalars as being of four constituent quark type in either case.

CONCLUSION

The main lesson from our preliminary treatment of a chiral model with mixing is perhaps that even though the M fields carry "chiral indices" it is not easy to assign an unambiguous quark substructure. On the other hand there is a great potentiality for learning more about non-perturbative QCD from further study of the light scalars. Such features as scalar mixing (including the possibility of mixing with glueballs for the I=0 states), four quark condensates and excited pseudoscalars may eventually get correlated with each other and with the experimental data on the scattering of light pseudoscalars.

ACKNOWLEDGMENTS

The work of D.B., S.M., S.N. and J.S. has been supported in part by the US DOE under contract DE-FG-02-85ER40231. The work of A.H.F. has been supported by grants from the State of New York/UUP Professional Development Committee, and the 2001 Faculty Summer Grant from the SUNY Institute of Technology.

REFERENCES

1. D. Black, A.H. Fariborz, S. Moussa, S. Nasri and J. Schechter, Phys. Rev. D **64**, 014031 (2001).
2. Review of Particle Physics, Euro. Phys. J. C **15**, 1 (2000).
3. R.L. Jaffe, Phys. Rev. D **15**, 267 (1977).
4. J. Weinstein and N. Isgur, Phys. Rev. D **41**, 2236 (1990).
5. N.A. Törnqvist, Z. Phys. C **68**, 647 (1995); E. van Beveren et al, Z. Phys. C **30**, 615 (1986).
6. V. Elias, A.H. Fariborz, Fang Shi and T.G. Steele, Nucl. Phys. A **633**, 279 (1998); Fang Shi, T.G. Steele, V. Elias, K.B. Sprague, Ying Xue and A.H. Fariborz, Nucl. Phys. A **671**, 416 (2000).
7. F.Sannino and J. Schechter, Phys. Rev. D **52**, 96 (1995); M. Harada, F. Sannino and J. Schechter, Phys. Rev. D **54**, 1991 (1996); Phys. Rev. Lett. **78**, 1603 (1997).
8. D. Black, A.H. Fariborz, F. Sannino and J. Schechter, Phys. Rev. D **58**, 054012 (1998); Phys. Rev. D **59**, 074026 (1999).
9. A.H. Fariborz and J. Schechter, Phys. Rev. D **60**, 034002 (1999).
10. D. Black, A.H. Fariborz and J. Schechter, Phys. Rev. D **61**, 074030 (2000).
11. D. Black, A.H. Fariborz and J. Schechter, Phys. Rev. D **61**, 074001 (2000).
12. D. Black, these proceedings.

FIELD THEORY I

Cutoff dependence of the Casimir effect

C. R. Hagen

*Department of Physics and Astronomy, University of Rochester
Rochester, N.Y. 14627*

Abstract. The problem of calculating the Casimir force on two conducting planes by means of the stress tensor is examined. The evaluation of this quantity is carried out using an explicit regularization procedure which has its origin in the underlying (2+1) dimensional Poincaré invariance of the system. The force between the planes is found to depend on the ratio of two independent cutoff parameters, thereby rendering any prediction for the Casimir effect an explicit function of the particular calculational scheme employed. Similar results are shown to obtain in the case of the conducting sphere.

In 1948 Casimir [1] first predicted that two infinite parallel plates in vacuum would attract each other. This remarkable result has its origin in the zero point energy of the electromagnetic field. While the latter is highly divergent, the change associated with this quantity for specific plate configurations has been found in numerous calculations to be finite as well as cutoff dependent and thus in principle observable. Early work to detect this small effect [2] was characterized by relatively large experimental uncertainties which left the issue in some doubt. More recent efforts [3] have provided quite remarkable data, but are based on a different geometry from that of Casimir. Since a rigorous theoretical calculation has never been carried out for the latter configuration, there remains room for skepticism as to whether the Casimir effect is as well established as is frequently asserted.

The most elementary calculation of the Casimir effect between two parallel conducting planes located at $z = 0$ and $z = a$ employs a mode summation in the framework of a regularization which depends only on the frequency $\omega_k = [\mathbf{k}^2 + (\frac{n\pi}{a})^2]^{\frac{1}{2}}$ where $n = 0, 1, 2, \ldots$. Upon combining the result obtained with the corresponding result for the interval $a \leq z \leq L$ where $L \gg a$ is the z-coordinate of a third conducting plane, a finite cutoff independent result

$F/A = \pi^2/240a^4$ [4] is obtained for the Casimir pressure on the plate at $z = a$.

A considerably more elegant approach to this problem is that of Brown and Maclay [5] who employ an image method to calculate $\langle 0|T^{\mu\nu}(x)|0\rangle$. Thus they showed that the photon propagator in the presence of conducting planes at $z = 0$ and $z = a$ could be expressed in terms of an infinite sum over the usual (i.e., $-\infty < z < \infty$) photon propagator with the z-coordinate of each term in the sum displaced by an even multiple of a. Since the stress tensor for the electromagnetic case is given by

$$T^{\mu\nu}(x) = F^{\mu\alpha}F^\nu_\alpha - \frac{1}{4}g^{\mu\nu}F^{\alpha\beta}F_{\alpha\beta} \tag{1}$$

where

$$F^{\mu\nu}(x) = \partial^\mu A^\nu(x) - \partial^\nu A^\mu(x),$$

it follows that upon taking appropriate derivatives with respect to the propagator arguments x and x' and invoking the limit $x \to x'$ a formal expression can be obtained for the vacuum expectation value of the stress tensor. On the basis of covariance arguments together with the divergence and trace free property of $T^{\mu\nu}(x)$ it was then found in ref. 5 that

$$\langle 0|T^{\mu\nu}(x)|0\rangle = (\frac{1}{4}g^{\mu\nu} - \hat{z}^\mu \hat{z}^\nu)(\frac{1}{2\pi^2 a^4}) \sum_{n=1}^\infty n^{-4} \tag{2}$$

where $\hat{z}^\mu$ is the unit vector (0,0,1,0) in the z-direction normal to the conducting planes.

However, there is some reason to question whether this approach has adequately dealt with the divergences which invariably occur in Casimir calculations. One notes in particular that the result (2) is obtained only after an obviously singular $n = 0$ term has been dropped from the sum which occurs in that equation. While one can argue as in [5] that such an a-independent term can be freely omitted since it is merely the usual subtraction of the large a result, it is well to note that the *entire* sum over n is required for a demonstration that the propagator satisfies correct boundary conditions at $z = 0, a$. Moreover, as is shown in this work, an appropriately regularized form of (2) does not necessarily allow a separation into cutoff dependent terms and a-dependent terms, in contrast with the result found in [5]. Of still greater import is the fact that more general regularizations than those usually considered in this calculation lead to an explicit cutoff dependence of the Casimir stress, a circumstance which would seem to deny its physical significance.

To establish the above claims one reverts from the image approach to one based on expansion of the Green's function in terms of orthogonal functions

[6]. To this end one notes that the free field propagator in the radiation gauge can be written as

$$G^{ij}(\mathbf{x} - \mathbf{x}', z, z', t - t') = \sum_{n\lambda} \int \frac{d\mathbf{k}d\omega}{(2\pi)^3} e^{-i\omega(t-t')}$$
$$\times \frac{A^i_{n\lambda}(\mathbf{k}, z) A^{j*}_{n\lambda}(\mathbf{k}, z')}{k^2 - \omega^2 + (n\pi/a)^2 - i\epsilon} e^{i\mathbf{k}\cdot(\mathbf{x}-\mathbf{x}')} \quad (3)$$

where $\lambda = 1, 2$ refers to the polarization, and spatial coordinates orthogonal to the z-direction are denoted by a boldface notation. The eigenfunctions $A^i_{n\lambda}(\mathbf{k}, z)$ satisfy the equation

$$\left[\frac{\partial^2}{\partial z^2} + (n\pi/a)^2\right] A^i_{n\lambda}(\mathbf{k}, z) = 0$$

and are given explicitly by

$$A^i_{n1}(\mathbf{k}, z) = \frac{\overline{k}_i}{|\mathbf{k}|} \left(\frac{2}{a}\right)^{\frac{1}{2}} \sin(n\pi z/a) \quad (4)$$

and

$$A^i_{n2}(\mathbf{k}, z) = \frac{1}{|\mathbf{k}|\omega_k} \left(\hat{\mathbf{z}}^i \omega_k^2 + \hat{\mathbf{z}} \cdot \nabla\nabla^i\right) \left(\frac{2}{a}\right)^{\frac{1}{2}} \cos(n\pi z/a) \quad (5)$$

where $\omega_k^2 = \mathbf{k}^2 + (n\pi/a)^2$ and $\overline{k}_i \equiv \epsilon^{ij} k_j$ with ϵ^{ij} being the usual alternating symbol. (In the $n = 0$ case the rhs of (5) must be multiplied by a factor of $2^{-\frac{1}{2}}$.) It is important to note that each eigenfunction $A^i_{n\lambda}(\mathbf{k}, z)$ satisfies the boundary conditions $\hat{\mathbf{z}} \times \mathbf{E} = \hat{\mathbf{z}} \cdot \mathbf{B} = 0$ at $z = 0, a$. This means that it is possible to introduce a regularization such that contributions from large values of $|\mathbf{k}|$ and/or n are reduced without destroying the validity of the boundary conditions. This stands in marked contrast with the image method which has no mechanism for the consistent suppression of the contributions of higher order reflections.

In order to determine the regularization appropriate to this calculation one should ideally make reference to the underlying symmetry. Since the latter consists of the reflection $z \to a - z$ and the (2+1) dimensional Poincaré group, it is natural to seek to classify regularization schemes according to representations of the latter. The usual cutoff method for this problem invokes a parameter which damps out the large ω_k contributions, an approach which makes no reference to the underlying Lorentz invariance. A far more appropriate technique is to generalize this to a cutoff based on a vector σ^μ in (2+1) dimensions as well as a scalar cutoff Σ which can be used to suppress large values of the (2+1) dimensional invariant $E^2 - \mathbf{P}^2$ where E and $\mathbf{P}$ are respectively the energy and momentum operators associated with this

(2+1) dimensional subspace. Clearly, the credibility of the Casimir effect requires that the result be independent of the relative importance of these two competing cutoffs.

The calculation proceeds by noting that since the limit $x \to x'$ is to be taken symmetrically at some point, it is appropriate to use only the imaginary part of the propagator. An appropriately regularized version of this function can be inferred from Eq.(3) to be [7]

$$\Im \ G^{ij}_{\sigma,\Sigma}(x,x') = \pi \sum_{n\lambda} \int \frac{d^3k}{(2\pi)^3} \delta(k^2 + (n\pi/a)^2)$$
$$\times A^i_{n\lambda}(\mathbf{k},z) A^{j*}_{n\lambda}(\mathbf{k},z') e^{ik^\mu(x-x')_\mu} e^{\sigma_\mu k^\mu \epsilon(k^0)} e^{\Sigma(-k^2)^{\frac{1}{2}}} \quad (6)$$

where $\epsilon(k^0)$ is the alternating function and a summation convention has been introduced in the Lorentz invariant subspace. Note that since both σ^μ and k^μ are three vectors in that space, they satisfy the orthogonality conditions $\hat{z}^\mu \sigma_\mu = \hat{z}^\mu k_\mu = 0$ [8]. In addition it is clearly necessary to impose $\bar{\sigma}^2 \equiv -\sigma^\mu \sigma_\mu > 0$ and $\sigma^0 > 0$ in order that this propagator exist. It will subsequently be found that its existence also requires $\Sigma < \bar{\sigma}$.

To proceed one uses the regularization (6) and the form of the stress tensor (1). When used in conjunction with the eigenfunctions (4) and (5) the vacuum expectation value of the regularized stress tensor can be determined. With some effort this is found by straightforward calculation to yield the coordinate independent result

$$\langle 0 | T^{\mu\nu} | 0 \rangle = \frac{2\pi}{a} \sum_{n=0}^{\infty}{}' \int \frac{d^3k}{(2\pi)^3} \delta(k^2 + (n\pi/a)^2) e^{\sigma_\mu k^\mu \epsilon(k^0)}$$
$$\times e^{\Sigma n\pi/a} \left[k^\mu k^\nu + \hat{z}^\mu \hat{z}^\nu (n\pi/a)^2 \right]$$

which is manifestly both symmetric and traceless. The prime on the summation denotes the fact that the $n = 0$ term must be multiplied by $\frac{1}{2}$ as a consequence of the normalization of $A^i_{02}(\mathbf{k},\mathbf{z})$. It can be more usefully written as

$$\langle 0 | T^{\mu\nu} | 0 \rangle = \frac{1}{a} \sum_{n=0}^{\infty}{}' e^{\Sigma n\pi/a}$$
$$\times \left(\frac{\partial}{\partial \sigma_\mu} \frac{\partial}{\partial \sigma_\nu} - \hat{z}^\mu \hat{z}^\nu \frac{\partial^2}{\partial \sigma^\alpha \partial \sigma_\alpha} \right) \Delta^{n\pi/a}(-i\sigma) \quad (7)$$

where $\Delta^{n\pi/a}(x)$ is the (2+1) dimensional function

$$\Delta^{n\pi/a}(x) = 2\pi \int \frac{d^3k}{(2\pi)^3} e^{ikx\epsilon(k^0)} \delta(k^2 + (n\pi/a)^2)$$

for a particle of mass $n\pi/a$. Since this is an $O(2,1)$ scalar, $\Delta^{n\pi/a}(-i\sigma)$ is a function of only the invariant $\bar{\sigma}$ which has the explicit form

$$\Delta^{n\pi/a}(-i\sigma) = \frac{1}{2\pi\bar{\sigma}} e^{-\bar{\sigma}n\pi/a}.$$

The insertion of this result into Eq.(7) clearly implies that the sum over n exists only for the case that $\Sigma < \bar{\sigma}$ as previously stated. Upon performing the summation over n it follows that

$$\langle 0|T^{\mu\nu}|0\rangle = \left(\frac{\partial^2}{\partial\sigma_\mu\partial\sigma_\nu} - \hat{z}^\mu\hat{z}^\nu\frac{\partial^2}{\partial\sigma^\alpha\partial\sigma_\alpha}\right)F(\bar{\sigma},\Sigma)$$

where

$$F(\bar{\sigma},\Sigma) = \frac{1}{4\pi a\bar{\sigma}}\coth\frac{(\bar{\sigma}-\Sigma)\pi}{2a}.$$

One now carries out the expansion of this expression discarding terms which give no contribution in the limit of vanishing cutoff, thereby obtaining

$$F(\bar{\sigma},\Sigma) \rightarrow \left[\frac{1}{2\pi^2}\frac{1}{\bar{\sigma}}\frac{1}{\bar{\sigma}-\Sigma} - \frac{\Sigma}{24a^2\bar{\sigma}} - \frac{(\bar{\sigma}-\Sigma)^3\pi^2}{1440\bar{\sigma}a^4}\right].$$

Upon performing the derivatives and rearranging terms there finally results [9, 10]

$$\langle 0 \mid T^{\mu\nu}|0\rangle = \left[g^{\mu\nu} + 3\frac{\sigma^\mu\sigma^\nu}{\bar{\sigma}^2} - \hat{z}^\mu\hat{z}^\nu\right]\left\{-\frac{\Sigma}{24a^2\bar{\sigma}^3} + \frac{(2\bar{\sigma}-\Sigma)(\bar{\sigma}-\Sigma)+\frac{2}{3}\bar{\sigma}^2}{2\pi^2\bar{\sigma}^3(\bar{\sigma}-\Sigma)^3} + \frac{\pi^2}{1440a^4}\frac{\Sigma}{\bar{\sigma}}\left(\frac{\Sigma^2}{\bar{\sigma}^2}-1\right)\right\}$$
$$+\left(\frac{1}{4}g^{\mu\nu} - \hat{z}^\mu\hat{z}^\nu\right)\left[\left(1-\frac{\Sigma}{\bar{\sigma}}\right)\frac{\pi^2}{180a^4} - \frac{4}{3\pi^2}\frac{1}{\bar{\sigma}(\bar{\sigma}-\Sigma)^3}\right].$$

If (following [5]) one subtracts the $a \rightarrow \infty$ result, this reduces to the more tractable form

$$\langle 0 \mid \overline{T}^{\mu\nu}|0\rangle = \left[g^{\mu\nu} + 3\frac{\sigma^\mu\sigma^\nu}{\bar{\sigma}^2} - \hat{z}^\mu\hat{z}^\nu\right]$$
$$\times\Sigma\left\{-\frac{1}{24a^2\bar{\sigma}^3} + \frac{\pi^2}{1440a^4\bar{\sigma}}\left(\frac{\Sigma^2}{\bar{\sigma}^2}-1\right)\right\}$$
$$+\left(\frac{1}{4}g^{\mu\nu} - \hat{z}^\mu\hat{z}^\nu\right)\left(1-\frac{\Sigma}{\bar{\sigma}}\right)\frac{\pi^2}{180a^4}$$

where an overbar notation has been used to denote this subtraction. It is noteworthy that even this removal of the large a result does not lead to regularization independent results, a fact which has been remarked upon earlier.

Of particular interest to Casimir calculations are the stress components $\langle 0|\overline{T}^{33}|0\rangle$ and the energy density per unit area $\mathcal{E} \equiv a\langle 0|\overline{T}^{00}|0\rangle$ which are given by

$$\langle 0|\overline{T}^{33}|0\rangle = -\frac{\pi^2}{240a^4}(1 - \frac{\Sigma}{\overline{\sigma}}) \tag{8}$$

and

$$\mathcal{E} = -\frac{\pi^2}{720a^3}\left\{1 - \frac{\Sigma}{\overline{\sigma}} - \frac{3\sigma_0^2 - \overline{\sigma}^2}{2\overline{\sigma}^3}\Sigma \right. \\ \left. \times \left[\frac{\Sigma^2}{\overline{\sigma}^2} - 1 - \frac{30a^2}{\pi^2\overline{\sigma}^2}\right]\right\} \tag{9}$$

respectively. It is striking that each of these terms retains a significant dependence on the cutoff details. In addition the usual relation assumed (as in [5]) to hold between $\mathcal{E}$ and the stress components, namely

$$\langle 0|\overline{T}^{33}|0\rangle = -\frac{\partial}{\partial a}\mathcal{E}, \tag{10}$$

is manifestly contradicted by Eqs.(8) and (9) in agreement with results found earlier in the context of the Casimir energy of a sphere [6]. It is significant that the relation (10) asserts a relationship between the vacuum stress $\langle 0|\overline{T}^{33}|0\rangle$ which transforms under $O(2,1)$ as a scalar while the right hand side transforms as the $\mu = \nu = 0$ component of a symmetric tensor under this group. Finally, note should be made of the fact that Eq.(9) predicts an additional Casimir force proportional to the divergent indeterminate form $\Sigma/a^2\overline{\sigma}^3$.

To reinforce the conclusions reached here in the case of parallel plates it is useful to consider also the case of the conducting sphere, the only other geometry in three dimensions which has proved amenable to exact calculation [11]. This case was first solved by Boyer [12] and subsequently verified by a number of authors [13-16]. Following reasoning similar to that of the parallel plate case note is made of the fact that the unbroken symmetry in this case consists of time translation and rotational invariance. Thus the natural cutoff parameters in this problem should refer to the energy and angular momentum. The former is the standard one and is well known to give cutoff independent results. It will be the goal here to examine the situation which occurs when a combination of these two is considered.

This is most economically achieved by reference to [16] which provides a useful separation of the Casimir energy into a finite part and one which requires regularization. Thus one writes for a sphere of radius a

$$E_c = E_{fin} + E_\sigma$$

where E_c, E_{fin}, and E_σ are respectively the total, the regularization independent part, and the formally divergent parts of the Casimir energy. The quantity E_σ is given by

$$E_\sigma = \frac{1}{4\pi a}\sum_{l=1}^\infty \Re e^{-i\phi}\int_0^\infty dy\, \exp(-i\nu\sigma y e^{-i\phi})y\frac{d}{dy}$$
$$\times (1+y^2 e^{-2i\phi})^{-3}$$

where $\nu = l + \frac{1}{2}$, $0 < \phi < \frac{\pi}{2}$, and σ is a dimensionless cutoff used to suppress the high frequency modes. Upon choosing a secondary cutoff of the form $e^{-\Sigma\nu}$ it is readily found that ΔE_σ (the *change* induced in E_σ in the limit of small cutoff) is given by

$$\Delta E_\sigma = -\frac{3\Sigma}{2\pi a\sigma^2}\int_0^\infty dy\,\frac{y^2}{(1+y^2)^4}\frac{1}{y^2+(\Sigma^2/\sigma^2)}.$$

This is evaluated to yield

$$\Delta E_\sigma = -\frac{3}{64a}\frac{\Sigma}{(\Sigma+\sigma)^4}[\Sigma^2 + 4\sigma\Sigma + 5\sigma^2],$$

a result which displays yet again the cutoff dependence of the Casimir effect for a more general choice of regularization. It may be noted that aside from confirming the vanishing of ΔE_σ for $\Sigma = 0$, this result shows that ΔE_σ goes as Σ^{-1} for $\sigma \to 0$ with intermediate values being obtained for finite Σ/σ.

In this work it has been shown that the Casimir effect is, prevailing opinion notwithstanding, highly dependent on the particular form of regularization employed for the extraction of the force. As remarked earlier as well as in ref.[6] the recent experiments which have seemed to many to provide the long awaited precision verification of this highly subtle effect are not based upon rigorous mathematical calculation. While the parallel plate Casimir experiment is fraught with difficulties beyond the ken of this author, it would seem that the successful completion of such experiments would be invaluable for purposes of setting to rest some of the issues which have been raised in this work.

Finally, it would be remiss not to mention in some way the very extensive work on the calculation of Casimir forces using the technique of zeta function regularization [17]. Historically, the successes of the Casimir approach in dealing with the parallel plate geometry and the sphere were obtained using conventional field theoretical subtraction procedures. Specifically, it was noted that only changes *relative* to the vacuum could be considered observable and it was therefore totally consistent to perform subtractions relative to the $a \to \infty$ vacuum. However, this step did not succeed in allowing one to obtain finite and observable results in more general applications. Eventually it was realized, however, that the application of zeta function regularization to such problems could yield finite results for some fairly general cases while at the same time agreeing with those obtained in the few instances in which more conventional subtractions could be successfully applied. This work makes no claim to having established any inconsistencies in the derivation of finite results for the Casimir effect when those efforts are based on the twin axioms of vacuum energy *and* zeta function regularization. Rather, the calculations presented here establish that the Casimir effect is generally cutoff dependent and hence incapable of being reliably determined whenever such calculations are performed using conventional (i.e., physically plausible) subtraction procedures.

ACKNOWLEDGMENTS

This work is supported in part by the U.S. Department of Energy Grant No.DE-FG02-91ER40685.This article appears by permission of Springer-Verlag, publishers of the European Physics Journal C (C. R. Hagen, "Cutoff dependence of the Casimir effect", Eur. Phys. J. **C19**, 677 (2001)).

REFERENCES

1. H. G. B. Casimir, Proc. K. Ned. Akad. Wet. **51**, 793 (1948).

2. M. J. Sparnaay, Physica **24**, 751 (1958).

3. S. K. Lamoreaux, Phys. Rev. Lett. **78**, 6 (1997); **83**, 3340 (1999); U. Mohideen and A. Roy, *ibid* **81**, 4549 (1998); **83**, 3341 (1999).

4. The units employed here are such that $\hbar = c = 1$ with the signature of the metric being (1,1,1,-1).

5. L. S. Brown and G. J. Maclay, Phys. Rev. **184**, 1272 (1969).

6. This approach has been used for the case of the sphere in C. R. Hagen, Phys. Rev. D **61**, 065005 (2000).

7. Equivalence to the usual regularization would require that $\sigma_i = \Sigma = 0$.

8. To do otherwise would also introduce complications associated with the fact that $\hat{z}^\mu P_\mu$ does not commute with E and $\mathbf{P}$.

9. It is of interest to note that from the metric $g^{\mu\nu}$ and the vectors σ^μ and $\hat{z}^\mu$ one can form three second rank tensors which subsequently reduce to two when the tracelessness condition is applied. The fact that there is no term of the form $(\sigma^\mu \hat{z}^\nu + \sigma^\nu \hat{z}^\mu)(a + b\hat{z}^\mu \sigma_\mu)$ is a consequence of the orthogonality condition $\hat{z}^\mu \sigma_\mu = 0$ and the invariance under $\hat{z}^\mu \to -\hat{z}^\mu$.

10. It should be noted here that the mode summation approach for parallel plate geometry has been carried out in the case $\sigma^\mu = -i\Lambda^{-1}\delta_0^\mu$, $\Sigma = 0$ by B. DeWitt, Phys. Reports **19C**, 295 (1975). In view of the fact that such a cutoff singles out the time axis it is not surprising that the result obtained there for the vacuum stress is the explicitly *noncovariant* form

$$\frac{\Lambda^4}{\pi^2}(g^{\mu\nu} + 4\delta_0^\mu\delta_0^\nu) + \frac{\pi^2}{720a^4}(g^{\mu\nu} - 4\hat{z}^\mu\hat{z}^\nu).$$

11. This leaves the case of one dimension as the only remaining example of a Casimir effect calculation which is regularization independent. In that application it could hardly be otherwise since the leading singularity is proportional to the square of an inverse cutoff parameter which is necessarily removed by subtraction of the $a \to \infty$ vacuum. In addition the next to leading term gives no effect since at most it could contribute a divergent a independent term to $\mathcal{E}$, thereby leaving a finite cutoff independent remainder to contribute to the force. In (3+1) dimensions, however, there are simply too many too many divergences, too many cutoff parameters, and too few physically reasonable subtractions to obtain a finite cutoff independent result.

12. T. H. Boyer, Phys. Rev. **174**, 1764 (1968).

13. B. Davies, J. Math. Phys. **13**, 1324 (1972).

14. R. Balian and B. Duplantier, Ann. Phys. (N.Y.) **112**, 165 (1978).

15. K. A. Milton, L. L. DeRaad, Jr., and J. Schwinger, Ann. Phys. **115**, 388 (1978).

16. M. E. Bowers and C. R. Hagen, Phys. Rev. D **59**, 025007 (1999).

17. E. Elizalde, S. D. Odintsov, A. Romeo, A. A. Bytsenko, and S. Zerbini, *Zeta Regularization Techniques with Applications* (World Scientific, Singapore, 1994).

Pair Production of Arbitrary Spin Particles with EDM and AMM and Vacuum Polarization

S. I. Kruglov

International Educational Centre, Toronto, Canada M3J 3G9

Abstract. The exact solutions of the wave equation for arbitrary spin particles with electric dipole and magnetic moments in the constant and uniform electromagnetic field were found. The differential probability of pair production of particles by an external electromagnetic field has been calculated on the basis of the exact solutions. We have also estimated the imaginary part of the constant and uniform electromagnetic field. The nonlinear corrections to the Maxwell Lagrangian have been calculated taking into account the vacuum polarization of arbitrary spin particles. The role of electric dipole and magnetic moments of arbitrary spin particles in instability of the vacuum is discussed.

The electric dipole moment (EDM) of particles violates the CP invariance and may be induced by the ϑ-term of the QCD vacuum. In the standard model (SM) the CP-violating interactions can be introduced by the Kobayashi-Maskawa matrix. The predicted EDM's of elementary particles are extremely small but some SUSY and multi-Higgs models may predict much stronger CP-violating effects [1]. It should be noted that in the framework of the QCD string theory mesons and baryons possess the EDM [2]. It is interesting to study the pair production probability and the vacuum polarization of particles because there is the vacuum instability of particles in electromagnetic fields. In particular, the vacuum of vector particles is non-stable in a magnetic field as there is a contribution to the negative part of the Callan-Symanzik β-function. The pair production probability of arbitrary spin particles with the gyromagnetic ratio g were studied in [3, 4]. Here we generalize this result on the case of the arbitrary spin particles with the EDM, and also study the vacuum polarization.

We proceed from the second order relativistic wave equation for arbitrary spin (s) particles with the EDM and AMM on the basis of the Lorentz representation $(0, s) \oplus (s, 0)$ of the wave function. The wave function in this approach has the minimal number $2(2s+1)$ of components, and obeys equations:

$$\left[\mathcal{D}_\mu^2 - m^2 - \frac{e}{2} \left(g F_{\mu\nu} - \sigma \widetilde{F}_{\mu\nu} \right) \Sigma_{\mu\nu}^{(\varepsilon)} \right] \Psi_\varepsilon(x) = 0, \tag{1}$$

where $\mathcal{D}_\mu = \partial_\mu - ieA_\mu$ is the covariant derivative, $\partial_\mu = \partial/\partial x_\mu$, $x_\mu = (\mathbf{x}, it)$, $F_{\mu\nu} = \partial_\mu A_\nu - \partial_\nu A_\mu$ is the electromagnetic field strength tensor, $\widetilde{F}_{\mu\nu} = (1/2)\varepsilon_{\mu\nu\alpha\beta} F_{\alpha\beta}$ is the dual tensor ($\varepsilon_{1234} = -i$). The numbers $\varepsilon = \pm 1$ correspond to the $(s, 0)$ and $(0, s)$ representations with the generators of the Lorenz group $\Sigma_{\mu\nu}^{(-)}$, $\Sigma_{\mu\nu}^{(+)}$, respectively. The gyromagnetic ratio $g = 1/s + \kappa$ (κ is the AMM of a particle) and the σ give the contribution to the magnetic moment $\mu = egs/(2m)$ and the EDM $\sigma/(2m)$ of arbitrary spin particles. The spin matrices S_k are connected with the generators $\Sigma_{\mu\nu}^{(\varepsilon)}$ by the relationships:

$$\Sigma_{ij}^{(\varepsilon)} = \varepsilon_{ijk} S_k, \qquad \Sigma_{4k}^{(\varepsilon)} = -i\varepsilon S_k,$$

$$[S_i, S_j] = i\varepsilon_{ijk} S_k, \qquad (S_1)^2 + (S_2)^2 + (S_3)^2 = s(s+1). \tag{2}$$

At the parity transformation, $\varepsilon \to -\varepsilon$, and the representation $(s,0)$ is converted into $(0,s)$. As a result Eqs.(1) (at $\varepsilon = \pm 1$) are invariant under the parity inversion if $\sigma = 0$. At $g = 1/s$ and $\sigma = 0$ we arrive at the approach [5].

Now we find the solutions to Eq.(1) for a particle in the field of uniform and constant electromagnetic fields. For simplicity we choose a coordinate system in which the electric $\mathbf{E}$ and magnetic $\mathbf{H}$ fields are parallel, so that $\mathbf{E} = \mathbf{n}E$, $\mathbf{H} = \mathbf{n}H$, $\mathbf{n} = (0,0,1)$ with the 4-vector potential $A_\mu = (0, x_1 H, -tE, 0)$. We get equations in the diagonal representation

$$\frac{1}{2}\Sigma_{\mu\nu}^{(+)} F_{\mu\nu} \Psi_+^{(s_z)}(x) = s_z X \Psi_+^{(s_z)}(x), \qquad \frac{1}{2}\Sigma_{\mu\nu}^{(+)} \widetilde{F}_{\mu\nu} \Psi_+^{(s_z)}(x) = s_z \widetilde{X} \Psi_+^{(s_z)}(x),$$

$$\frac{1}{2}\Sigma_{\mu\nu}^{(-)} F_{\mu\nu} \Psi_-^{(s_z)}(x) = s_z X^* \Psi_-^{(s_z)}(x), \qquad \frac{1}{2}\Sigma_{\mu\nu}^{(-)} \widetilde{F}_{\mu\nu} \Psi_-^{(s_z)}(x) = s_z \widetilde{X}^* \Psi_-^{(s_z)}(x), \tag{3}$$

where $X = H + iE$, $\widetilde{X} = -E + iH$, $X^* = H - iE$, $\widetilde{X}^* = -E - iH$ and the spin projection s_z is given by $s_z = \pm s, \pm(s-1), \cdots 0$ for bosons, and $s_z = \pm s, \pm(s-1), \cdots \pm 1/2$ for fermions. With the help of Eqs.(3), we can represent Eqs.(1) as

$$\left[\mathcal{D}_\mu^2 - m^2 - es_z\left(gX - \sigma\widetilde{X}\right)\right]\Psi_+^{(s_z)}(x) = 0,$$

$$\left[\mathcal{D}_\mu^2 - m^2 - es_z\left(gX^* - \sigma\widetilde{X}^*\right)\right]\Psi_-^{(s_z)}(x) = 0. \tag{4}$$

For every projection s_z Eqs.(4) are the Klein-Gordon type equations with the complex "effective" masses $m_{eff}^2 = m^2 + es_z\left(gX - \sigma\widetilde{X}\right)$. Solutions to Eq.(4) (for $\Psi_+^{(s_z)}(x)$) are given by [6]

$$\pm\Psi_+^{(s_z)}(x) = N_0 \exp\left\{i(p_2 x_2 + p_3 x_3) - \frac{\eta^2}{2}\right\} H_n(\eta) \pm \chi^{(s_z)}(\tau), \tag{5}$$

where $H_n(\eta)$ are the Hermite polynomials; $\eta = (p_2 - eHx_1)/\sqrt{eH}$, $\tau = \sqrt{eE}(t + p_3/(eE))$ (see [7, 8]), and

$$_+\chi^{(s_z)}(\tau) = D_\nu[-(1-i)\tau], \qquad _-\chi^{(s_z)}(\tau) = D_\nu[(1-i)\tau],$$

$$^+\chi^{(s_z)}(\tau) = D_{-(\nu+1)}[(1+i)\tau], \qquad ^-\chi^{(s_z)}(\tau) = D_{-(\nu+1)}[-(1+i)\tau]; \tag{6}$$

$D_\nu(x)$ are the parabolic-cylinder functions, $\nu = ik_{s_z}^2/(2eE) - 1/2$. Eigenvalues $k_{s_z}^2$ obey the requirement

$$k_{s_z}^2 - m^2 - es_z\left(gX - \sigma\widetilde{X}\right) = eH(2n+1), \tag{7}$$

where $n = 1, 2, \ldots$ is the principal quantum number. Four solutions (6) possess different asymptotic at $t \to \pm\infty$ [7].

The probability of pair production of arbitrary spin particles with EDM and AMM by constant and uniform electromagnetic fields may be obtained using the asymptotic form of the solutions (5) at $t \to \pm\infty$. The functions ${}^+_+\chi^{(s_z)}(\tau)$ have positive frequency and ${}^-_+\chi^{(s_z)}(\tau)$ have negative frequency when the time approaches $\pm\infty$. The momentum projections p_2, p_3 and the $k_{s_z}^2$ are the conserved numbers. The wave functions (5) satisfy the relationships [7]:

$$
\begin{aligned}
{}_+\Psi_+^{(s_z)}(x) &= c_{1ns_z}\, {}^+\Psi_+^{(s_z)}(x) + c_{2ns_z}\, {}^-\Psi_+^{(s_x)}(x), \\
{}_-\Psi_+^{(s_z)}(x) &= c^*_{2ns_z}\, {}^+\Psi_+^{(s_z)}(x) + c^*_{1ns_z}\, {}^-\Psi_+^{(s_z)}(x),
\end{aligned}
\qquad (8)
$$

where the variable c_{2ns_z} is

$$
c_{2ns_z} = \exp\left[-\frac{\pi}{2}(\lambda + i)\right], \qquad \lambda = \frac{m^2 + es_z\left(gX - \sigma\widetilde{X}\right) + eH(2n+1)}{eE}.
$$

Coefficients c_{1ns_z} and c_{2ns_z} obey the relations as follows $|c_{1ns_z}|^2 - |c_{2ns_z}|^2 = 1$ for bosons, and $|c_{1ns_z}|^2 + |c_{2ns_z}|^2 = 1$ for fermions. The value $|c_{2ns_z}|^2$ is the probability of pair production of arbitrary spin particles in the state with the principal quantum number n and the spin projection s_z throughout all space and during all time. According to the approach [7], the average number of particle pairs produced from a vacuum is given by

$$
\overline{N} = \sum_{n, s_z} |c_{2ns_z}|^2\, \frac{e^2 EHVT}{(2\pi)^2}, \qquad (9)
$$

where V is the normalization volume, T is the time of observation. Calculating the sum over the principal quantum number n and spin projections s_z (see [4]), we find the pair production probability per unit volume and per unit time

$$
I(E,H) = \frac{\overline{N}}{VT} = \frac{e^2 EH}{8\pi^2}\, \frac{\exp\left[-\pi m^2/(eE)\right]}{\sinh(\pi H/E)}\, \frac{\sinh\left[(s+1/2)\pi(\sigma + gH/E)\right]}{\sinh\left[(\pi/2)(\sigma + gH/E)\right]}. \qquad (10)
$$

We find from Eq.(10) that in the case $\sigma = g = 0$, the pair production of arbitrary spin particles is $(2s+1)$ times that for the scalar particle pair production due to the $(2s+1)$ physical degrees of freedom of the arbitrary spin field. At $\sigma = 0$ Eq. (10) converts into expressions derived in [4]. The intensity of the pair creation (10) is the generalization of the results [3, 4] on the case of arbitrary spin particles with the EDM and AMM. In the general case $\sigma \neq 0$, $g \neq 0$ there is also a pair production of arbitrary spin particles if $E = 0$, $H > m^2/e$ at $gs > 1$ (see [3]), i.e. there is instability of the vacuum in a magnetic field.

The imaginary part of the Lagrangian of the electromagnetic field can be obtained as

$$
\mathrm{Im}\mathcal{L} = \frac{1}{2}\int \sum_{n,s_z} \ln|c_{1ns_z}|^2\, \frac{e^2 EH}{(2\pi)^2}
$$

$$= \frac{e^2 EH}{16\pi^2} \sum_{n=1}^{\infty} \frac{\beta_n}{n} \exp\left(-\frac{\pi m^2 n}{eE}\right) \frac{\sinh[n(s+1/2)\pi(\sigma+gH/E)]}{\sinh(n\pi H/E)\sinh[n(\pi/2)(\sigma+gH/E)]}, \quad (11)$$

where $\beta_n = (-1)^{n-1}$ for integer spins, and $\beta_n = 1$ for half-integer spins. In accordance with [9] the first term ($n=1$) in (11) gives the half of the pair production probability per unit volume and per unit time. It should be noted that $\mathrm{Im}\,\mathcal{L}$ (11) and the pair production probability (10) do not depend on the scheme of renormalization as all divergences and the renormalizability are contained in $\mathrm{Re}\,\mathcal{L}$ [9].

Let us consider the problem of one-loop corrections to the Lagrange function of a constant and uniform electromagnetic field due to the field interaction with a vacuum of arbitrary spin particles with the EDM and AMM. This problem has been solved for fields of spins 0, 1/2 (at $\sigma = 0$, $g = 2$) in [9], and for spin one in [10]. The nonlinear corrections to the Maxwell Lagrangian due to vacuum polarization are connected with the cross-section of scattering photons by photons. Using the Schwinger method [9], we obtain the one-loop corrections to Lagrangian of a constant and uniform electromagnetic field[1]

$$\mathcal{L}^{(1)} = \frac{\varepsilon}{32\pi^2} \int_0^\infty d\tau\, \tau^{-3} \exp\left(-m^2\tau - l(\tau)\right) \mathrm{tr}\exp\left[\frac{e_0\tau}{2}\Sigma_{\mu\nu}\left(gF_{\mu\nu} - \sigma\tilde{F}_{\mu\nu}\right)\right], \quad (12)$$

$$\Sigma_{\mu\nu} = \Sigma_{\mu\nu}^{(+)} \oplus \Sigma_{\mu\nu}^{(-)}, \quad l(\tau) = \frac{1}{2}\mathrm{tr}\ln\left[(e_0 F\tau)^{-1}\sin(e_0 F\tau)\right],$$

where $\oplus$ is the direct sum, $\varepsilon = 1$ for bosons and $\varepsilon = -1$ for fermions due to different signs of loop integrals for bosons and fermions. The index 0 in Eq.(12) refers to the unrenormalized variables, so that e_0 is the bare electric charge. The Lagrangian $\mathcal{L}^{(1)}$, (12), is the effective nonlinear Lagrangian which is an integral over the proper time τ. Calculating the trace (tr) of the matrices we arrive at

$$\mathcal{L}^{(1)} = \frac{\varepsilon}{16\pi^2} \int_0^\infty d\tau\, \tau^{-3} \exp\left(-m^2\tau\right)$$
$$\times \left\{ \frac{(e_0\tau)^2 \mathcal{G}_0}{\mathrm{Im}\cosh(e_0\tau X_0)} \mathrm{Re}\, \frac{\sinh\left[(s+1/2)e_0\tau\left(gX_0 - \sigma\tilde{X}_0\right)\right]}{\sinh\left[(e_0\tau/2)\left(gX_0 - \sigma\tilde{X}_0\right)\right]} - (2s+1) \right\}, \quad (13)$$

where $X_0 = H_0 + iE_0$, $\mathcal{G}_0 = \mathbf{E}_0 \cdot \mathbf{H}_0$; $\mathbf{E}_0$, $\mathbf{H}_0$ are bare electric and magnetic fields, respectively. We subtracted the constant to have vanishing $\mathcal{L}^{(1)}$ when the electromagnetic fields approach zero, (see [9]). Setting $\sigma = 0$, $g = 2$, $\varepsilon = -1$ in Eq.(13) we come to the Schwinger Lagrangian [9]. The integral (13) is the nonlinear correction to Maxwell's Lagrangian due to the vacuum polarization of arbitrary spin particles with the EDM and AMM. The Lagrangian (13) has the quadratic term in the electromagnetic fields, and that renormalizes the Lagrangian of the free electromagnetic fields $\mathcal{L}^{(0)} = -\mathcal{F}_0 = \left(\mathbf{E}_0^2 - \mathbf{H}_0^2\right)/2$. Expanding Eq. (13) in weak electromagnetic fields and adding the Lagrangian of the free electromagnetic fields $\mathcal{L}^{(0)}$ we obtain the renormalized

---

[1] The factor ε was omitted in [4].

Lagrangian of electromagnetic fields that takes into account the vacuum polarization of arbitrary spin particles with the EDM and AMM:

$$\mathcal{L} = \mathcal{L}^{(0)} + \mathcal{L}^{(1)} = -\mathcal{F} + \frac{\varepsilon}{16\pi^2} \int_0^\infty d\tau \tau^{-3} \exp\left(-m^2\tau\right)$$

$$\times \left\{ \frac{(e\tau)^2 \mathcal{G}}{\text{Im}\cosh(e\tau X)} \text{Re} \frac{\sinh\left[(s+1/2)e\tau\left(gX - \sigma\widetilde{X}\right)\right]}{\sinh\left[(e\tau/2)\left(gX - \sigma\widetilde{X}\right)\right]} \right. \quad (14)$$

$$\left. - (2s+1) - \frac{(2s+1)(e\tau)^2 \mathcal{F}}{3} \left[s(s+1)\left(g^2 - \sigma^2\right) - 1\right] \right\},$$

where we renormalize fields and charges, $\mathcal{F} = Z_3^{-1}\mathcal{F}_0$, $e = Z_3^{1/2} e_0$. The renormalization constant is given by

$$Z_3^{-1} = 1 - \frac{\varepsilon e_0^2 (2s+1)\left[s(s+1)\left(g^2 - \sigma^2\right) - 1\right]}{48\pi^2} \int_{\tau_0}^\infty d\tau \tau^{-1} \exp\left(-m^2\tau\right). \quad (15)$$

If the electromagnetic fields **E**, **H** are absent, the Lagrangian (14) vanishes. The renormalization constant Z_3^{-1} diverges logarithmically when the cutoff factor τ_0 at the lower limit in the integral (15) approaches zero ($\tau_0 \to 0$). It follows from Eq.(15) that if the inequality $\varepsilon\left[s(s+1)\left(g^2 - \sigma^2\right) - 1\right] > 0$ is valid the renormalization constant of the charge $Z_3^{1/2} > 1$. This case indicates asymptotic freedom in the field [11, 12], and is not realized in QED because $g = 2$, $\sigma = 0$ and $\varepsilon = -1$. However, for boson fields, when $\varepsilon = 1$, asymptotic freedom occurs for the small value of the EDM σ. The asymptotically free behavior in the boson fields is due to the AMM, and the EDM of particles suppresses the phenomena of asymptotic freedom.

Eq. (15) allows us to obtain the Callan-Zymanzik β-function:

$$\beta = -\frac{\varepsilon e_0^2 (2s+1)\left[s(s+1)\left(g^2 - \sigma^2\right) - 1\right]}{48\pi^2}. \quad (16)$$

Asymptotic freedom occurs if the β-function is negative ($\beta < 0$). The AMM and spin (s) of bosons assures asymptotic freedom and instability of the vacuum in a magnetic field. However this discussion concerns only the renormalizable theory. We imply that the scheme considered is an effective theory. Expanding Eq. (14) in the weak electromagnetic fields and using the equality $\int_0^\infty d\tau \tau \exp\left(-m^2\tau\right) = 1/m^4$ we obtain after renormalization the Maxwell Lagrangian with the nonlinear corrections:

$$\mathcal{L} = \frac{1}{2}\left(\mathbf{E}^2 - \mathbf{H}^2\right) - \frac{2s(s+1)\sigma g}{s(s+1)(g^2 - \sigma^2) - 1}(\mathcal{G} - \mathcal{G}_0) + \frac{\varepsilon \alpha^2 (2s+1)}{90m^4}$$

$$\times \left\{ \left[s(s+1)(3s^2 + 3s - 1)\left(g^4 - 6g^2\sigma^2 + \sigma^4\right) - 10s(s+1)(g^2 - \sigma^2) + 7\right] \mathcal{F}^2 \right. \quad (17)$$

$$\left. + \left[s(s+1)(3s^2 + 3s - 1)\left(6g^2\sigma^2 - g^4 - \sigma^4\right) + 1\right] \mathcal{G}^2 \right.$$

$$+4s(s+1)\sigma g\left[2(3s^2+3s-1)\left(g^2-\sigma^2\right)-5\right]\mathcal{G}\mathcal{F}\bigg\},$$

where $\alpha = e^2/(4\pi)$. The second and last terms in Eq.(17) indicate CP - violation due to the EDM of a particle. We can consider the second term in Eq.(17) as anomaly for a particle with the EDM because such a quadratic in fields term does not present in the bare Lagrangian $\mathcal{L}^{(0)}$. The Lagrangian (17) is the Heisenberg-Euler type Lagrangian [13] for the case of arbitrary spin particles with the EDM and AMM. For the case of QED at $s = 1/2$, $g = 2$, $\varepsilon = -1$, Eq.(17) becomes the Schwinger Lagrangian [9]. Eq.(14) allows us to find the limit at $eE/m^2 \rightarrow \infty$ and $eH/m^2 \rightarrow \infty$ if we ignore the dependence of the AMM and EDM on the strong external electromagnetic fields that is some approximation.

The pair-production probability (10), and the effective Lagrangian for electromagnetic fields (14) which takes into account the polarization of the vacuum, are the generalization of the Schwinger result on the case of the theory of particles with arbitrary spins in the external electric and magnetic fields. It follows from Eq.(10) that there is a pair production of particles by a purely magnetic field in the case of $H > H_0$ ($H_0 = m^2/e$ is the critical value of a magnetic field), $gs > 1$ assuring asymptotic freedom and instability of the vacuum in a magnetic field. The presence of the EDM of a particle does not lead to instability of the vacuum in a magnetic field but it suppresses the phenomena of asymptotic freedom. In the presence of the magnetic field the probability decreases for scalar particles and increases for higher spin particles. The intensity of pair production of arbitrary spin particles (10) does not depend on the renormalization scheme as all divergences and the renormalizability are contained in Re$\mathcal{L}$. It is interesting to investigate quantum processes with CP-violation as they may give a sensitive probe for New Physics.

REFERENCES

1. Barr, S. M., and Marciano, W. J., "Electric Dipole Moments," in *CP Violation*, edited by C. Jarlskog, World Scientific, Singapore, 1989, pp. 455-499.
2. Kruglov, S. I., *Phys. Lett.* **B390**, 283-286 (1997); **B397**, 283-286 (1997); *Phys. Rev.* **D60**, 116009 (1999); preprint hep-ph/9910514.
3. Marinov, M. S., and Popov, V. S., *Sov. J. Nucl. Phys.* **15**, 702-709 (1972) [*Yad. Fiz.* **15**, 1271-1285 (1972)].
4. Kruglov, S. I., *Int. J. Theor. Phys.* **40**, 511-532 (2001); *Symmetry and Electromagnetic Interaction of Fields with Multi-Spin*, Nova Science Publishers, Inc., Huntington, New York, 2001, pp. 182-189.
5. Hurley, W. J., *Phys. Rev.* **D4**, 3605-3616 (1971); **D10**, 1185-1200 (1974); *Phys. Rev. Lett.* **29**, 1475-1477 (1972).
6. Bateman, H., and Erdelyi, A., *Higher Transcendental Functions*, New York, McGraw-Hill, 1953.
7. Nikishov, A. I., *Sov. Phys.-JETP* **30**, 660-662 (1970); *Nucl. Phys.* **B21**, 346-358 (1970).
8. Kruglov, S. I., *J. Phys. G: Nucl. Phys.* **21**, 1643-1656 (1995).
9. Schwinger, J., *Phys. Rev.* **82**, 664-679 (1951).
10. Vanyashin, V. S., and Terent'ev, M. V., *Sov. Phys.-JETP* **21**, 375-380 (1965).
11. Gross, D. J., and Wilczek, F., *Phys. Rev. Lett.* **30**, 1343-1346 (1973).
12. Politzer, D. H., *Phys. Rev. Lett.* **30**, 1346-49 (1973).
13. Heisenberg, W., and Euler, H., *Z. Physik* **98**, 714 (1936).

Analyzing the 't Hooft Model on a Light-Cone Lattice

Joel S. Rozowsky

Department of Physics, Syracuse University, Syracuse NY 13244

Abstract. We study the 't Hooft model (large N_c QCD in 2 space-time dimensions) using an improved approach to digitizing the sum of gauge theory Feynman diagrams based on light-cone gauge discretized lattice. Our purpose is to test the new formalism in a solvable case, with the hope to learn how it might be usefully applied to the physically interesting case of 4 dimensional QCD.

In this proceeding we report on recent a recent paper with Thorn [1]. Last year, with Bering and Thorn we proposed [2] a new method to digitize the sum of planar diagrams selected by 't Hooft's $N_c \to \infty$ limit of $SU(N_c)$ gauge theories [3]. The proposal, based on the light-cone or infinite momentum frame description of the dynamics, involved discretization of both the p^+ carried by each line of the diagram and the propagation time $\tau = ix^+$, as in [4, 5, 6]. But the main advantage of the new version was a coherent prescription for resolving most of the ambiguities due to $p^+ = 0$ divergences that typically plague the light-cone description.

We hope that our formalism will eventually allow an improved understanding of QCD in 4 dimensional space-time. But here, we merely wish to test the proposal in the context of the well-understood case of large N_c gauge theories in two space-time dimensions, namely the 't Hooft model [7]. Our purpose is not to unearth new aspects of the model, but rather to see how its well known properties can be obtained from our new discretization.

The physical content of the 't Hooft model boils down to an integral equation, essentially a Bethe-Salpeter equation [8], that determines the mass spectrum of $q\bar{q}$ mesons. The reason the limit $N_c \to \infty$ reduces to ladder diagrams (albeit with self-energy corrected quark propagators), is that the 2 dimensional gluon is not dynamical (there are no transverse polarizations). Thus, as with any axial gauge, the light-cone gauge $A_- = 0$ eliminates all gluon self-interactions, so A_+ can be integrated out inducing an instantaneous Coulomb potential. But the 't Hooft limit $N_c \to \infty$ further eliminates all quark loops and all non-planar diagrams, leaving only the planar self energy corrections to the quark propagator, and the ladder bare gluon exchanges (Coulomb interaction) between quark anti-quark lines in the singlet $q\bar{q}$ channel. In light-cone parameters the Bethe-Salpeter equation summing these ladder $q\bar{q}$ diagrams simplifies to the single variable 't Hooft integral equation [7].

$$\mathcal{M}^2\varphi(x) = \left(\frac{1}{x} + \frac{1}{1-x}\right)\mu^2\varphi(x) - \frac{g_s^2 N_c}{2\pi} P \int_0^1 dy \frac{\varphi(y) - \varphi(x)}{(y-x)^2}, \qquad (1)$$

where the integral is understood to be evaluated by the principal value prescription. The variable x is the fraction carried by the quark of the total P^+ of the system (the anti-quark carries P^+ fraction $1-x$). Also $\mathcal{M}$ is the mass of the meson bound state and φ satisfies the boundary conditions, $\varphi(0) = \varphi(1) = 0$.

Since the new formalism discretizes $\tau \equiv ix^+ = ka$ in addition to $p^+ = lm$, the corresponding simplifications lead to an equation that is not a straightforward discretization of this integral equation. In particular, the continuum limit can be taken in different ways depending on the ratio $T_0 = m/a$ (which would be infinite for continuous τ), and we want to explore to what extent these different continuum limits lead to the same physics. We shall find that some care must be taken with the setup of the discrete τ dynamics in order for this to be true. Indeed, a numerical study shows that the most simple-minded treatment leads to a ground state that becomes unstable at moderate 't Hooft coupling even with relatively small $P^+/m \equiv M$ unless the ratio $a/m = 1/T_0$ is tuned to be sufficiently small (perhaps infinitesimal for large M). If this feature were robust, it would cast doubt on any potential utility of the discretization of τ.

To overcome this difficulty, we find it necessary to veto some of the "densest" discretized Feynman diagrams: a quark must be forbidden to emit 2 gluons at immediately successive time steps, with a similar veto on two successive absorptions. With this simple veto (which is prescribed locally in time), we shall show that the continuum limit reduces to the 't Hooft model provided only that the total P^+ of the $q\bar{q}$ system is large compared to the discretization unit m. In particular it is not necessary that the ratio T_0 be large. Keeping T_0 finite in the continuum limit leads to the 't Hooft equation with a non-trivial renormalization of the coupling. Because of this effect, it turns out that the effective (renormalized) coupling is small for both large and small bare coupling, reminiscent of strong/weak coupling duality. The strong coupling limit favors the densest diagrams, so vetoing some of the densest ones has a dramatic effect on the strong coupling behavior of the theory. This possibility was anticipated and discussed in [2] in connection with the nature of the fishnet diagrams in higher dimensional space-time.

The Lagrange density for $SU(N_c)$ gauge fields coupled to quarks in the fundamental representation is given by

$$\mathcal{L} = -\frac{1}{4}\mathrm{Tr} F^{\mu\nu} F_{\mu\nu} + \bar{q}[i\gamma \cdot (\partial - igA) - \mu_0]q, \qquad (2)$$

where $F_{\mu\nu} = \partial_\mu A_\nu - \partial_\nu A_\mu - ig[A_\mu, A_\nu]$. We remind the reader that the normalization of gauge fields appropriate for matrix fields and dictated by the gluon kinetic term differs by a factor $1/\sqrt{2}$ from the more standard one:

$$-\frac{1}{4}\sum_a F_a^{\mu\nu} F_{a\mu\nu} = -\frac{1}{2}\mathrm{Tr} F_s^{\mu\nu} F_{s\mu\nu},$$

with $F_s \equiv \sum_a \frac{\lambda_a}{2} F_a$. Thus $A_s = A/\sqrt{2}$, and we conclude that $g = g_s/\sqrt{2}$. In 2 space-time dimensions we choose the representation of γ matrices for which the light-like components are

$$\gamma^+ = \sqrt{2}\begin{pmatrix} 0 & 1 \\ 0 & 0 \end{pmatrix} \qquad \gamma^- = \sqrt{2}\begin{pmatrix} 0 & 0 \\ 1 & 0 \end{pmatrix}. \qquad (3)$$

With this choice the field equation for the upper component of the quark spinor does not involve the "time" derivative and is an equation of constraint relating the upper component, q_1, to the lower component, q_2. Working in light-cone gauge ($A_- = A^+ = 0$), we can eliminate the upper component in favor of the lower component yielding the light-cone gauge Lagrange density

$$\mathcal{L} = +\frac{1}{2}\text{Tr}(\partial_- A_+)^2 + i\psi^\dagger \left[\partial_+ - igA_+ + \frac{\mu_0^2}{2\partial_-}\right]\psi, \tag{4}$$

where $\psi = 2^{1/4} q_2$.

Our discretization of Feynman diagrams is based on the x^+ representation of each bare propagator

$$D(p^+, x^+) = \int \frac{dp^-}{2\pi} \tilde{D}(p^+, p^-) e^{-ix^+ p^-}. \tag{5}$$

Performing the p^- integral gives the following Feynman rules for the continuum theory

$$\begin{aligned}
D_\psi(p^+, x^+) &= e^{-ix^+ \mu_0^2/2p^+} \to e^{-\tau \mu_0^2/2p^+} \\
D_A(p^+, x^+) &= i\frac{\delta(x^+)}{p^{+2}} \to -\frac{\delta(\tau)}{p^{+2}} \\
V_{\psi^\dagger \psi A} &= ig \to g,
\end{aligned} \tag{6}$$

where the arrows indicate the rules to use with imaginary time.

One way to digitize the 't Hooft equation (1) is to put the variables x, y on a grid, which amounts to discrete light-cone quantization [9, 4], where one discretizes the amount of P^+ each line of the ladder diagram carries in quanta of m

$$p^+ = lm \qquad l = 1, 2, 3, \ldots.$$

One can then focus on a state of the system of interest (in our case a $q\bar{q}$ system) with total $P^+ = Mm$. The continuum theory is recovered by taking the combined limits $m \to 0$ and $M \to \infty$ while keeping $P^+ = Mm$ fixed. Following [4, 2], in addition to discretizing the p^+ of each particle, we also discretize imaginary light-cone time, $\tau = ix^+ = ka$ ($k = 1, 2, 3, \ldots$). This discretization (which also serves as an ultraviolet cutoff) allows the continuum limit to be reached by keeping $T_0 \equiv m/a$ fixed and taking both $m, a \to 0$ and $M \to \infty$ simultaneously. Actually, since the physics of the discretized model depends only on the ratio m/a, the continuum limit is nothing but the large M limit. The conventional continuous time DLCQ approach (see [10] and references therein) corresponds to the special case $T_0 \to \infty$.

Discretization of the quark propagator poses no difficulty. However, for the instantaneous interaction induced by integrating out A_+, we shall use

$$D_A(Mm, -ia) = -\frac{T_0}{M^2}. \tag{7}$$

The instantaneous gluon only propagates one discretized time-step. In [2] we had a more generalized instanteous propagator which is not necessary in this context (this corresponds to $f_1 = 1, f_{k>1} = 0$ in [2]).

The Feynman rules can be simplified if we absorb the negative sign from the antiquark propagator into the corresponding vertex factor. We define new parameters

$$\alpha \equiv e^{-\mu_0^2/2T_0} \quad \text{and} \quad \kappa \equiv \sqrt{\frac{g^2 N_c}{2\pi T_0}}. \tag{8}$$

We also recall that in 't Hooft's large N_c limit every additional pair of cubic vertices in the ladder sum corresponds to a completed color index loop, which produces a factor N_c. Thus we shall also absorb a factor of $\sqrt{N_c}$ into each vertex. Simply put, all terms in the ladder sum are only dependent on the 't Hooft coupling $g^2 N_c$. The simplified Feynman rules are presented in Fig. 1.

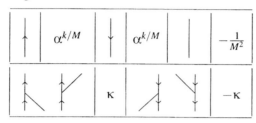

FIGURE 1. Simplified discretized Feynman Rules for 't Hooft model.

One way to proceed is generate a transfer matrix which evolves the $q\bar{q}$ system one time-step at a time. This is not very efficient as it leads to very large matrices when evaluated numerically. A more logical way to proceed is to generate a Bethe-Salpeter type equation which evolves the system from one ladder exchange to the next. When one does this with the naive Feynman rules one encounters solutions which becomes complex at relatively small coupling κ^2 (as M increases). See [1] for more details.

One way to avoid these unwanted solutions is to slightly modify the discretized Feynman rules so that the rung will attach to the same lines whichever way the exchanged gluon propagates. As seen in Fig. 2, the asymmetry stems from the possibility of consecutive gluon emissions (absorptions) on immediately successive time steps. If this possibility is disallowed, the basic exchange rung can be taken to be the sum of the two different exchanges. In addition to removing unwanted solutions this veto rule also leads to simpler equations, with a more transparent continuum limit. As we shall see shortly, it also produces a more physical strong coupling behavior a discretization without a veto.

The Bethe-Salpeter equation for the discretized 't Hooft model, with the veto imposed is given by

$$\Psi(l) = \sum_{r=1}^{M-l-1} \frac{\kappa^2}{r^2} u\alpha^{1/l+1/(M-l-r)} \mathcal{D}_{q\bar{q}}(l+r) \Psi(l+r)$$

$$+ \sum_{r=1}^{l-1} \frac{\kappa^2}{r^2} u\alpha^{1/(l-r)+1/(M-l)} \mathcal{D}_{q\bar{q}}(l-r) \Psi(l-r), \tag{9}$$

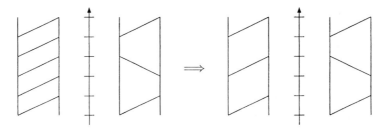

FIGURE 2. Asymmetry in the densest configuration of exchanges in the same sense and opposite sense. The double arrow indicates implementing the veto.

where

$$\mathcal{D}_{q\bar{q}}(l) = \frac{u\alpha^{M/l(M-l)}\left(1 - u^2\kappa^4 \Sigma_l' \Sigma_{M-l}'\right)}{\left(1 - u^2\kappa^4 \Sigma_l' \Sigma_{M-l}'\right)^2 - u\alpha^{M/l(M-l)}\left(1 - u\kappa^2 \Sigma_{M-l}'\right)\left(1 - u\kappa^2 \Sigma_l'\right)}, \qquad (10)$$

with $\Sigma_l' \equiv \alpha^{1/(M-l)} \sum_{r=1}^{l-1} \alpha^{1/(l-r)}/r^2$ and $u \equiv 1/t = e^{aE}$.

By imposing the veto we have reduced the rank of the eigenvalue problem from $2(M-2)$ to $M-1$. With the veto, this discretized equation is much easier to analyze in the continuum limit $M \to \infty$. We obtain the continuum 't Hooft equation

$$\left[\mathcal{M}^2 - \mu^2 \frac{1}{x(1-x)}\right]\Phi(x) = \frac{g_{\text{eff}}^2 N_c}{\pi} P \int_0^1 dy \frac{\Phi(y) - \Phi(x)}{(y-x)^2}. \qquad (11)$$

Comparing with Eq. 1, we see that the only effect on the continuum limit of keeping T_0 finite is a finite renormalization of the gauge coupling g^2, and a coupling constant dependent shift in μ^2. Since α is a free parameter, we can access all positive values of μ^2 by tuning it. Where $\kappa^2 = 6\eta(1 + \eta + 2\eta^2)/(1 - \eta^2)(1 + \eta)\pi^2$, is given in terms of a parameter η. The effective coupling in the 't Hooft equation

$$\frac{g_{\text{eff}}^2 N_c}{\pi} = \frac{12\eta(1-\eta^2)(1+\eta)T_0}{\pi^2(1+\eta+7\eta^2-\eta^3)}, \qquad (12)$$

and the renormalized mass parameter

$$\mu^2 = \mu_0^2 + \frac{12\eta^2(3+\eta^2)T_0}{\pi^2(1+\eta+7\eta^2-\eta^3)}, \qquad (13)$$

where we have used $\alpha = e^{-\mu_0^2/2T_0}$.

As a check, note that the continuous time limit corresponds to $T_0 \to \infty$ or $\kappa^2 \to 0$, whence $\eta \to 0$. Then the effective coupling Eq. 12 goes to $12T_0\eta/\pi^2 = 2T_0\kappa^2 = g_s^2 N_c/2\pi$ as it should. Next, with discrete time, we see that, in order to have real energy and κ ($\kappa^2 > 0$), we must place the restriction $0 < \eta < 1$. Small κ corresponds to small η, and large κ corresponds to η near unity. Interestingly, we note that the effective coupling in the 't Hooft equation is small in *both* the small and large κ regimes.

It is easy to understand the small effective coupling at large κ in terms of our discrete time Feynman diagrams. With discrete time, $\kappa^2 \to \infty$ causes the diagrams with a maximal number of powers of κ^2 per time step to dominate.

Now we turn to a numerical analysis of our discretized dynamics in order to understand how the continuum limit is approached in practice. We can write this equation as an eigenvalue problem by rescaling Ψ and isolating the eigenvalue t as a function of $\chi \equiv u\kappa^2$. The resulting eigenvalue problem to solve is

$$t\Phi(l) = \frac{\alpha_l}{\left(1-\chi^2 \Sigma'_l \Sigma'_{M-l}\right)} \left[\frac{(1-\chi\Sigma'_l)(1-\chi\Sigma'_{M-l})}{(1-\chi^2 \Sigma'_l \Sigma'_{M-l})} \Phi(l) + \chi \sum_{r=1}^{l-1} \frac{\alpha^{1/r+1/(M-l)}}{(l-r)^2} \Phi(r) \right.$$
$$\left. + \chi \sum_{r=l+1}^{M-1} \frac{\alpha^{1/l+1/(M-r)}}{(l-r)^2} \Phi(r) \right]. \quad (14)$$

We used numerical procedures in MAPLE and MATLAB to find the eigenvalues $t_n(\chi)$ of the matrix on the right hand side of this equation as a function of χ. The value of κ^2 is different for each t_n since $\kappa^2 = \chi t_n$. However by varying $0 \leq \chi \leq \infty$ we can generate the real solutions, t_n, for all κ^2.

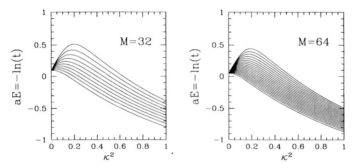

FIGURE 3. Plots of the lowest lying energy eigenstates of the Bethe-Salpeter equation with the veto for $M = 32, 64$.

The problem of contamination of the lowest lying states by complex solutions has been solved by our veto prescription (unlike the case without the veto, see [1]).

The lowest lying states remain intact for all coupling κ^2, see Fig. 3. When we analyze Eq. 14 for increasing M (see Fig. 3 for $M = 32, 64$) we see that the number of low lying states that remain uncrossed for all couplings increases with increasing M. We also see that the spacing between these states decreases as M increases.

Our results for large values of M (we were able to compute for values of $M \leq 4096$ which is hopefully sufficiently close to the continuum limit) can be compared to those of 't Hooft [7]. This detailed study (see [1]) agrees with the results of 't Hooft.

In this proceeding we have explored the efficacy of the discretization of large N_c QCD proposed in [2] by applying it to the well-understood 't Hooft model. For a smooth continuum limit over the whole range of bare coupling κ, we had to introduce a refinement of the discrete time gluon emission vertex. This amounted to insisting that after an emission, at least 2 time steps had to intervene before the next emission, with a similar restriction on consecutive absorptions. In contrast, an absorption is allowed to immediately follow an emission and *vice versa*. With this refinement in place we found that the continuum 't Hooft equation describes the mass spectrum for all real κ. However, the parameters that occur in the equation are renormalized from their bare values, as summarized in Eqs. 12, 13.

An amusing outcome of this renormalization phenomenon is that the effective coupling goes to zero in both the small and large κ limits. Perhaps this feature is a version of weak/strong coupling duality, much celebrated in recent developments in string/M theory. However, we must concede that 2 dimensional QCD may be too trivial to expect anything other than the usual continuum theory to emerge from any continuum limit. Another caveat against attributing much significance to this "duality" phenomenon, is that the physics of the continuum limit really only depends on the ratio $\mu^2/N_c g^2$. This is because one can always choose the effective coupling as the fundamental unit of energy. Then the theories at different coupling but with the same value of this ratio (0 for example) are physically identical: any differences in description can be removed by a change of units.

At any rate, we conclude that the discretization of [2] can be meaningfully applied to QCD in 2 space-time dimensions, with some intriguing hints about the nature of weak/strong coupling duality. An obvious and important limitation of the 2 dimensional case, however, is that the gluon has no dynamical degrees of freedom. Thus there is no opportunity for the P^+ of the system to be shared amongst an infinite number of gluons. This must occur for the fishnet diagrams to be relevant, and is allowed in higher dimensional space-time. The next step is to study the three dimensional case, the simplest gauge theory where fishnet diagrams can be relevant.

ACKNOWLEDGEMENTS

We acknowledge Charles Thorn with whom this work was done. We also recognize support from the following DOE grants: DE-FG02-97ER-41029 and DE-FG02-85ER40237.

REFERENCES

1. J. S. Rozowsky and C. B. Thorn, *Phys. Rev.* **D63** (2001) 085006, hep-ph/0007019.
2. K. Bering, J. S. Rozowsky and C. B. Thorn, *Phys. Rev.* **D61** (2000) 045007, hep-th/9909141.
3. G. 't Hooft, *Nucl. Phys.* **B72** (1974) 461.
4. C. B. Thorn, *Phys. Lett.* **70B** (1977) 85; *Phys. Rev.* **D17** (1978) 1073.
5. R. Giles, L. McLerran, and C. B. Thorn, *Phys. Rev.* **D17** (1978) 2058.
6. R. Brower, R. Giles, and C. Thorn, *Phys. Rev.* **D18** (1978) 484.
7. G. 't Hooft, *Nucl. Phys.* **B75** (1974) 461.
8. E. E. Salpeter and H. A. Bethe, *Phys. Rev.* **84** (1951) 1232.
9. T. Maskawa and K. Yamawaki, Prog. Theor. Phys. **56** (1976) 270; A. Casher, *Phys. Rev.* **D14** (1976) 452.
10. S. J. Brodsky, H-C. Pauli, and S. J. Pinsky, *Phys. Rept* **301**, (1998) 299; hep-ph/9705477.

Fuzzy Non-Trivial Gauge Configurations

Badis Ydri

Physics Department, Syracuse University, Syracuse, N.Y, 13244-1130, U.S.A.

Abstract. In this talk we will report on few results of discrete physics on the fuzzy sphere. In particular non-trivial field configurations such as monopoles and solitons are constructed on fuzzy S^2 using the language of K-theory, i.e projectors. As we will show, these configurations are intrinsically finite dimensional matrix models. The corresponding monopole charges and soliton winding numbers are also found using the formalism of noncommutative geometry and cyclic cohomology.

Fuzzy physics is aimed to be an alternative method to approach discrete physics. Problems of lattice physics especially those with topological roots are all avoided on fuzzy spaces. For example, chiral anomaly, Fermion doubling and the discretization of non-trivial topological field configurations were all formulated consistently on the fuzzy sphere [see [1] and the extensive list of references therein]. The paradigm of fuzzy physics is "discretization by quantization", namely given a space, we treat it as a phase space and then quantize it. This requires the existence of a symplectic structure on this space. One such class of spaces which admit symplectic forms are the co-adjoint orbits, for example $\mathbf{CP}^1 = \mathbf{S}^2$, $\mathbf{CP}^2$, $\mathbf{CP}^3$ and so on. Their quantization to obtain their fuzzy counterparts is done explicitly in [2, 1]. Here we will only summarize the important results for $\mathbf{S}^2$ which are needed for the purpose of this paper.

FUZZY S^2

Fuzzy $\mathbf{S}^2$ or $\mathbf{S}_F^2$ is the algebra $\mathbf{A} = Mat_{2l+1}$ of $(2l+1) \times (2l+1)$ matrices which is generated by the operators n_i^F, $i = 1, 2, 3$, which are defined by

$$n_i^F = \frac{L_i}{\sqrt{l(l+1)}}. \tag{1}$$

L_i's satisfy $[L_i, L_j] = i\varepsilon_{ijk}L_k$ and $\sum_{i=1}^{3} L_i^2 = l(l+1)$ respectively, where l is a positive integer. In other words, L_i's are the generators of the IRR l of $SU(2)$. A general element $\hat{f}$ of $\mathbf{A}$ admits an expansion, in terms of n_i^F's, of the form $\hat{f}(\vec{n}^F) = \sum_{i_1,\ldots,i_k} f_{i_1,\ldots,i_k} n_{i_1}^F \ldots n_{i_k}^F$, which will terminate by the nature of the operators n_i^F's. The continuum limit is defined by $l \longrightarrow \infty$. In such a limit the fuzzy coordinates n_i^F's tend, by definition, to the commutative coordinates n_i's [by inspection the commutators of the fuzzy coordinates among each others vanish at $l \longrightarrow \infty$, but from the Casimir equation above we must have $\sum_{i=1}^{3} n_i^2 = 1$]. Furthermore, the noncommutative algebra at this limit becomes the commutative algebra of functions on continuum $\mathbf{S}^2$, namely $\mathbf{A} \longrightarrow \mathcal{A}$, where a general element f of $\mathcal{A}$ will admit the expansion $f(\vec{n}) = \Sigma_{i_1,\ldots,i_k} f_{i_1,\ldots,i_k} n_{i_1} \ldots n_{i_k}$.

Viewing $\mathbf{S}^2$ as a submanifold of $\mathbf{R}^3$, one can check the following basic identity[3]

$$\mathcal{D}_2 = \mathcal{D}_3|_{r=\rho} + \frac{i\gamma^3}{\rho}. \qquad (2)$$

$\gamma^a = \sigma_a$, $a = 1, 2, 3$, are the flat gamma matrices in 3-dimensions. $\mathcal{D}_2$, $\mathcal{D}_3$ are the Dirac operators on $\mathbf{S}^2$ and $\mathbf{R}^3$ respectively. $\mathcal{D}_3|_{r=\rho}$ is the restriction of the Dirac operator on $\mathbf{R}^3$ to the sphere $r = \rho$, where ρ is the radius of the sphere, namely $\Sigma_{a=1}^3 x_a^2 = \rho^2$ for any $\vec{x} \in \mathbf{S}^2$. The Clifford algebra on $\mathbf{S}^2$ is two dimensional and therefore at each point $\vec{n} = \vec{x}/\rho$ one has only two independents gamma matrices, they can be taken to be γ^1 and γ^2. γ^3 should then be identified with the chirality operator $\gamma = \vec{\sigma}.\vec{n}$ on $\mathbf{S}^2$.

Next by using the canonical Dirac operator $\mathcal{D}_3 = -i\sigma_a \partial_a$ in (2) one can derive the two following equivalent expressions for the Dirac operator $\mathcal{D}_2$ on $\mathbf{S}_2$:

$$\begin{aligned}\mathcal{D}_{2g} &= \frac{1}{\rho}(\vec{\sigma}\vec{L} + 1) \\ \mathcal{D}_{2w} &= -\frac{1}{\rho}\varepsilon_{ijk}\sigma_i n_j \mathcal{J}_k. \end{aligned} \qquad (3)$$

$\mathcal{L}_k = -i\varepsilon_{kij} x_i \partial_j$ is the orbital angular momentum and $\mathcal{J}_k = \mathcal{L}_k + \frac{\sigma_k}{2}$ is the total angular momentum. g and w in (3) stands for Grosse-Klimčík-Prešnajder [3] and Watamuras Dirac operators [4] respectively. It is not difficult to check that $\mathcal{D}_{2w} = i\gamma \mathcal{D}_{2g} = \mathcal{D}_3|_{r=\rho} + \frac{i\gamma}{\rho}$ which means that $\mathcal{D}_{2w}$ and $\mathcal{D}_{2g}$ are related by a unitary transformation and therefore are equivalent. The spectrum of these Dirac operators is trivially derived to be given by $\pm \frac{1}{\rho}(j + \frac{1}{2})$ where j is the eigenvalue of $\vec{\mathcal{J}}$, i.e $\vec{\mathcal{J}}^2 = j(j+1)$ and $j = 1/2, 3/2, \ldots$.

The fuzzy versions of the Dirac operators (3) are taken to be

$$\begin{aligned} D_{2g} &= \frac{1}{\rho}(\vec{\sigma}.ad\vec{L} + 1) \\ D_{2w} &= \frac{1}{\rho}\varepsilon_{ijk}\sigma_i n_j^F L_k^R. \end{aligned} \qquad (4)$$

$ad\vec{L} = \vec{L}^L - \vec{L}^R$ is the fuzzy derivation which annihilates the identity matrix in $\mathbf{A}$ as the classical derivation $\vec{L}$ annihilates the constant function in $\mathcal{A}$. $\vec{L}^L$ and $-\vec{L}^R$ are the generators of the IRR l of $SU(2)$ which act on the left and on the right of the algebra $\mathbf{A}$ respectively, i.e $\vec{L}_i^L f = \vec{L} f$ and $-L_i^R f = -f L_i$ for any $f \in \mathbf{A}$. From this definition one can see that AdL_i provide the generators of the adjoint action of $SU(2)$ on $\mathbf{A}$, namely $Ad\vec{L}(f) = [\vec{L}, f]$ for any $f \in \mathbf{A}$.

These two fuzzy Dirac operators are not unitarily equivalent anymore. This can be checked by computing their spectra. The spectrum of D_{2g} is exactly that of the continuum only cut-off at the top total angular momentum $j = 2l + \frac{1}{2}$. In other words the spectrum of D_{2g} is equal to $\{\pm\frac{1}{\rho}(j+\frac{1}{2}), j = \frac{1}{2}, \frac{3}{2}, \ldots 2l - \frac{1}{2}\}$ and $\tilde{D}_{2g}(j) = \frac{1}{\rho}(j+\frac{1}{2})$

for $j = 2l + \frac{1}{2}$. The spectrum of D_{2w} is, however, highly deformed as compared to the continuum spectrum especially for large values of j. It is given by $D_{2w}(j) = \pm \frac{1}{\rho}(j + \frac{1}{2})\sqrt{1 + \frac{1-(j+1/2)^2}{4l(l+1)}}$. From these results it is obvious that D_{2g} is superior to D_{2w} as an approximation to the continuum.

In the same way one can find the fuzzy chirality operator Γ by the simple replacement $\vec{n} \longrightarrow \vec{n}^F$ in $\gamma = \vec{\sigma}.\vec{n}$ and insisting on the result to have the following properties : 1)$\Gamma^2 = 1$, $\Gamma^+ = \Gamma$ and $[\Gamma, f] = 0$ for all $f \in \mathbf{A}$. One then finds[4]

$$\Gamma = \frac{1}{l + \frac{1}{2}}(-\vec{\sigma}\vec{L}^R + \frac{1}{2}). \tag{5}$$

Interestingly enough this fuzzy chirality operator anticommutes with D_{2w} and not with D_{2g} so D_{2w} is a better approximation to the continuum than D_{2g} from this respect. This is also clear from the spectra above, in the spectrum of D_{2g} the top angular momentum is not paired to anything and therefore D_{2g} does not admit a chirality operator.

FUZZY NON-TRIVIAL GAUGE CONFIGURATIONS

Classical Monopoles

Monopoles are one of the most fundamental non trivial configurations in field theory. The wave functions of a particle of charge q in the field of a monopole p, which is at rest at $r = 0$, are known to be given by the expansion [2]

$$\psi^{(N)}(r,g) = \sum_{j,m} c_m^j(r) < j, m|D^{(j)}(g)|j, -\frac{N}{2} >, \tag{6}$$

where $D^{(j)} : g \longrightarrow D^{(j)}(g)$ is the j IRR of $g \in SU(2)$. The integer N is related to q and p by the Dirac quantization condition : $N = \frac{qp}{2\pi}$. r is the radial coordinate of the relative position $\vec{x}$ of the system, the angular variables of $\vec{x}$ are defined through the element $g \in SU(2)$ by $\vec{\tau}.\vec{n} = g\tau_3 g^{-1}$, $\vec{n} = \vec{x}/r$. It is also a known result that the precise mathematical structure underlying this physical system is that of a $U(1)$ principal fiber bundle $SU(2) \longrightarrow \mathbf{S}^2$. In other words for a fixed $r = \rho$, the particle q moves on a sphere $\mathbf{S}^2$ and its wave functions (6) are precisely elements of $S(\mathbf{S}^2, SU(2))$, namely sections of a $U(1)$ bundle over $\mathbf{S}^2$. They have the equivariance property

$$\psi^{(N)}(\rho, ge^{i\theta\frac{\tau_3}{2}}) = e^{-i\theta\frac{N}{2}}\psi^{(N)}(\rho, g), \tag{7}$$

i.e they are not really functions on $\mathbf{S}^2$ but rather functions on $SU(2)$ because they clearly depend on the specific point on the $U(1)$ fiber. In this paper, we will only consider the case $N = \pm 1$. The case $|N| \neq 1$ being similar and is treated in great detail in [1, 5].

An alternative description of monopoles can be given in terms of K-theory and projective modules. It is based on the Serre-Swan's theorem [7, 8] which states that there is a complete equivalence between vector bundles over a compact manifold $\mathbf{M}$

and projective modules over the algebra $C(\mathbf{M})$ of smooth functions on $\mathbf{M}$. Projective modules are constructed from $C(\mathbf{M})^n = C(\mathbf{M}) \otimes \mathbf{C}^n$ where n is some integer by the application of a certain projector p in $\mathcal{M}_n(C(\mathbf{M}))$, i.e the algebra of $n \times n$ matrices with entries in $C(\mathbf{M})$.

In our case $\mathbf{M} = \mathbf{S}^2$ and $C(\mathbf{M}) = \mathcal{A} \equiv$ the algebra of smooth functions on $\mathbf{S}^2$. For a monopole system with winding number $N = \pm 1$, the appropriate projective module will be constructed from $\mathcal{A}^2 = \mathcal{A} \otimes \mathbf{C}^2$. It is $\mathcal{P}^{(\pm 1)} \mathcal{A}^2$ where $\mathcal{P}^{(\pm 1)}$ is the projector

$$\mathcal{P}^{(\pm 1)} = \frac{1 \pm \vec{\tau}.\vec{n}}{2}. \tag{8}$$

It is clearly an element of $\mathcal{M}_2(\mathcal{A})$ and satisfies $\mathcal{P}^{(\pm 1)2} = \mathcal{P}^{(\pm 1)}$ and $\mathcal{P}^{(\pm 1)+} = \mathcal{P}^{(\pm 1)}$. $\mathcal{P}^{(\pm 1)} \mathcal{A}^2$ describes a monopole system with $N = \pm 1$ as one can directly check by computing its Chern character as follows

$$\pm 1 = \frac{1}{2\pi i} \int Tr \mathcal{P}^{(\pm 1)} d\mathcal{P}^{(\pm 1)} \wedge d\mathcal{P}^{(\pm 1)}. \tag{9}$$

On the contrary to the space of sections $\mathcal{S}(\mathbf{S}^2, SU(2))$, elements of $\mathcal{P}^{(\pm 1)} \mathcal{A}^2$ are by construction invariant under the action $g \longrightarrow g \exp(i\theta \frac{\tau_3}{2})$. The other advantage of $\mathcal{P}^{(\pm 1)} \mathcal{A}^2$ as compared to $\mathcal{S}(\mathbf{S}^2, SU(2))$ is the fact that its fuzzification is much more straight forward.

On The Equivalence of $\mathcal{P}^{(\pm 1)} \mathcal{A}^2$ and $\mathcal{S}(\mathbf{S}^2, SU(2))$

Before we start the fuzzification of $\mathcal{P}^{(\pm 1)} \mathcal{A}^2$, let us first comment on the relation between the wave functions $\psi^{(\pm 1)}$ given in equation (6) and those belonging to $\mathcal{P}^{(\pm 1)} \mathcal{A}^2$. The projector $\mathcal{P}^{(\pm 1)}$ can be rewritten as $\mathcal{P}^{(\pm 1)} = D^{(\frac{1}{2})} \frac{1 \pm \tau_3}{2} D^{(\frac{1}{2})+}(g)$ where $D^{(\frac{1}{2})} : g \longrightarrow D^{(\frac{1}{2})}(g) = g$ is the $\frac{1}{2}$ IRR of $SU(2)$. Hence $\mathcal{P}^{(\pm 1)} D^{(\frac{1}{2})}(g) | \pm > = D^{(\frac{1}{2})}(g) \frac{1 \pm \tau_3}{2} | \pm > = D^{(\frac{1}{2})}(g) | \pm >$, where $| \pm >$ are defined by $\tau_3 | \pm > = \pm | \pm >$. In the same way one can show that $\mathcal{P}^{(\pm 1)} D^{(\frac{1}{2})}(g) | \mp > = 0$. This last result means that

$$\mathcal{P}^{(\pm 1)} = D^{(\frac{1}{2})}(g) | \pm > < \pm | D^{(\frac{1}{2})+}(g) \tag{10}$$

$< \pm | D^{(\frac{1}{2})+}(g)$ defines a map from $\mathcal{P}^{(\pm 1)} \mathcal{A}^2$ into $\mathcal{S}(\mathbf{S}^2, SU(2))$ as follows

$$< \pm | D^{(\frac{1}{2})+}(g) \; : \; |\psi> \longrightarrow < \pm | D^{(\frac{1}{2})+}(g) | \psi > = \psi^{(\pm 1)}(\rho, g). \tag{11}$$

$< \pm | D^{(\frac{1}{2})+}(g) | \psi >$ has the correct transformation law (7) under $g \longrightarrow g \exp(i\theta \frac{\tau_3}{2})$ as one can check by using the basic equivariance property

$$D^{(\frac{1}{2})}(g e^{i\theta \frac{\tau_3}{2}}) | \pm > = e^{\pm i \frac{\theta}{2}} D^{(\frac{1}{2})}(g) | \pm >. \tag{12}$$

In the same way $D^{(\frac{1}{2})}(g)|\pm>$ defines a map, $S(\mathbf{S}^2, SU(2)) \longrightarrow \mathcal{P}^{(\pm 1)}\mathcal{A}^2$, which takes the wave functions $\psi^{(\pm 1)}$ to the two components elements $\psi^{(\pm 1)}D^{(\frac{1}{2})}(g)|\pm>$ of $\mathcal{P}^{(\pm 1)}\mathcal{A}^2$. Under $g \longrightarrow g\exp(i\theta\frac{\tau_3}{2})$, the two phases coming from $\psi^{(\pm 1)}$ and $D^{(\frac{1}{2})}(g)|\pm>$ cancel exactly so that their product is a function over $\mathbf{S}^2$.

Fuzzy Monopoles

Towards fuzzification one rewrites the winding number (9) in the form

$$\pm 1 = -\frac{1}{4\pi} \int d(\cos\theta) \wedge d\phi \, \mathrm{Tr}\, \gamma \mathcal{P}^{(\pm 1)} [\mathcal{D}, \mathcal{P}^{(\pm 1)}] [\mathcal{D}, \mathcal{P}^{(\pm 1)}](\vec{n})$$
$$= -Tr_\omega \left(\frac{1}{|\mathcal{D}|^2} \gamma \mathcal{P}^{(\pm 1)} [\mathcal{D}, \mathcal{P}^{(\pm 1)}] [\mathcal{D}, \mathcal{P}^{(\pm 1)}] \right). \tag{13}$$

The first line is trivial to show starting from (9), whereas the second line is essentially Connes trace theorem [7]. $|\mathcal{D}| =$ positive square root of $\mathcal{D}^\dagger \mathcal{D}$ while Tr_ω is the Dixmier trace [7, 9, 10]. In the fuzzy setting, this Dixmier trace will be replaced by the ordinary trace because the algebra of functions on fuzzy $\mathbf{S}_F^2$ is finite dimensional.

$\mathcal{D}$ in (13) is either $\mathcal{D}_{2g}$ or $\mathcal{D}_{2w}$ which are given in equation (3). They both give the same answer ± 1. The fuzzy analogues of $\mathcal{D}_{2g}$ and $\mathcal{D}_{2w}$ are respectively D_{2g} and D_{2w} given by equation (4). These latter operators were shown to be different and therefore one has to decide which one should we take as our fuzzy Dirac operator. D_{2g} does not admit as it stands a chirality operator and therefore its use in the computation of winding numbers requires more care which is done in [1, 6]. D_{2w} admits the fuzzy chirality operator (5) which will be used instead of the continuum chirality $\gamma = \vec{\sigma}.\vec{n}$. However D_{2w} has a zero eigenvalue for $j = 2l + \frac{1}{2}$ so it must be regularized for its inverse in (13) to make sense. This will be understood but not done explicitly in this paper, a careful treatment is given in [1, 5].

Finally the projector $\mathcal{P}^{(\pm 1)}$ will be replaced by a fuzzy projector $p^{(\pm 1)}$ which we will now find. We proceed like we did in finding the chirality operator Γ, we replace $\vec{n}$ in (8) by $\vec{n}^F = \vec{L}^L/\sqrt{l(l+1)}$ and insist on the result to have the properties $p^{(\pm 1)2} = p^{(\pm 1)}$ and $p^{(\pm 1)+} = p^{(\pm 1)}$. We also require this projector to commute with the chirality operator Γ, the answer for winding number $N = +1$ turns out to be $p^{(+1)} = \frac{1}{2} + \frac{1}{2l+1}[\vec{\tau}.\vec{L}^L + \frac{1}{2}]$. This can be rewritten in the following useful form

$$p^{(+1)} = \frac{\vec{K}^{(1)2} - (l-\frac{1}{2})(l+\frac{1}{2})}{(l+\frac{1}{2})(l+\frac{3}{2}) - (l-\frac{1}{2})(l+\frac{1}{2})}, \tag{14}$$

where $\vec{K}^{(1)} = \vec{L}^L + \frac{\vec{\tau}}{2}$. This allows us to see immediately that $p^{(+1)}$ is the projector on the subspace with the maximum eigenvalue $l + \frac{1}{2}$. Similarly, the projector $p^{(-1)}$ will correspond to the subspace with minimum eigenvalue $l - \frac{1}{2}$, namely

$$p^{(-1)} = \frac{\vec{K}^{(1)2} - (l+\frac{1}{2})(l+\frac{3}{2})}{(l-\frac{1}{2})(l+\frac{1}{2}) - (l+\frac{1}{2})(l+\frac{3}{2})}. \tag{15}$$

By construction (14) as well as (15) have the correct continuum limit (8), and they are in the algebra $\mathcal{M}_2(\mathbf{A})$ where $\mathbf{A}$ is the fuzzy algebra on fuzzy $\mathbf{S}_F^2$, i.e $2(2l+1) \times 2(2l+1)$ matrices. Fuzzy monopoles with winding number ± 1 are then described by the projective modules $p^{(\pm 1)} \mathbf{A}^2$.

If one include spin, then $\mathbf{A}^2$ should be enlarged to $\mathbf{A}^4$. It is on this space that the Dirac operator D_{2w} as well as the chirality operator Γ are acting. In the fuzzy the left and right actions of the algebra $\mathbf{A}$ on $\mathbf{A}$ are not the same. The left action is generated by L_i^L whereas the right action is generated by $-L_i^R$ so that we are effectively working with the algebra $\mathbf{A}^L \otimes \mathbf{A}^R$. A representation Π of this algebra is provided by $\Pi(\alpha) = \alpha \otimes \mathbf{1}_{2\times 2}$ for any $\alpha \in \mathbf{A}^L \otimes \mathbf{A}^R$. It acts on the Hilbert space $\mathbf{A}^4 \oplus \mathbf{A}^4$.

With all these considerations, one might as well think that one must naively replace $Tr_\omega \longrightarrow Tr$, $\gamma \longrightarrow \Gamma$, $\mathcal{D} \longrightarrow D_{2w}$ and $\mathcal{P}^{(\pm 1)} \longrightarrow p^{(\pm 1)}$ in (13) to get its fuzzy version. This is not totally correct since the correct discrete version of (13) turns out to be

$$c(\pm 1) = -Tr \varepsilon P^{(\pm 1)}[F_{2w}, P^{(\pm 1)}][F_{2w}, P^{(\pm 1)}], \qquad (16)$$

with

$$\mathbf{F}_{2w} = \begin{pmatrix} 0 & \frac{D_{2w}}{|D_{2w}|} \\ \frac{D_{2w}}{|D_{2w}|} & 0 \end{pmatrix}, \quad \varepsilon = \begin{pmatrix} \Gamma & 0 \\ 0 & \Gamma \end{pmatrix}. \qquad (17)$$

and

$$P^{(\pm 1)} = \begin{pmatrix} \frac{1+\Gamma}{2} p^{(\pm)} & 0 \\ 0 & \frac{1-\Gamma}{2} p^{(\pm)} \end{pmatrix}. \qquad (18)$$

[For a complete proof see [1] or [5]]. For $p^{(+1)}$ one finds that $c(+1) = +1 + [2(2l+1) + 1]$ while for $p^{(-)}$ we find $c(-1) = -1 + [2(2l) + 1]$. They are both wrong if compared to (13)!

The correct answer is obtained by recognizing that $c(\pm 1)$ is nothing but the index of the operator

$$\hat{f}^{(+)} = \frac{1-\Gamma}{2} p^{(\pm 1)} \frac{D_{2w}}{|D_{2w}|} p^{(\pm 1)} \frac{1+\Gamma}{2}. \qquad (19)$$

This index counts the number of zero modes of $\hat{f}^{(+)}$. The proof starts by remarking that, by construction, only the matrix elements $< p^{(\pm 1)} U_- | \hat{f}^{(+)} | p^{(\pm 1)} U_+ >$ where $U_\pm = \frac{1\pm\Gamma}{2} \mathbf{A}^4$, exist and therefore $\hat{f}^{(+)}$ is a mapping from $\hat{V}_+ = p^{(\pm 1)} U_+$ to $\hat{V}_- = p^{(\pm 1)} U_-$. Hence $Index \hat{f}^{(+)} = dim \hat{V}_+ - dim \hat{V}_-$.

Since one can write the chirality operator Γ in the form $\Gamma = \frac{1}{l+1/2}\left[j(j+1) - (l+1/2)^2\right]$ where j is the eigenvalue of $(-\vec{L}^R + \frac{\vec{\sigma}}{2})^2$, $j = l \pm 1/2$ for which $\Gamma|_{j=l\pm 1/2} = \pm 1$ defines the subspace $U_\pm$ with dimension $2(l \pm 1/2) + 1$. On the other hand, for $p^{(+1)}$ which projects down to the subspace with maximum eigenvalue $k_{max} = l + \frac{1}{2}$ of the operator $\vec{K}^{(1)} = \vec{L} + \frac{\vec{\tau}}{2}$, $\hat{V}_\pm$ has dimension $[2(l\pm 1/2) + 1][2(l + 1/2) + 1]$ and so the index is $Index \hat{f}^{(+)} = c(+1) = 2(2l+2)$. This result signals the existence of zero modes of the operator $\hat{f}^{(+)}$. Indeed for $\Gamma = +1$ one must couple $l + \frac{1}{2}$ to $l + \frac{1}{2}$ and obtain $j = 2l+1, 2l,..0$, whereas for $\Gamma = -1$ we couple $l + \frac{1}{2}$ to $l - \frac{1}{2}$ and obtain

$j = 2l, ..., 1$. j here denotes the total angular momentum $\vec{J} = \vec{L}^L - \vec{L}^R + \frac{\vec{\sigma}}{2} + \frac{\vec{\tau}}{2}$. Clearly the eigenvalues $j^{(+1)} = 2l + 1$ and 0 in $\hat{V}_+$ are not paired to anything. The extra piece in $c(+1)$ is therefore exactly equal to the number of the top zero modes, namely $2j^{(+1)} + 1 = 2(2l + 1) + 1$. These modes do not exist in the continuum and therefore they are of no physical relevance and must be projected out. This can be achieved by replacing the projector $p^{(+1)}$ by a corrected projector $\pi^{(+1)} = p^{(+1)}[1 - \pi^{(j^{(+1)})}]$ where $\pi^{(j^{(+1)})}$ projects out the top eigenvalue $j^{(+1)}$, it can be easily written down explicitly. Putting $\pi^{(+1)}$ in (16) gives exactly $c(+1) = +1$ which is the correct answer.

The same analysis goes for $p^{(-1)}$. Indeed if we replace it by the corrected projector $\pi^{(-1)} = p^{(-1)}[1 - \pi^{(j^{(-1)})}]$ where $\pi^{(j^{(-1)})}$ projects out the top eigenvalue $j^{(-1)} = 2l$, then equation (16) will give exactly $c(-1) = -1$ which is what we want.

CONCLUSION

It was shown in this article that topological quantities can be precisely and strictly defined in the discrete setting by using the methods of noncommutative geometry and fuzzy physics.

REFERENCES

1. Badis Ydri, *Fuzzy Physics*, a thesis which will be submitted in partial fulfillment of the requirements for the degree of Ph.D in physics, syracuse university.
2. A.P.Balachandran, G.Marmo, B-S.Skagerstan and A.Stern, *Classical Topology and Quantum States*, World Scientific, Singapore, 1991.
3. H. Grosse and P. Prešnajder, *Lett.Math.Phys.* **33**, 171 (1995). H. Grosse, C. Klimčík and P. Prešnajder,*Commun.Math.Phys.* **178**,507-526 (1996); **185**, 155 (1997);H. Grosse and P. Prešnajder, *Lett.Math.Phys.* **46**, 61 (1998) and ESI preprint,1999. H. Grosse, C. Klimčík, and P. Prešnajder,*Comm.Math.Phys.* **180**, 429 (1996),hep-th/9602115.H. Grosse, C. Klimčík, and P. Prešnajder, in *Les Houches Summer School on Theoretical Physics*, 1995,hep-th/9603071.
4. U. Carow-Watamura and S. Watamura,hep-th/9605003,*Comm.Math.Phys.* **183**, 365 (1997), hep-th/9801195.
5. S. Baez, A. P. Balachandran, S. Vaidya and B. Ydri,hep-th/9811169 and Comm.Math.Phys.**208**,787(2000).
6. A. P. Balachandran, T. R. Govindarajan and B. Ydri, *SU-4240-712,IMSc-99/10/36* and hep-th/9911087;A. P. Balachandran, T. R. Govindarajan and B. Ydri, *SU-4240-712,IMSc-99/10/36* and hep-th/0006216 and *Mod.Phys.Lett.*A**15**, 1279 (2000);
7. A. Connes,*Noncommutative Geometry*,Academic Press, London, 1994;
8. G. Landi ."Deconstructing Monopoles and Instantons". math-ph/9812004.
9. G. Landi,*An Introduction to Noncommutative spaces And Their Geometries*, Springer-Verlag, Berlin, 1997.hep-th/9701078.
10. J. C. Varilly and J. M. Gracia-Bondia.*J. Geom. Phys.*, 12:223–301, 1993.J. C. Varilly.*An Introduction to Noncommutative Geometry.physics/9709045.*

Generalized Coherent State Approach to Star Products and Applications to the Fuzzy Sphere

G. Alexanian*, A. Pinzul† and A. Stern†

Department of Physics, Syracuse University, Syracuse NY 13244.
†*Department of Physics, University of Alabama, Tuscaloosa AL 35487*

Abstract. We construct a star product associated with an arbitrary two dimensional Poisson structure using generalized coherent states on the complex plane. From our approach one easily recovers the star product for the fuzzy torus, and also one for the fuzzy sphere. For the latter we need to define the 'fuzzy' stereographic projection to the plane and the fuzzy sphere integration measure, which in the commutative limit reduce to the usual formulae for the sphere.

The star product is an important tool for deformation quantization and noncommutative geometry. The most well studied star product is often referred to as the Moyal star product [2],[3]. (For a nice review see [4].) It allows for a quantum mechanical description on phase space. In recent times it has found application in the string theory approach to noncommutative geometry. It is of particular importance for the fuzzy torus and has a simple form when acting on the periodic modes. Another star product due to Grosse and Presnajder[5] is applicable to the fuzzy sphere[6],[7],[9],[8],[10],[11], and is also of current interest in string theory[12]. This star product is constructed from coherent states on S^2 and is generalizable to arbitrary coset manifolds.[13] By relying on coherent states, the property of associativity is assured. The only other requirement on the star product is that it has a proper commutative limit. For this one assumes it to be a function of a parameter, say $\hbar$, which can be Taylor expanded about $\hbar = 0$. At zeroth order the star product reduces to the ordinary product, and at first order the star (or Moyal) commutator should be proportional to the Poisson bracket. The relevant issue is to find the star product associated with a given Poisson manifold. In this regard, a nontrivial constructive approach was given by Kontsevich[14] which is applicable for arbitrary Poisson structures.

The approach discribed in this talk is along the lines of Berezin quantization[15],[13], and relies on generalized coherent states on the complex plane developed by Man'ko, Marmo, Sudarshan and Zaccaria[16]. As usual, the stereographic projection has a coordinate singularity, which we choose at the north pole, in the commutative limit. However, away from $j \to \infty$, we can argue that there is no coordinate singularity, simply because the quantum mechanical probability of being at the north pole is zero. Since the fuzzy sphere is a (finite dimensional) matrix model, the corresponding field theory must be absent of any divergences, in contrast to the case of field theories on the fuzzy torus. We plan to pursue this in future works.

For our star product we introduce generalized coherent states $|\zeta>$, with the label ζ corresponding to a point on the complex plane. They are assumed to form an (overcomplete) basis for Hilbert space H. As is usual for coherent states, they are unit vectors $<\zeta|\zeta>=1$ and satisfy the completeness relation

$$1 = \int d\mu(\zeta,\bar{\zeta}) |\zeta><\zeta|, \qquad (1)$$

where $d\mu(\zeta,\bar{\zeta})$ is the appropriate measure on the complex plane, and the bar denotes complex conjugation. We also assume the existence of another basis for H, and in terms of this basis the states $|\zeta>$ are expressible in a power series in ζ times some overall normalization.

To every operator A on Hilbert space H one can associate a function $\mathcal{A}(\zeta,\bar{\zeta})$ on the complex plane according to $\mathcal{A}(\zeta,\bar{\zeta}) =<\zeta|A|\zeta>$. An associative product for two such functions is then defined by

$$\mathcal{A}(\zeta,\bar{\zeta}) \star \mathcal{B}(\zeta,\bar{\zeta}) =<\zeta|AB|\zeta>= \int d\mu(\eta,\bar{\eta}) <\zeta|A|\eta><\eta|B|\zeta> \qquad (2)$$

If $|\zeta>$ ($<\zeta|$) is, up to a normalization factor, analytic (anti-analytic) in ζ, then the ratio $<\eta|A|\zeta>/<\eta|\zeta>$ is analytic in ζ and anti-analytic in η. Furthermore, it can be obtained from $\mathcal{A}(\zeta,\bar{\zeta})$ by acting with the translation operator twice

$$e^{-\zeta\frac{\partial}{\partial\eta}} e^{\eta\frac{\partial}{\partial\zeta}} \mathcal{A}(\zeta,\bar{\zeta}) = e^{-\zeta\frac{\partial}{\partial\eta}} \frac{<\zeta|A|\zeta+\eta>}{<\zeta|\zeta+\eta>} = \frac{<\zeta|A|\eta>}{<\zeta|\eta>} \qquad (3)$$

Alternatively, we can write $e^{-\zeta\frac{\partial}{\partial\eta}} e^{\eta\frac{\partial}{\partial\zeta}}$ (acting on η-independent functions) as an *ordered exponentials*

$$:\exp(\eta-\zeta)\frac{\overrightarrow{\partial}}{\partial\zeta}:, \quad :\exp\frac{\overleftarrow{\partial}}{\partial\zeta}(\eta-\zeta):$$

where the derivatives with the right (left) arrow are ordered to the right (left) in each term in the Taylor expansion, and they also act to the right (left). Substituting into (2), we can then write the product on functions of ζ and $\bar{\zeta}$ according to

$$\star = \int d\mu(\eta,\bar{\eta}) \; :\exp\frac{\overleftarrow{\partial}}{\partial\zeta}(\eta-\zeta): \; |<\zeta|\eta>|^2 \; :\exp(\bar{\eta}-\bar{\zeta})\frac{\overrightarrow{\partial}}{\partial\bar{\zeta}}: \qquad (4)$$

The product is thus determined once we know $d\mu(\zeta,\bar{\zeta})$ and $<\zeta|\eta>$.

The product (4) has the property that in general it is not symmetric. It reduces to the ordinary product if the function on the right is analytic in ζ and the function on the left is anti-analytic in ζ. If we have that the states $|\zeta>$ are eigenvectors of some operator $\tilde{a}$ with eigenvalues ζ

$$\tilde{a}|\zeta>=\zeta|\zeta>, \qquad (5)$$

then the product (4) between two analytic (or two anti-analytic) functions also reduces to the ordinary product. For this we only need $<\zeta|\tilde{a}^\dagger = \bar{\zeta}<\zeta|$. To get nontrivial results we can take the function on the right to be anti-analytic in ζ and the function on the left to be analytic in ζ

$$\mathcal{A}(\zeta) \star \mathcal{B}(\bar{\zeta}) =<\zeta|\mathcal{A}(\tilde{a}) \; \mathcal{B}(\tilde{a}^\dagger)|\zeta>= \mathcal{A}(\zeta) \mathcal{B}(\bar{\zeta}) + <\zeta|[\mathcal{A}(\tilde{a}), \mathcal{B}(\tilde{a}^\dagger)]|\zeta> \qquad (6)$$

which can be evaluated once we know the commutation relations for $\tilde{\mathbf{a}}$ and $\tilde{\mathbf{a}}^\dagger$. Say the commutation relations are of the form

$$[\tilde{\mathbf{a}}, \tilde{\mathbf{a}}^\dagger] = F(\tilde{\mathbf{a}}\tilde{\mathbf{a}}^\dagger) \qquad (7)$$

for some function F, and F can be Taylor expanded in some (commuting) parameter $\hbar$, where the lowest order term is linear in $\hbar$. For the "classical limit" defined as $\hbar \to 0$, one demands that $\mathcal{A}(\zeta, \bar{\zeta}) \star \mathcal{B}(\zeta, \bar{\zeta}) \to \mathcal{A}(\zeta, \bar{\zeta}) \mathcal{B}(\zeta, \bar{\zeta})$ and $[\mathcal{A}, \mathcal{B}]_\star \to O(\hbar)$ and the coefficient on the right hand side of $\star$-commutator is identified with the Poisson bracket. [For the star product of an analytic function with an anti-analytic function, these two conditions reduce to one: $\mathcal{A}(\zeta) \star \mathcal{B}(\bar{\zeta}) \to \mathcal{A}(\zeta) \mathcal{B}(\bar{\zeta}) + O(\hbar)$.

From the standard coherent states it is easy to recover the Moyal[2], [3] (or actually the Voros[17]) star product. * Here we identify $\tilde{\mathbf{a}}$ and $\tilde{\mathbf{a}}^\dagger$ with the standard lowering and raising operators for the harmonic oscillator $\mathbf{a}$ and $\mathbf{a}^\dagger$, satisfying $[\mathbf{a}, \mathbf{a}^\dagger] = 1$. (For the moment we supress $\hbar$.) It is also easy to perform the integral in (4) in this case. The scalar product squared is $|<\zeta|\eta>|^2 = e^{-|\zeta-\eta|^2}$ and the measure is $d\mu(\eta, \bar{\eta}) = \frac{1}{\pi} d\eta_R \, d\eta_I$, η_R and η_I being the real and imaginary parts of η. The integrand in (4) is then a Gaussian:

$$\star = \int \frac{d\eta_R \, d\eta_I}{\pi} : \exp \frac{\overleftarrow{\partial}}{\partial \zeta}(\eta - \zeta) : \exp(-|\zeta - \eta|^2) : \exp(\bar{\eta} - \bar{\zeta}) \frac{\overrightarrow{\partial}}{\partial \bar{\zeta}} := \exp \frac{\overleftarrow{\partial}}{\partial \zeta} \frac{\overrightarrow{\partial}}{\partial \bar{\zeta}} \qquad (8)$$

Note that the ordering of the exponential function can be dropped after the change of integration variables.

A more general class of coherent states on the complex plane was given in [16]. These coherent states provide a more convenient basis than the standard coherent states when studying functions of operators $\tilde{\mathbf{a}}$ and $\tilde{\mathbf{a}}^\dagger$. Now we assume such operators satisfy the general commutation relations (7), while the coherent states satisfy (5). The expectation values of $\tilde{\mathbf{a}}$ and $\tilde{\mathbf{a}}^\dagger$ for the coherent state $|\zeta>$ are ζ and $\bar{\zeta}$, respectively, and the star product (4) can be directly applied to functions of these variables.

The procedure of [16] requires the existence of a map from the usual harmonic oscillator algebra generated by annihilation and creation operators $\mathbf{a}$ and $\mathbf{a}^\dagger$, satisfying $[\mathbf{a}, \mathbf{a}^\dagger] = 1$, to the algebra generated by $\tilde{\mathbf{a}}$ and $\tilde{\mathbf{a}}^\dagger$. The map is expressed in the form $\tilde{\mathbf{a}} = f(\mathbf{n}+1) \, \mathbf{a}$, $\mathbf{n}$ being the number operator $\mathbf{n} = \mathbf{a}^\dagger \mathbf{a}$, and the function f being determined from F. In this section we regard $f(\mathbf{n})$ as a nonsingular function, while we adapt the formalism to a singular function in the section 3. Following [16] we restrict to real functions, as only the real part of f determines F. We can introduce the Hilbert space H spanned by orthonormal states $|n>$, $n = 0, 1, 2, ...$, with $\mathbf{a}|0> = 0$ and $\mathbf{n}|n> = n|n>$. Following [16] one can construct the analogue of the standard coherent states according to:

$$|\zeta> = N(|\zeta|^2)^{-\frac{1}{2}} \exp\{\zeta f(\mathbf{n})^{-1} \mathbf{a}^\dagger\} f(\mathbf{n})^{-1} |0> = N(|\zeta|^2)^{-\frac{1}{2}} \sum_{n=0}^{\infty} \frac{\zeta^n}{\sqrt{n!}[f(n)]!} |n>, \qquad (9)$$

---

* After posting an earlier version of this article on the hep-th archive, we were informed of a similar discussion in [18].

where $[f(n)]! = f(n)f(n-1)...f(0)$. These states are diagonal in $\tilde{a}$, rather than a, with associated eigenvalues ζ as in (5). Requiring them to be of unit norm fixes $N(|\zeta|^2)$,

$$N(x) = \sum_{n=0}^{\infty} \frac{x^n}{n! \, ([f(n)]!)^2} , \qquad (10)$$

which reduces to the exponential function for standard coherent states. As with the standard coherent states, the states (9) are not orthonormal. Expressing $<\eta|\zeta>$ through $N(x)$ and using completeness relation one can show that integration measure should satisfy [1]

$$N(\bar{\zeta}\lambda) = \int d\mu(\eta,\bar{\eta}) \, \frac{N(\bar{\zeta}\eta) \, N(\bar{\eta}\lambda)}{N(|\eta|^2)} , \qquad (11)$$

for arbitrary complex coordinates ζ and λ. For the $d\mu(\zeta,\bar{\zeta}) = ih(|\zeta|^2) \, d\zeta \wedge d\bar{\zeta}$, one can solve equation (11) by the means of inverse Mellin transformation [1].

It is possible to compute the star product for a class of function without using a specific expression for measure. For example, $1 \star 1 = 1$, which follows from (11). One can show that functions

$$Z_{nm} = \frac{\bar{\zeta}^n \zeta^m}{N(|\zeta|^2)}, \quad n,m = 0,1,2,3,..., \qquad (12)$$

with the property $\bar{Z}_{nm} = Z_{mn}$, form a closed algebra:

$$Z_{nm} \star Z_{rs} = m! \, ([f(m)]!)^2 \, \delta_{m,r} \, Z_{ns} \qquad (13)$$

With a small modification of the above procedure we can write down the star product for the fuzzy sphere. The modification is necessary because we will no longer have (5), except in the commutative limit. We associate deformed annihilation and creation operators, $\tilde{a}$ and $\tilde{a}^\dagger$, with the operator analogue of the stereographic coordinates of a sphere. Its algebra now leads to a highest weight state $|2j>$ and therefore finite $(2j+1)$ dimensional representations, which is another departure from the treatment of the previous section. It requires that we terminate the series in (9) and what follows, and quantities computed previously now depend on j. The infinite series is recovered when $j \to \infty$, which is the commutative limit of the fuzzy sphere.

We now recall that the stereographic projection of a sphere of radius 1, $x_i x_i = 1$, $i = 1,2,3$, to the complex plane which maps the north pole to infinity is given by $z = (x_1 - ix_2)/(1-x_3)$. To obtain the algebra of the fuzzy sphere one promotes the coordinates x_i tooperators $\mathbf{x}_i$'s, satisfying commutation relations:

$$[\mathbf{x}_i, \mathbf{x}_j] = i\alpha \, \varepsilon_{ijk} \mathbf{x}_k , \qquad (14)$$

as well as $\mathbf{x}_i \mathbf{x}_i = 1$, where α is a parameter which vanishes in the commutative limit and 1 now denotes the unit operator. For $\alpha = \frac{1}{\sqrt{j(j+1)}}$, $j = \frac{1}{2}, 1, \frac{3}{2}, ...$, $\mathbf{x}_i$ has finite dimensional representations, which are simply given by $\mathbf{x}_i = \alpha \mathbf{J}_i$, $\mathbf{J}_i$ being the angular momentum matrices. To define an operator analogue of the stereographic projection of operators $\mathbf{x}_i$, we need to choose an ordering in the definition of operators $\mathbf{z}$ and $\mathbf{z}^\dagger$. We define the following deformation map of the algebra:

$$z = (x_1 - ix_2)(1 - x_3)^{-1}, \qquad z^\dagger = (1 - x_3)^{-1}(x_1 + ix_2) \qquad (15)$$

From the commutation relations (14) one gets $[z, \chi^{-1}] = -(\alpha/2)z$, where $\chi^{-1} = \frac{1}{2}(1 - x_3)$. It then follows that χ^{-1} commutes with $|z|^2 = zz^\dagger$ and

$$[z, z^\dagger] = \alpha\chi\left(1 + |z|^2 - \frac{1}{2}\chi(1 + \frac{\alpha}{2}|z|^2)\right). \qquad (16)$$

This is the analogue of eq. (7), the right hand side corresponding to the function F. For all finite dimensional matrix representations of the fuzzy sphere, χ^{-1} is represented by a nonsingular matrix, and the above equation makes sense. More generally, χ^{-1} is a nonsingular operator. Since it is a hermitian operator, and it should be possible to write it in terms of $|z|^2$. To get its form start with the identity $x_i x_i = 1$ and applying commutation relations above we get $\chi = 2/\alpha + 1/\xi - 2/\alpha\sqrt{1/\xi + (\alpha/(2\xi))^2}$, where $\xi = 1 + \alpha|z|^2$ The sign choice is so that χ reduces to $1 + |z|^2$ in the limit $j \to \infty$. Analyzing the eigenvalues of the right-hand side of this equation it follows that χ is an invertable operator. There is thus a $1 - 1$ correspondence between representations of the algebra generated by z and $z^\dagger$ and the algebra of the fuzzy sphere.

More generally, we denote the states of an irreducible representation Γ^j as usual by $|j, m>$, $j = \frac{1}{2}, 1, \frac{3}{2}, ..., m = -j, -j+1, ..., j$. The states span vector space H^j. Then $|z|^2|j, m> = \lambda_{j,m}|j, m>$, with

$$\lambda_{j,m} = \frac{j(j+1) - m(m+1)}{(\sqrt{j(j+1)} - m - 1)^2}. \qquad (17)$$

As $j \to \infty$, $\lambda_{j,m}$ ranges between 0 and $8j + 4$.

We next define the map from the harmonic oscillator algebra. This is clearly a singular map since H^j is finite dimensional and the Hilbert space H for the latter is not. For irreducible representation Γ^j, we can restrict the map to act on the finite dimensional subspace of H spanned by the first $2j + 1$ states $|n>$, $n = 0, 1, 2, ..., 2j$. More precisely, we identify $|j, m>$ in H^j with $|j + m>$ of H, and the map is applied to this subspace. Because the map depends on j we include a j subscript

$$z = f_j(n + 1)\, a \qquad (18)$$

From (16) and () the eigenvalues of $f_j(n)^2$ in H^j are $\lambda_{j,n-j-1}/n$. Therefore,

$$f_j(n) = \frac{\sqrt{2j - n + 1}}{\sqrt{j(j+1)} + j - n} \qquad (19)$$

It is zero when acting on $|2j+1>$ and hence $z^\dagger|2j> = 0$. It is ill-defined for harmonic oscillator states with $n > 2j + 1$. The map (18) is similar to that of Holstein and Primakoff[19], who instead go from the angular momentum algebra to the oscillator algebra.

We now construct coherent states, as before, with a linear combination of **n** eigenstates. Only here we need to truncate the series at $n = 2j$:

$$|\zeta, j> = N_j(|\zeta|^2)^{-\frac{1}{2}} \sum_{n=0}^{2j} \frac{\zeta^n}{\sqrt{n!\,[f_j(n)]!}} |n> . \tag{20}$$

In the paper [1] we have shown that both the integration measure $d\mu_j(\zeta, \bar{\zeta})$ and this normalization function $N(x)_j$ can be expressed as product of hypergeometric functions, and thus the star product (4) for the fuzzy sphere can now be given as an integral of hypergeometric functions.

For finite j the coherent states (20) are not diagonal in **z**. (**z** has only zero eigenvalues for all finite j.) Instead,

$$\mathbf{z}|\zeta, j> = \zeta|\zeta, j> - \frac{N_j(|\zeta|^2)^{-\frac{1}{2}} \zeta^{2j+1}}{\sqrt{(2j)!\,[f_j(2j)]!}} |2j> . \tag{21}$$

So as indicated earlier, we don't have the analogue of (5). On the other hand, $|\zeta, j>$ tend to **z** eigenstates in the commutative limit $j \to \infty$. For this we need the asymptotic behavior of $N_j(x)$. The result for $x \equiv |\zeta|^2 \ll j$ is the following behavior of the measure for large j in terms of these variables (see [1]):

$$d\mu_j(\zeta, \bar{\zeta}) \sim \frac{j}{\pi} \frac{i\,d\zeta \wedge d\bar{\zeta}}{(1+|\zeta|^2)^2} , \quad x \ll j . \tag{22}$$

So we recover the usual measure for S^2. (We can rescale the coordinates to absorb the j factor.)

We now compute some star products. For the variables ζ and $\bar{\zeta}$:

$$\begin{aligned}
\bar{\zeta} \star \zeta &= |\zeta|^2, \quad \zeta \star \zeta = \zeta^2 + \frac{Z_{2j,2j+2}}{(2j)!\,[f_j(2j)!]^2}, \quad \bar{\zeta} \star \bar{\zeta} = \bar{\zeta}^2 + \frac{Z_{2j+2,2j}}{(2j)!\,[f_j(2j)!]^2}, \\
\zeta \star \bar{\zeta} &= |\zeta|^2 + <\zeta|[\mathbf{z}, \mathbf{z}^\dagger]|\zeta> - \frac{Z_{2j+2,2j+2}}{(2j)!\,[f_j(2j)!]^2}
\end{aligned} \tag{23}$$

The second and third equations do not agree with what we have seen before - i.e. absence of corrections for the purely holomorphic or anti-holomorphic functions. These correction terms are due to the fact that the coherent states are not eigenvectors of **z**, except in the commutative limit. Actually, rather than ζ and $\bar{\zeta}$, a more usefull set of variables are the 'fuzzy' stereographic coordinates z_F and $\bar{z}_F$ defined by $z_F = <\zeta|\mathbf{z}|\zeta>$ and $\bar{z}_F = <\zeta|\mathbf{z}^\dagger|\zeta>$. Using (21), one can compute how they are related to ζ and $\bar{\zeta}$ [1]. Also, z_F and $\bar{z}_F$ tend to ζ and $\bar{\zeta}$ in the commutative limit, and just as with the latter variables, the star products of z_F and $\bar{z}_F$ reduce to the ordinary products in the limit. For the star commutator of z_F with $\bar{z}_F$ we can use the definition

$$z_F \star \bar{z}_F - \bar{z}_F \star z_F = <\zeta|[\mathbf{z}, \mathbf{z}^\dagger]|\zeta> \tag{24}$$

For $j \to \infty$ the right hand side reduces to $\frac{1}{2j}(1 + |z_F|^2)^2$, corresponding to the Poisson bracket of z_F and $\bar{z}_F$ [1].

For finite j, Z_{nm}, $n, m = 0, 1, 2, \ldots 2j$ generate a $(2j+1)^2$ dimensional algebra given by (13). This algebra is isomorphic to the algebra of $(2j+1) \times (2j+1)$ matrices associated with the jth representation of the fuzzy sphere. The latter are generated by $(2j+1) \times (2j+1)$ matrix representations for $\mathbf{z}$ and $\mathbf{z}^\dagger$ [1].

Finally, we return to the stereographic projection. For any j we can write it (or more precisely, the inverse stereographic projection) in terms of $N_j(x)$. Here we invert (15) to solve for the three dependent coordinates $(x_i)_F = <\zeta|\mathbf{x}_i|\zeta>$ of the fuzzy sphere. After using $\mathbf{x}_3 = \alpha(\mathbf{n} - j)$, it identically follows that $\sum (x_i)_F \star (x_i)_F = 1$. We can check that the standard inverse stereographic projection is recovered in the commutative limit [1].

ACKNOWLEDGEMENT

We are gratefull for valuable discussions with A. P. Balachandran, G. Karatheodoris, D. O'Connor and C. Zachos (who also informed us of numerous references). This work was supported by the joint NSF-CONACyT grant E120.0462/2000. G. A. was supported in part by the U.S. Department of Energy under contract number DE-FG02-85ER40231, and A. P. and A. S. were supported in part by the U.S. Department of Energy under contract number DE-FG05-84ER40141.

REFERENCES

1. G. Alexanian, A. Pinzul and A. Stern, Nucl.Phys. **B600** 531-547 (2001).
2. H. Groenewold, Physica (Amsterdam) **12**, 405 (1946).
3. J. Moyal, Proc. Camb. Phil. Soc. **45**, 99 (1949).
4. C. Zachos, J. Math. Phys. **41**, 5129 (2000); hep-th/0008010, to be published in the proceedings of NATO Advanced Research Workshop on Integrable Hierarchies and Modern Physical Theories (NATO ARW - UIC 2000), Chicago, Illinois, 22-26 Jul 2000.
5. H. Grosse and P. Presnajder, Lett.Math.Phys. **28**, 239 (1993); P. Presnajder, hep-th/9912050 .
6. J. Madore, Class. and Quant. Geom. **9**, 69 (1992); *An Introduction to Noncommutative Differential Geometry and its Applications*, Cambridge University Press, Cambridge, 1995.
7. U. Carow-Watamura, S. Watamura, Commun. Math. Phys. **183**, 365 (1997); Int. J. Mod. Phys. **A13**, 3235 (1998); Commun. Math. Phys. **212**, 395 (2000).
8. S. Baez, A.P. Balachandran, B. Idri, S. Vaidya, Commun.Math.Phys. **208**, 787 (2000).
9. A.P. Balachandran, T.R. Govindarajan, B. Ydri, hep-th/9911087; hep-th/0006216.
10. A.P. Balachandran, S. Vaidya, hep-th/9910129.
11. A.P. Balachandran, X. Martin, D.O'Connor, hep-th/0007030.
12. A. Alekseev, A. Recknagel, V. Schomerus, JHEP 0005:010,2000; P.-M. Ho, hep-th/0010165.
13. A. Perelomov, *Generalized Coherent States and Their Applications* (Springer-Verlag, Berlin, 1986); J.R. Klauder and B.-S. Skagerstam *Coherent States* (World Scientific, Singapore, 1985).
14. M. Kontsevich, q-alg/9709040.
15. F.A. Berezin, Commun.Math.Phys. **40** 153 (1975).
16. V.I. Man'ko, G. Marmo, E.C.G. Sudarshan, F. Zaccaria, Physica Scripta **55**, 528 (1997).
17. F. Bayen, in *Group Theoretical Methods in Physics*, ed. E. Beiglböck , et. al. [Lect. Notes Phys. **94**, 260 (1979)]; A. Voros, Phys. Rev. **A 40**, 6814 (1989).
18. N. Read, Phys.Rev. **B58**, 16262 (1998).
19. T. Holstein and H. Primakoff, Phys.Rev. **58**, 1098 (1940).

FIELD THEORY II

Spontaneous CPT Violation in Confined QED

E.J. Ferrer[*1], V. de la Incera[1], and A. Romeo[2]

[1]*Department of Physics, State University of New York at Fredonia, Houghton Hall 118, Fredonia, NY 14063, USA*
[2]*Institute for Space Studies of Catalonia, CSIC, Edif. Nexus, Gran Capita 2-4, 08034, Barcelona, Spain*

Abstract. Symmetry breaking effects induced by untwisted fermions in QED in a nonsimply connected spacetime with topology $S^1 \times R^3$ is investigated. It is found that the discrete CPT symmetry of the theory is spontaneously broken by the appearance of a constant vacuum expectation value of the electromagnetic potential along the direction of space periodicity. The constant potential is shown to be gauge nonequivalent to zero in the nonsimply connected spacetime under consideration. Due to the symmetry breaking, one of the electromagnetic modes of propagation is massive with a mass that depends on the inverse of the compactification length. As a result the system exhibits a sort of topological directional superconductivity. A second electromagnetic mode is superluminal, with a superluminal velocity that increases logarithmically with the compactification radius.

Symmetry breaking in non-trivial space-time, where the curvature, as well as the topology, can play an important role, has received much attention in the last years because of its possible applications in cosmology. In the present work we discuss specifically the relevance of the nontrivial space topology for the magnetic response in confined QED. This result could find applications in high-T_c superconductivity where, as it is well known, the system is confined to a quasi-two-dimensional space.

A characteristic of quantum field theory in space-time with non-trivial topology is the possible existence of nonequivalent types of fields with the same spin [1]. In particular, for a fermion system in space-time which is locally flat but with topology $S^1 \times R^3$ (i.e. a Minkowskian space with one of the spatial dimensions compactified in a circle S^1 of finite length a), the nontrivial topology is transferred into periodic boundary conditions for untwisted fermions or antiperiodic boundary conditions for twisted fermions

$$\psi(t, x, y, z - a/2) = \pm \psi(t, x, y, z + a/2) \qquad (1)$$

Henceforth we restrict our analysis to the untwisted fermions case. The results for twisted fermions can be easily read off the results at finite temperature, since in the Euclidean space the two theories are basically the same after the interchange of the four-space subindexes $3 \leftrightarrow 4$.

When untwisted fermions are considered, the effect of vacuum polarization upon photon propagation yields a tachyonic mass for the third component of the photon

---

[*] Conference speaker

field [2]. Tachyonic modes in a quantum theory are an indication of symmetry breaking, since the vacuum under consideration is not really the physical one. In this talk we will show that the stable vacuum configuration is given by a constant electromagnetic potential A_3 along the compactified dimension, which spontaneously breaks the discrete CPT symmetry of the theory. As discussed below, even though such a vacuum configuration has $F_{\mu\nu}=0$, it cannot be gauged to zero, because the gauge transformation that would be needed does not respect the periodicity of the function space in the $S^1 \times R^3$ domain. This is a sort of Aharonov-Bohm effect that takes place due to the non-simply connected topological structure of the considered space-time.

The existence of constant gauge potentials with physical meaning are known in statistics, where A_0 cannot be gauged away by reasons similar to those pointed out above [3]. The minimum of the potential in the statistical case is however at $A_0 = 0$, since only twisted fermions are allowed in statistics. On the other hand, in the many-particle electroweak theory, a non-trivial constant vacuum A_0 is induced by the fermion density and cannot be gauged away [4]. There, in contrast to the system we are considering, an additional parameter to the temperature (a leptonic and/or baryonic chemical potential) is needed to trigger the non-trivial constant minimum for A_0.

Let us consider the QED action in a space-time domain with compactified dimension of length a in the OZ-direction

$$S = \int_{-a/2}^{a/2} dx_3 \int_{-\infty}^{\infty} dx_0 d^2 x_\perp \left[-\frac{1}{4} F^2_{\mu\nu} + \overline{\psi}(i\gamma\cdot\partial - e\gamma\cdot A - m)\psi \right] \quad (2)$$

To find the physical vacuum that stabilizes QED with untwisted fermions, we propose the following ansatz for the vacuum solution: $\overline{A}_\nu = \Lambda \delta_{\nu 3}$, with Λ an arbitrary constant that will be determined from the minimum equation of the effective potential. Notice that a constant electromagnetic potential in the direction of the periodicity ($\nu=3$ in our case), although seemingly physically equivalent to the pure vacuum $\overline{A}_\nu = 0$, since both have $F_{\mu\nu}=0$, cannot be gauged to zero. The reason is that due to the periodicity of the fields in the $S^1 \times R^3$ space, the gauge transformations $A_\mu \rightarrow A_\mu - \frac{1}{e}\partial_\mu \alpha$ are restricted to those satisfying $\alpha(x_3+a) = \alpha(x_3) + 2l\pi$, $l \in Z$ [3]. However, the gauge transformation $\alpha(x) = e\Lambda(x \cdot n)$ connecting the constant field configuration with zero does not satisfy the required periodicity condition unless Λ satisfies $\Lambda = 2l\pi/ea$, $l \in Z$.

After summing in the discrete momentum corresponding to the fermion periodic boundary condition, $p_3 = 2n\pi/a$ ($n = 0, \pm 1, \pm 2,...$) the one-loop effective potential for fermion fields taken on the background of the vacuum configuration Λ is given by

$$V(\Lambda) = -(2\pi)^{-3} \int_{-\infty}^{\infty} d^3 \hat{p} \left[\varepsilon_p + 2a^{-1} \operatorname{Re}\ln\left(1 - e^{-a(\varepsilon_p - ie\Lambda)}\right) \right] \quad (3)$$

where $\hat{p}_\mu = (p_0, p_1, p_2, 0)$ and $\varepsilon_p = \sqrt{\hat{p}^2 + m^2}$. The extremum solution $\partial V / \partial \Lambda = 0$ is $\Lambda = l\pi/ea$. Nevertheless, the minimum condition $\partial^2 V / \partial \Lambda^2 > 0$ is only satisfied by the subset

$$\Lambda_{min} = \frac{(2l+1)\pi}{ea}, \quad l \in Z \qquad (4)$$

The solution (4) is similar to the one found in Ref. [5] in the case of massless QED with periodic fermions on a circle (QED with $S^1 \times R^1$ topology).

The elements in the set of minimum solutions (4), are all connected by allowed large gauge transformations ($\alpha(x_3 + a) = \alpha(x_3) + 2l\pi$). It should be pointed out, however, that the solutions (4) are gauge nonequivalent to the trivial vacuum $\Lambda = 0$, since none of them satisfies that $\Lambda = 2l\pi/ea$.

Substituting with the minimum solution (4) in Eq. (3) we obtain in the $am \ll 1$ approximation that the effective potential of untwisted fermions reduces to $V(\Lambda_{min}) = -7\pi^2/360a^4$, which coincides with the result reported for twisted fermions in Ref. [2]. Thus, the vacuum energy of both classes of fermions coincides if the corresponding correct vacuum solution is used.

As discussed above, the appearance of the constant vacuum solution $\overline{A}_\nu = \Lambda_{min} \delta_{\nu 3}$ at the one-loop level must be associated with symmetry breaking. It is easy to corroborate that if the theory is expanded around one of the minimum solutions (4), the gauge field in the action (2) can be, as usual, shifted by $A_\mu \to A_\mu - \Lambda_{min}\delta_{\mu 3}$. Thus, a new term $e\Lambda_{min}\psi\gamma_3\overline{\psi}$ appears in the action, violating the CPT symmetry

$$x_\mu = -x_\mu, \quad A_\mu(x) = -A_\mu(-x), \quad \psi(x) = -i\psi(-x)\gamma_5\gamma_0 \qquad (5)$$

of the original action. We stress that the vacuum solution (4) breaks the CPT invariance by breaking each discrete symmetry separately, but maintaining CP or any other product of two symmetries intact. From our results we conclude that the spatial compactification leads perturbatively to CPT symmetry breaking with vacuum configuration given by a constant electromagnetic potential whose amplitude increases with the decreasing of the compactification radius as $1/a$.

Let us find the electromagnetic modes' masses for photons propagating perpendicularly to the compactified direction OZ on the background of the new vacuum (4). With this end, the general structure of the electromagnetic field Green's function in the $S^1 \times R^3$ space should be considered

$$\Delta_{\mu\nu}(k) = P\left(g_{\mu\nu} - \frac{k_\mu k_\nu}{k^2}\right) + Q\left(\frac{k_\mu k_\nu}{k^2} - \frac{k_\mu n_\nu + n_\mu k_\nu}{k \cdot n} + \frac{k^2 n_\mu n_\nu}{(k \cdot n)^2}\right) + \frac{\alpha}{k^4} k_\mu k_\nu \qquad (6)$$

Here, due to the breaking of Lorentz invariance, in addition to the usual tensor structures k_μ and $g_{\mu\nu}$, a spacelike unit vector $n_\mu = (0,0,0,1)$, pointing in the direction of the periodicity, must be introduced. In Eq. (6) α is a gauge fixing parameter corresponding to the gauge condition $\frac{1}{\alpha}\partial_\mu A^\mu = 0$, and the coefficients P and Q are given by

$$P = \frac{1}{k^2 + \Pi_0}, \qquad Q = \frac{\Pi_1}{(k^2 + \Pi_0)\{k^2 + \Pi_0 - \Pi_1[k^2/(k\cdot n)^2 + 1]\}} \qquad (7)$$

The parameters Π_0 and Π_1, are the coefficients of the polarization operator $\Pi_{\mu\nu}$, whose general structure is

$$\Pi_{\mu\nu}(k) = \Pi_0 \left(g_{\mu\nu} - \frac{k_\mu k_\nu}{k^2} \right) + \Pi_1 \left(\frac{k_\mu k_\nu}{k^2} - \frac{k_\mu n_\nu + n_\mu k_\nu}{k\cdot n} + \frac{k^2 n_\mu n_\nu}{(k\cdot n)^2} \right) \qquad (8)$$

The coefficient Π_1 is absent in the usual flat space case with trivial topology. It arises here because of the explicit breaking of the Lorentz invariance in the $S^1 \times R^3$ space-time. The situation is similar to that in statistical quantum field theory [6], where the presence of a medium breaks the Lorentz invariance giving rise to the compactification of the time variable. The role of n_μ is played by the four-velocity of the medium u_μ in the finite temperature case.

From (8) it is easy to see that the polarization operator coefficients can be expressed in terms of the two independent tensor components Π_{33} and Π_{00}, through the relations

$$\Pi_0 = \frac{k^2}{\kappa}\left(k_3^2 k_0^2 \Pi_{33} - \hat{k}^4 \Pi_{00}\right) \qquad (9)$$

$$\Pi_1 = \frac{k^2 k_3^2}{\kappa}\left(\vec{k}^2 \Pi_{33} - \hat{k}^2 \Pi_{00}\right) \qquad (10)$$

where we are introducing the new notation $\kappa = \hat{k}^2(\vec{k}^2\hat{k}^2 - k_0^2 k_3^2)$.

The electromagnetic modes' masses are found from the poles of the Green's function (6). From (7), (9) and (10) we have that for photons propagating in a direction perpendicular to the direction of periodicity ($k_3=0$) the dispersion relations are

$$k_0^2 - k_\perp^2 - (\hat{k}^2/k_\perp^2)\Pi_{00} = 0 \qquad (11)$$

$$k_0^2 - k_\perp^2 - \Pi_{33} = 0 \tag{12}$$

Let us find the solutions of Eqs. (11)-(12) at the one-loop level. With this end we need to calculate the one-loop polarization operator components Π_{00} and Π_{33} for untwisted fermions. Considering the free propagator for untwisted fermions on the minimum solution (4)

$$\widetilde{G}(x - x') = \frac{1}{(2\pi)^3 a} \sum_{p_3} \int d^3 \hat{p} \exp[ip.(x-x')] G(\widetilde{p}) \tag{13}$$

where

$$G(\widetilde{p}) = \frac{\widetilde{p} \cdot \gamma - m}{\hat{p}^2 - m^2 + i\varepsilon}, \qquad \widetilde{p}_\mu = (p_0, p_\perp, p_3 - e\Lambda) \tag{14}$$

the corresponding one-loop polarization operator is given by

$$\Pi_{\mu\nu}(k) = \frac{-4ie^2}{(2\pi)^3 a} \sum_{p_3} \int d^3 \hat{p} \, \frac{\widetilde{p}_\mu(\widetilde{p}_\nu - k_\nu) + \widetilde{p}_\nu(\widetilde{p}_\mu - k_\mu) - \widetilde{p} \cdot (\widetilde{p} - k) g_{\mu\nu} + m^2 g_{\mu\nu}}{(\widetilde{p}^2 - m^2)[(\widetilde{p} - k)^2 - m^2]} \tag{15}$$

In the $a|\hat{k}| \ll am \ll 1$ limit, we obtain

$$\Pi_{00}(k_3 = 0, k_0 = 0, k_\perp \approx 0) \cong \frac{e^2}{3\pi^2} k_\perp^2 [\tfrac{1}{2}\ln\xi + O(\xi^\circ)] + O(k_\perp^4) \tag{16}$$

$$\Pi_{33}(k_3 = 0, k_0 = 0, k_\perp \approx 0) \cong \frac{e^2}{a^2}[\tfrac{1}{3} + O(\xi^2)] + O(k_\perp^2) \tag{17}$$

where $\xi = am/2\pi \ll 1$. Using the results (16) and (17) in the dispersion equations (11)-(12), we find that the electromagnetic modes propagate on the nontrivial vacuum with masses $M_1^2 = (\hat{k}^2/k_\perp^2)\Pi_{00} = 0$ and $M_2^2 = \Pi_{33} = e^2/3a^2 > 0$. It is interesting to notice that if we make the change $a \to \beta$ (β the inverse temperature) in M_2^2, then we would formally obtain the well-known result in statistical QED [6] of the Debye screening mass.

We then conclude that in the new vacuum state no tachyon is present. Instead, a massive electromagnetic mode, with a topological mass that depends on the inverse of the compactification radius a, arises. When the compactification length is taken to infinity ($a \to \infty$), the flat-space QED theory is regained with zero photon mass.

Defining the photon velocity for each propagation mode with dispersion equations (11)-(12) by using the formula $v(\mathbf{k}) = \partial k_0 / \partial |\mathbf{k}|$ *, we find that within the considered approximation the transverse and longitudinal modes propagate perpendicularly to the compactified direction with velocities

$$v_\perp^T = 1 - \frac{e^2}{12\pi^2} \ln \xi, \qquad (18)$$

$$v_\perp^L = 1 - [M_2^2 / 2k_\perp^2], \qquad (19)$$

It is worth to mention that the modifications found for the two velocities, $v_\perp^T$ and $v_\perp^L$, have different origins [7]. The modification of the longitudinal velocity $v_\perp^L$ is due to the appearance of the topological mass M_2; while the transverse superluminal velocity $v_\perp^T$ (note that $v_\perp^T < c$ because $\xi < 1$ in the used approximation) appears as a consequence of a genuine variation of the refraction index in the considered space-time. Modifications of the photon velocity in non-trivial vacua have been previously reported in the literature [8]. For compactification lengths in agreement with the used approximation, $a < 1/m \approx 10^3$ fm., the transverse velocity $v_\perp^T$ is about 0.1% larger than the light velocity in trivial space-time.

The existence of a massive electromagnetic mode in confined QED for untwisted fermions has also implications for the magnetic response of the system. To understand this, let us consider the modified Maxwell equation in linear response theory

$$[\Box \delta_{\mu\nu} - \lambda_M^{-2} \delta_{3\mu} \delta_{3\nu}] A_\nu = e J_\mu \qquad (20)$$

taken in the Lorentz gauge $\partial_\mu A^\mu = 0$. In Eq. (20), we introduced the magnetic length $\lambda_M = 1/M_2$. Considering an external static and constant current I flowing inside the confined space along the OZ-axis, the current density in Eq. (20) becomes $eJ_3(x) = I\delta^2(x_\perp)$. Then, the induced potential, which is a solution of (20) with periodic boundary conditions in the OZ-direction, will be

$$A_3(x) = -\frac{I}{2\pi} K_0[|x_\perp|/\lambda_M], \qquad (21)$$

where K_0 denotes a modified Bessel function of the third kind. In the $|x_\perp| \gg \lambda_M$ limit we obtain for the corresponding magnetic field

---

* In the considered low-frequency approximation the group and phase velocities coincide and can be found by the same proposed formula.

$$\mathbf{B}(r) = -\frac{I}{2\sqrt{2\pi}} \frac{\exp(-r/\lambda_M)}{\sqrt{r\lambda_M}} \hat{\theta}, \qquad (22)$$

with $r = |x_\perp|$ and $\hat{\theta}$ denoting the azimuth-angle unit vector in cylindrical coordinates. From (22) we see that an azimuthal magnetic field will be screened along the radial direction, in a distance equal to the inverse of the topological mass M_2. When the separation between the two infinite parallel plates decreases, the screening effect increases.

Our results indicate that confined QED exhibits a sort of topological directional superconducting behavior, with a Meissner effect taking place for magnetic fields induced by electric currents flowing in the direction of the compactified dimension.

ACKNOWLEDGMENTS

It is a pleasure for two of the authors (EJF and VI) to express their gratitude to the Institute for Spatial Studies of Catalonia for the warm hospitality extended to them during the time this work was completed, and in particular to Dr. E. Elizalde for facilitating this collaboration. Also the authors would like to thank Dr. V.P. Gusynin and Dr. M.A. Shifman for helpful discussions. This work has been supported in part by US-Spain Science and Technology CCCT-6-151 grant, by NSF grant PHY-0070986 (EJF and VI), and by NSF POWRE grant PHY-9973708 (VI).

REFERENCES

1. Isham, C.J., Proc. R. Soc. A **362**, 383 (1978); **364**, 591 (1978); Avis, S.J. and Isham, C.J., Proc. R. Soc. A **363**, 581 (1978); Banach, R., J. Phys. A: Math. Gen. **14**, 901 (1981).
2. Ford, L.H., Phys. Rev. D **21**, 933 (1980); **22**, 3003 (1980).
3. Batakis, N. and Lazarides, G., Phys. Rev. D **18**, 4710 (1978); Polyakov, A.M., Phys. Lett. **72**, 477 (1978); Affleck, I., Nucl. Phys. B **162**, 461 (1980); Gross, D., Pisarski, R. and Yaffe, L., Rev. Mod. Phys. **53**, 43 (1981); Weiss, N., Phys. Rev. D **24**, 475 (1981); D **25**, 2667 (1982); Actor, A., Phys. Rev. D **27**, 2548 (1983); Ann. Phys. **159**, 445 (1985); J. Phys. A **33**, 4585 (2000).
4. Linde, A.D., Phys. Lett. B **86**, 39 (1979); Ferrer, E.J., de la Incera, V. and Shabad, A.E., Phys. Lett. B **185** (1987) 407; Nucl. Phys. B **309** (1988) 120.
5. Hetrick, J.E., and Hosotani, Y., Phys. Rev. D **38**, 2621 (1988).
6. Fradkin, E.S., Proceedings of Quantum Field Theory and Hydrodynamics, P. N. Lebedev Physical Institute, Vol. **29**, 7 (1965, Moscow Nauka) (Engl. Transl., 1967 New York: Consultant Bureau).
7. Ferrer, E.J., de la Incera, V., and Romeo, A., hep-ph/0107229 (to appear in Phys. Lett. B).
8. Scharnhost, K., Phys. Lett. B **236**, 354 (1990); Barton, G., Phys. Lett. B **237**, 559 (1990); Tarrach, R., Phys. Lett. B **133**, 259 (1983); Latorre, J.I., Pascual, P., Tarrach, R., Nucl. Phys. B **437**, 60 (1995); Scharnhost, K., Annalen Phys. **7**, 700 (1998); Dittrich, W., and Gies, H., Phys. Rev. D **58**, 025004 (1998).

Issues on Radiatively Induced Lorentz and CPT Violation in Quantum Electrodynamics

W.F. Chen

Department of Mathematics and Statistics, University of Guelph
Guelph, Ontario, Canada N1G 2W1

Abstract. Various ambiguous results on radiatively induced Lorentz and CPT violation in quantum electrodynamics with a modified fermionic sector are reviewed and explanations for this ambiguity in the literature are commentated. Furthermore, joint between stringent limit from astrophysical observation and theoretical prediction on Lorenz and CPT violation is discussed.

INTRODUCTION

Lorentz symmetry is algebraic foundation of the theory of special relativity. Nearly one hundred years the theory of special relativity keeps the status as a cornerstone of modern physics and has been supporting by numerous high energy physics experiments and astrophysical observation. However, physics is a science born out of experimental observation. With the availability of higher precision experimental or observational data, it is conceivable that even the most fundamental principles may someday have to be modified or even abandoned. It is partly in this spirit that an investigation on the possible breaking of Lorentz symmetry is not fantastic.

A straightforward reason of considering Lorentz and CPT violation in quantum electrodynamics (QED) was from astrophysical observation. A lopsided analysis on the polarized electromagnetic radiation emitted by distant radio galaxies revealed that the universe may present cosmological anisotropy [1] and further suggested that this chiral effect can be well described at lower derivative expansion by a modified classical electrodynamics proposed a decade ago [2],

$$S = \int d^4x \left(-\frac{1}{4} F_{\mu\nu} F^{\mu\nu} + \frac{1}{2} \varepsilon^{\mu\nu\lambda\rho} k_\mu F_{\nu\lambda} A_\rho \right). \tag{1}$$

The first term of (1) is the familiar Maxwell term, and the second one is called the Chern-Simons-like (CS) term,

$$\mathcal{L}_{CS} = \frac{1}{2} \varepsilon^{\mu\nu\lambda\rho} k_\mu F_{\nu\lambda} A_\rho, \tag{2}$$

which explicitly violate Lorentz and discrete CPT symmetries since k_μ is certain background constant vector in four-dimensional space-time. Despite that a more rigorous analysis on the astrophysical observation data has excluded the polarization effect of electromagnetic wave in propagating from the distant radio sources [3], this still stimulates an effort to explore a possible Lorentz and CPT violating mechanism theoretically.

Furthermore, a $SU(3) \times SU(2) \times U(1)$ standard model with explicit Lorentz and CPT violating extension had been constructed and hence a quantitative physical theory of studying Lorentz and CPT violation was furnished [4]. Predictions on the possible Lorentz and CPT violation from this extended Standard Model can be tested by high-precision measurements in numerous existing experiments and possibly in next generation accelerator [5]. In addition, the spontaneous breaking of Lorentz symmetry is actually a natural consequence of string theory since it generally involves interactions that make a Lorentz tensor get non-zero vacuum expectation value [6].

In this talk I shall concentrate on a typical quantum field theory problem, namely, whether the CS term in Eq. (2) can be induced from quantum correction with a modified fermionic sector

$$\mathcal{L}_{\text{fermion}} = \overline{\psi}(i\slashed{\partial} - e\slashed{A} - \slashed{b}\gamma_5 - m)\psi, \tag{3}$$

where b_μ is a constant prescribed four-vector. The new introduced gauge invariant interaction term between constant vector b_μ and axial vector current $j_\mu^5(x) = \overline{\psi}\gamma_\mu\gamma_5\psi$ violates Lorenz and CPT symmetries explicitly, since b_μ picks up a fixed direction in space-time. If the CS term can be induced from the radiative correction with the coefficient $k_\mu \propto b_\mu$, a constraint on $\mathcal{L}_{CS}$ from astrophysical observation will restrict a possible Lorentz and CPT violation in the fermionic sector.

In section 2 we shall review various results on radiatively induced CS term, and then analyze the possible origin for this ambiguity in section 3. Finally we summarize and discuss the joint of astrophysical observation with theoretical prediction on Lorentz and CPT violation and some relevant problems.

CONTROVERSIAL AND AMBIGUOUS RESULTS ON RADIATIVELY INDUCED CHERN-SIMONS-LIKE TERM

As a general procedure, the quantum effective action can be obtained by integrating out fermionic fields,

$$\begin{aligned} e^{i\Gamma[A,b]} &= \int D\overline{\psi}D\psi e^{i\int d^4 x \overline{\psi}(i\slashed{\partial} - e\slashed{A} - \slashed{b}\gamma_5 - m)\psi} \\ &= \det(i\slashed{\partial} - e\slashed{A} - \slashed{b}\gamma_5 - m); \\ \Gamma[A,b] &= -i\text{Tr}\ln(i\slashed{\partial} - e\slashed{A} - \slashed{b}\gamma_5 - m). \end{aligned} \tag{4}$$

The radiatively induced Chern-Simons term will be b-linear and parity-odd part of above effective action.

It is well known that $\Gamma[A,b]$ or equivalently the relevant fermionic determinant cannot be evaluated exactly. A perturbative expansion or certain approximation must be utilized and a use of regularization scheme must be made in the calculation. At first sight, the evaluation of CS term in the quantum effective action is a simple quantum field theory problem. However, the concrete calculation turned out to be rather non-trivial and the result presented remarkable ambiguities: distinct relations between k_μ and b_μ can yield depending on concrete calculation schemes. The various ambiguous results are listed in the following:

-
$$k_\mu = 0. \tag{5}$$

Coleman and Glashow argued first that the CS term cannot be generated [7]. They considered that the axial vector current $j_\mu^5(x)$ should keep gauge invariant in the quantum theory at any momentum or equivalently at any space-time point. Since $\langle j_\mu^5(x) \rangle = \delta \mathcal{L}(x)/\delta b_\mu$, this hypothesis is actually equivalent to the requirement that the Lagrangian density corresponding to the quantum effective action should be gauge invariant. Thus, based on this requirement, the CS term cannot be generated since its Lagrangian density is explicitly not invariant under gauge transformation $A_\mu \to A_\mu + \partial_\mu \Lambda$. Furthermore, Bonneau studied the renormalization of an extended QED including the CS term (2) and the modified fermionic sector (3). He found that Ward identities and the renormalization conditions determine inevitably the absence of CS term from quantum correction [8]. In fact, when deriving Ward identities, Bonneau introduced external source fields for the axial vector current and the CS term, so the Ward identities he derived actually impose gauge invariance on Lagrangian density and hence coincide with the "no-go theorem" requirement argued by Coleman and Glashow [7]. Recently, Adam and Klinkhamer put forward an independent line of reasoning the vanishing of radiatively induced CS term instead of gauge symmetry [9]. They argued that if the perturbative expansion near $b^2 = 0$ is valid and further the gauge field A_μ is regarded as a quantized dynamical field rather than an external background field, the presence of CS term with a purely time-like coefficient may violate the causality principle of a quantum field theory. This argument thus excluded the radiative induction of CS term. In addition, explicit perturbative calculations in Pauli-Villars regularization [4] and dimensional regularization [8] showed that k_μ should vanish.

-
$$k_\mu = \frac{3e^2}{16\pi^2} b_\mu. \tag{6}$$

However, Jackiw and Kostelecký thought that since $j_\mu^5(x)$ only couples with a constant 4-vector b_μ, it is true to require only that $j_\mu(x)$ with zero-momentum (i.e. $\int d^4x\, j_\mu^5(x)$) is gauge invariant at quantum level. Since $\langle \int d^4x\, j_\mu^5(x) \rangle = \delta \Gamma / \delta b_\mu$, this statement is equivalent to the requirement that the quantum effective action should be gauge invariant. Thus the dynamical generation of CS term can escape from above "no-go theorem" conjectured by Coleman and Glashow, since the action of CS term is gauge invariant. Based on this observation, Jackiw and Kostelecký calculated the b-linear part of the one-loop vacuum polarization tensor with b_μ-exact propagator. They ingeniously manipulated the linear divergent term in the loop momentum integration by writing it into a finite term plus an external momentum derivative term, and then arrived at the finite result (6). This result was actually obtained earlier by Chung and Oh in calculating the one-loop effective action (4) in terms of dimensional regularizatoion plus derivative expansion [11]. Furthermore, Pérez-Victoria [12] and Chung [13] proved through an explicit calculation on the parity-odd part of one-loop vacuum polarization tensor at zero external momentum that the result (6) stands to any order of b_μ.

$$k_\mu = Cb_\mu, \quad C \text{ being an arbitrary constant.} \tag{7}$$

It was suggested by Jackiw that the perturbative ambiguity for radiatively induced CS term can be revealed quantitatively in a new developed regularization method called differential regularization [14]. This new calculation technique works for a Euclidean field theory in coordinate space. Its invention is based on the observation that the short-distance singularity representing the UV divergence of a primitively divergent Feynman diagram prevent the amplitude from having a Fourier transform into momentum space. So one can regulate such an amplitude by writing its singular term as a derivative of another function, which has a well defined Fourier transform, then performing Fourier transform into momentum space through partial integration and discarding the surface term. In this way one can directly get a renormalized amplitude. The great advantages of this regularization over conventional regularization schemes lie in that it does not modify the original classical action and in particular, it does not impose or violate gauge symmetry in the process of calculation. Only at the end of calculation, one can get the preferred symmetry by an appropriate choice on indefinite renormalization scales. This is the reason why differential regularization can quantitatively show the perturbative ambiguity. The b-linear part of one-loop vacuum polarization tensor in differential regularization reads [15]

$$\Pi_{\mu\nu}^{(b)}(x) = \frac{1}{2\pi^4} b_\lambda \varepsilon_{\lambda\mu\nu\alpha} \left\{ 4\pi^2 \ln \frac{M_1}{M_2} \frac{\partial}{\partial x_\rho} \delta^{(4)}(x) + m^3 \frac{\partial}{\partial x_\rho} \left[\frac{K_1(mx)K_0(mx)}{x} \right] \right\}. \tag{8}$$

The corresponding Fourier transform is thus

$$\Pi_{\mu\nu}^{(b)}(p) = \int d^4 x e^{-ip\cdot x} \Pi_{\mu\nu}(x) = \frac{2}{\pi^2} b_\lambda \varepsilon_{\lambda\mu\nu\rho} i p_\rho \left[\ln \frac{M_1}{M_2} \right.$$
$$\left. + \frac{m}{4p\sqrt{1+p^2/(4m^2)}} \ln \frac{\sqrt{1+p^2/(4m^2)}+p/(2m)}{\sqrt{1+p^2/(4m^2)}-p/(2m)} \right], \tag{9}$$

and the CS term is relevant to above polarization tensor at low-energy limit,

$$\left.\Pi_{\mu\nu}^{(b)}(p)\right|_{p^2=0} = \frac{2i}{\pi^2} \varepsilon_{\rho\mu\nu\lambda} b_\rho p_\lambda \left(\ln \frac{M_1}{M_2} + \frac{1}{4} \right). \tag{10}$$

Since M_1 and M_2 are two arbitrary renormalization scales, the above result shows that the relation between k_μ and b_μ is completely arbitrary. Furthermore, the same conclusion was drawn by Chung [16] through an analysis on the non-invariance of path integral measure under axial vector gauge transformation as Fujikawa's method [17] of evaluating chiral anomaly, and then an explicit calculation on the b-linear part of the vacuum polarization tensor with the action manifesting the non-invariance under axial vector gauge transformation.

$$k_\mu = \frac{e^2}{8\pi^2} b_\mu \qquad (11)$$

It was found by Chan [18], when adopting the covariant derivative expansion [19] to evaluate the anomalous contribution to the effective action (4), that due to the noncommutativity of the operators ∂ and $A_\mu(x)$ there arises a non-Feynman diagram contribution to the b-linear part of the vacuum polarization tensor. This additional term looks quite exotic from the viewpoint of perturbation theory, and it seems to represent a somehow non-perturbative contribution. As a consequence, the result (6) was modified to $e^2/(8\pi^2)b_\mu$. The basic idea of covariant derivative expansion is to express local quantum effective Lagrangian in powers of gauge covariant derivative $\Pi_\mu = i\partial_\mu - eA_\mu$ rather than in powers of $i\partial_\mu$ and A_μ separately [19]. The remarkable difference between (6) and (11) and the feature of covariant derivative expansion motivated us to re-calculate the b-linear part of the effective action (4) in the technique of Schwinger's constant field approximation [20] since this method shares the same feature as covariant derivative expansion. The essence of this method is converting the calculation of an effective action as (4) into solving a harmonic oscillator problem in non-relativistic quantum mechanics. We found that if the following trace condition is satisfied,

$$\lim_{x' \to x} \frac{(x-x')_\mu (x-x')_\nu}{(x-x')^2} = \lim_{x' \to x} \frac{1}{\int d^4 x'} \int d^4 x' \frac{(x-x')_\mu (x-x')_\nu}{(x-x')^2} = \frac{1}{4} g_{\mu\nu}, \qquad (12)$$

then the same result as in covariant derivative expansion can be reproduced [21],

$$\langle J_\mu(x) \rangle = \frac{\delta \Gamma_{CS}}{\delta A_\mu(x)} = \frac{e^2}{4\pi^2} \left\{ \exp\left[-ie \int_{x'}^{x} dy^\rho A_\rho(y) \right] \frac{m K_1([-m^2(x-x')^2]^{1/2}}{[-m^2(x-x')^2]^{1/2}} \right.$$

$$\left. \times \, (x-x')_\mu (x-x')_\rho b_\nu \varepsilon^{\rho\nu\alpha\beta} F_{\alpha\beta} \right\} \bigg|_{x' \to x} = -\frac{e^2}{16\pi^2} \varepsilon^{\mu\nu\lambda\rho} b_\nu F_{\lambda\rho}, \qquad (13)$$

$$\Gamma_{CS} = \frac{e^2}{16\pi^2} \int d^4 x \, \varepsilon^{\mu\nu\lambda\rho} b_\mu A_\nu F_{\lambda\rho}. \qquad (14)$$

However, it should be emphasized that the limit given in Eq. (12) has a potential ambiguity and the general result will be that $\lim_{x \to 0} x_\mu x_\nu / x^2 = C g_{\mu\nu}$, thus the induced CS term is actually ambiguous. Furthermore, this ambiguous result was confirmed by Chungs to any order of b_μ in the same method [22].

ORIGIN OF AMBIGUITY

Four finite but entirely different results on radiatively induced CS term are shown in last section. Two convincing explanations for this ambiguity were proposed by Sitenko [23] and Pérez-Victoria [24]. The former emphasized the calculation technique cause, while the latter indicated a theoretical origin.

The explanation proposed by Sitenko [23] concerns with two different formulations of the quantum effective action (4) adopted by Chaichian et al [21] and Chung et al [11],

$$\Gamma^{(1)} = -i\text{Tr}\ln(i\slashed{\partial} - \slashed{A} - m) + i\int_0^1 dz\,\text{Tr}\left[\left(i\slashed{\partial} - \slashed{A} - z\gamma^5\slashed{b} - m\right)^{-1}\gamma^5\slashed{b}\right], \quad (15)$$

$$\Gamma^{(2)} = -i\text{Tr}\ln\left(i\slashed{\partial} - \gamma^5\slashed{b} - m\right) + i\int_0^1 dz\,\text{Tr}\left[\left(i\slashed{\partial} - z\slashed{A} - \gamma^5\slashed{b} - m\right)^{-1}\slashed{A}\right]. \quad (16)$$

Their b-linear sectors read, respectively,

$$i\int_0^1 dz\,\text{Tr}\left[\left(i\slashed{\partial} - \slashed{A} - z\gamma^5\slashed{b} - m\right)^{-1}\gamma^5\slashed{b}\right]_{(b)} = \frac{1}{16\pi^2}b_\mu\int d^4x\,\varepsilon^{\mu\nu\lambda\rho}F_{\nu\lambda}A_\rho$$

$$+ \frac{1}{8\pi^2}b_\mu I_\nu\int d^4x\,\varepsilon^{\mu\nu\lambda\rho}F_{\lambda\rho} + \frac{1}{8\pi^2}I_{\alpha\beta}\left(g^{\beta\nu}\varepsilon^{\mu\alpha\rho\lambda} + g^{\beta\lambda}\varepsilon^{\mu\nu\rho\alpha}\right)b_\mu\int d^4x\,A_\nu\partial_\rho A_\lambda, \quad (17)$$

$$i\int_0^1 dz\,\text{Tr}\left[\left(i\slashed{\partial} - z\slashed{A} - \gamma^5\slashed{b} - m\right)^{-1}\slashed{A}\right]_{(b)} = \frac{3}{32\pi^2}b_\mu\int d^4x\,\varepsilon^{\mu\nu\lambda\rho}F_{\nu\lambda}A_\rho$$

$$+ \frac{1}{8\pi^2}I_{\alpha\beta}\left(g^{\beta\nu}\varepsilon^{\mu\alpha\rho\lambda} + g^{\beta\lambda}\varepsilon^{\mu\nu\rho\alpha} + g^{\beta\rho}\varepsilon^{\mu\nu\rho\alpha}\right)b_\mu\int d^4x\,A_\nu\partial_\rho A_\lambda, \quad (18)$$

where

$$I_\mu \equiv \frac{1}{i\pi^2}\int d^4k\,\frac{k_\mu}{(k^2-m^2)^2},$$

$$I_{\mu\nu} \equiv \frac{1}{i\pi^2}\int d^4k\,\frac{4k_\mu k_\nu - k^2 g_{\mu\nu}}{(k^2-m^2)^3} = \frac{1}{2}g_{\mu\nu} - \frac{1}{i\pi^2}\int d^4k\,\frac{\partial}{\partial k^\mu}\left[\frac{k_\nu}{(k^2-m^2)^2}\right] \quad (19)$$

are two momentum integrals with superficially linear and logarithmic divergence, respectively. Due to the explicit breaking of Lorentz symmetry, one cannot naively put $I_\mu = 0$ and use the formula $\int d^n k\,k_\mu k_\nu f(k^2) = 1/n\,g_{\mu\nu}\int d^n k\,k^2 f(k^2)$. The approaches of regularization schemes manipulating these two divergent integrals lead to above finite ambiguities on the radiatively induced CS term [23]:

- If a regularization scheme defines $I_\mu = 0$ and imposes the trace condition $g^{\mu\nu}I_{\mu\nu} = 0$, this actually yields $I_{\mu\nu} = 0$ due to the fact $I_{\mu\nu} \propto g_{\mu\nu}$, the b-linear parts of $\Gamma^{(1)}$ and $\Gamma^{(2)}$ will yield the results obtained by Jackiw et al and Chan et al, respectively,

$$\Gamma^{(1)}_{\text{CS}} = \frac{1}{16\pi^2}b_\mu\int d^4x\,\varepsilon^{\mu\nu\lambda\rho}F_{\nu\lambda}A_\rho; \quad \Gamma^{(2)}_{\text{CS}} = \frac{3}{32\pi^2}b_\mu\int d^4x\,\varepsilon^{\mu\nu\lambda\rho}F_{\nu\lambda}A_\rho. \quad (20)$$

- In a regularization scheme defining $I_\mu = 0$ and $I_{\mu\nu} = 1/2 g_{\mu\nu}$ (i.e. omitting the surface term of $I_{\mu\nu}$), the conclusion argued by Coleman and Glashow will be reproduced,

$$\Gamma^{(1)}_{\text{CS}} = \Gamma^{(2)}_{\text{CS}} = 0. \quad (21)$$

- If defining the one-loop quantum effective action as an arbitrary combination of (15) and (16) and choosing a regularization scheme imposing $I_\mu = I_{\mu\nu} = 0$, one can get the result obtained by Chen in differential regularization,

$$\Gamma_{CS} = (1-c)\Gamma_{CS}^{(1)} + c\Gamma_{CS}^{(2)} = \frac{2+c}{32\pi^2}b_\mu \int d^4x \varepsilon_{\mu\nu\lambda\rho}F_{\nu\lambda}A_\rho \equiv Cb_\mu \int d^4x \varepsilon_{\mu\nu\lambda\rho}F_{\nu\lambda}A_\rho. \quad (22)$$

The above explanation shows that the origin for the finite ambiguous CS term lies in the inequivalence between two formulations (15) and (16) of the quantum effective action (4) and the ambiguity of a regularization method in manipulating logrithmically and linearly divergent loop momentum integrals.

Pérez-Victoria [24] put forward another explanation through revealing a relation between the radiatively induced CS coefficient and triangle chiral anomaly via an intermediate model having spontaneous breaking of Lorentz and CPT symmetries. The fermionic sector of this model takes following form [24],

$$\mathcal{L}'_{\text{fermion}} = \overline{\psi}\left(i\slashed{D} - m + \frac{\tilde{b}}{\Lambda}\gamma_5\slashed{\partial}\phi + ic\gamma_5\phi + \frac{d}{\Lambda}\phi^2\right)\psi, \quad (23)$$

where ϕ is a real pseudoscalar field (i.e., an axion), Λ is certain large scale and $\tilde{b}$, c and d are indefinite parameters. In the choice that

$$c = d = 0, \quad \langle\phi\rangle = \frac{\Lambda}{\tilde{b}}b_\mu x^\mu, \quad (24)$$

the above model will restore the fermionic Lagrangian (3).

As initially argued by Coleman and Glashow [7], the radiatively induced CS coefficient $\tilde{k}$ in the model (23) can be detected by evaluating the quantum vertex $\Gamma_{\mu\nu}(p,q)$ composed of one axion and two photons [24], i.e., the 1PI part of the correlation function $\langle A_\mu(p)A_\nu(q)\phi(-p-q)\rangle$,

$$\Gamma^{\mu\nu}(p,q) = \varepsilon^{\mu\nu\lambda\rho}p_\lambda q_\rho C(p,q), \quad \tilde{k} = -\frac{\Lambda}{2}C(0,0). \quad (25)$$

The Lagrangian (23) shows that the $\tilde{b}$-linear part of $\Gamma_{\mu\nu}(p,q)$ is explicitly related to the triangle amplitude $V^{\mu\nu\rho}(p,q) = \langle j^\mu(p)j^\nu(q)j_5^\rho(-p-q)\rangle$,

$$\Gamma^{\mu\nu}_{\tilde{b}}(p,q) = \frac{\tilde{b}}{\Lambda}e^2(p_\rho + q_\rho)V^{\mu\nu\rho}(p,q). \quad (26)$$

It is well known that $V^{\mu\nu\rho}$ satisfies the celebrated anomalous Ward identity,

$$(p_\rho + q_\rho)V^{\mu\nu\rho}(p,q) = 2imV^{\mu\nu}(p,q) + \varepsilon^{\mu\nu\lambda\rho}p_\lambda q_\rho \mathcal{A}, \quad (27)$$

where $\mathcal{A}$ is the chiral anomaly coefficient. Further, the tensor structure of the canonical term $V^{\mu\nu}(p,q)$, which comes from an explicit breaking of chiral symmetry by fermionic mass term, takes the following form,

$$V^{\mu\nu}(p,q) = \langle j^\mu(p)j^\nu(q)j_5(-p-q)\rangle = \varepsilon^{\mu\nu\lambda\rho}p_\lambda q_\rho V(p,q), \quad j_5 \equiv \overline{\psi}\gamma_5\psi. \quad (28)$$

Eqs. (25)-(28) establish a relation among the form factor $C_{\widetilde{b}}(p,q)$ of the $\widetilde{b}$-linear sector of $\Gamma^{\mu\nu}(p,q)$, the form factor $V(p,q)$ of the canonical term in the anomalous Ward identity (27) and the anomaly coefficient [24],

$$C_{\widetilde{b}} = \frac{\widetilde{b}}{\Lambda}e^2\left[2mV(p,q) + \mathcal{A}\right]. \qquad (29)$$

The canonical term $V^{\mu\nu}(p,q)$ in the anomalous Ward (27) identity is finite and unambiguous. A comparison between $\langle j^\mu(p)j^\nu(q)j_5(-p-q)\rangle$ and $\Gamma_{\mu\nu}(p,q)$ shows that $V^{\mu\nu}(p,q)$ is actually equal to the c-linear part of $\Gamma^{\mu\nu}(p,q)$. It turned out that the c-linear part of $\Gamma^{\mu\nu}(p,q)$ could be easily calculated since it is convergent [24],

$$\widetilde{k}_c = -\frac{ce^2\Lambda}{2}V(0,0) = \frac{ce^2\Lambda}{8\pi^2 m}. \qquad (30)$$

An insertion of Eqs. (25) and (30) into (29) immediately lead to a relation among $k_{\widetilde{b}}$, i.e. the coefficient of $\widetilde{b}$-linear part of $\Gamma^{\mu\nu}(p,q)$ defined on photon mass-shell, the parameter $\widetilde{b}$ and the anomaly coefficient $\mathcal{A}$,

$$\widetilde{k}_{\widetilde{b}} = e^2\widetilde{b}\left(\frac{1}{4\pi^2} - \frac{1}{2}\mathcal{A}\right). \qquad (31)$$

Upon choosing the parameters shown in (24), one can find a relation between radiatively induced CS coefficient of the model (3) and chiral anomaly coefficent [24],

$$k_\mu = e^2 b_\mu \left(\frac{1}{4\pi^2} - \frac{1}{2}\mathcal{A}\right). \qquad (32)$$

The origin of the ambiguity on radiatively induced CS coefficient is thus revealed since the chiral anomaly coefficient is ambiguous and regularization dependent. This fact was explicitly and quantitatively shown in differential regualrization [14, 25].

SUMMARY AND DISCUSSION

The issue on radiatively induced Lorentz and CPT violation in quantum electrodynamics is reviewed and it has not been completely settled down yet. In this talk we emphasize how the CS term is induced from quantum correction. The physical effects it causes in classical electrodynamics were described by Jackiw [26]. Since the present astrophysical observation data has excluded the physical consequence of the CS term, thus no matter how this k_μ arises, either as a radiative correction induced from the fermionic sector or as free parameter set up by hand, it must vanish. Here let us discuss how to "input" this ambiguous quantum correction to compare with astrophysical observation. There are two approaches in the literature to include this CS term. One is starting from the conventional QED plus an explicit Lorentz and CPT violating in the fermionic sector, i.e.

$$\mathcal{L} = -\frac{1}{4}F_{\mu\nu}F^{\mu\nu} + \overline{\psi}(i\slashed{\partial} - e\slashed{A} - m)\psi - \overline{\psi}\slashed{b}\gamma_5\psi = \mathcal{L}_{\text{QED}} - \mathcal{L}_b, \qquad (33)$$

and the CS term (2) will be induced from quantum correction with the coefficient $k_\mu \propto e^2 b_\mu$. The other one is introducing the CS term (2) at classical level,

$$\widetilde{\mathcal{L}} = \mathcal{L}_{\text{QED}} - \mathcal{L}_b + \mathcal{L}_{\text{CS}}, \tag{34}$$

and the coefficient of CS term is a free parameter. The *radiatively induced* CS terms calculated in these two models have different meanings from the viewpoint of renormalization theory, despite that the processes of calculating CS term and the results are identical. In the framework described by $\mathcal{L}$, the induced CS term can only be considered as a radiative correction, while in the later model $\widetilde{\mathcal{L}}$, depending on the renormalization condition, the induced CS term can be cancelled by a finite counterterm and keep the classical parameter k_μ as the renormalized one. According to the perturbation theory of a renormalizable quantum field theory, a quantum correction calculated in certain regularization scheme, now matter how it is, finite or infinite, has no physical meaning before a renormalization procedure is implemented. Only when a renormalization condition is assigned, the quantum correction is decomposed into two parts, one part will be cancelled by certain counterterm and absorbed into the classical Lagrangian to redefine the various parameters such as mass or coupling constant, the other part is the radiative correction and reflects the observable quantum effects. Based on this fact, it can be easily seen that in the first model the induced CS term cannot be cancelled by introducing a counterterm since its has no counterpart in the classical Lagrangian, hence it can only belong to the radiative correction like chiral anomaly and anomalous magnetic moment etc. However, the induced CS term, despite of being a radiative correction, has an essential difference with anomalous magnetic moment and chiral anomaly: it can not be ambiguously fixed by the principle of a quantum field theory itself. Anomalous magnetic moment can be uniquely evaluated by a gauge invariant regularization scheme. Chiral anomaly, despite that it is ambiguous, can be determined if vector gauge symmetry is required. Whereas for the induced CS term, it seems that gauge symmetry cannot dominate it [24]. The ambiguity of the induced CS term make the theory (33) awkward, since this means that the theory cannot make a definite prediction on the quantum phenomena. If we recall the explanation by Sitenko [23] on the origin of this ambiguity, this specific example seems to imply an inkling that a quantum field theory, as the most successful framework of describing subatomic physics up to now, has an intrinsic deficiency in certain specific situation such as Lorentz symmetry breaking [27]. This speculation might be comprehensible, since a relativistic quantum field theory was born out of a combination between quantum theory and the theory of special relativity. If Lorentz symmetry, the algebraic soul of special relativity, collapsed, what could one expect from a relativistic quantum field model! Of course, it may not be so serious, since there exists a possible way out for the model (33), namely, considering the contribution to the CS coefficient from all the fermion species in the Standard Model [11], $k_\mu \propto \sum_i e_i^2 b_\mu^i$, the index i summing over all the leptons and quarks of the Standard Model. If b_μ^i come from the vacuum expectation value $\langle A_\mu^5 \rangle$ of an axial vector gauge field, then the induced CS coefficient may vanish according to the anomaly cancellation in the Standard Model. However, as indicated by Chung et al [11], it is also possible that b_μ may not be related to the vacuum expectation value of an axial vector field.

In contrast, the second setting (34) is more appealing in discussing the physical effects of CS term. Since CS term is put by hand at the classical level, one can introduce a finite counterterm to define renormalized CS coefficient [8, 24]

$$k^\mu_{ren} = k^\mu_{bare} + k^\mu_{quant} + k^\mu_{counter}, \qquad (35)$$

and then input $k^\mu_{ren} = 0$ to yield to the astrophysical observation data. However, in this case, both k^μ_{ren} and b_μ are regarded as independent measurable parameters, k^μ_{ren} has nothing with b_μ, and hence the astrophysical observation data on the vanishing of k_μ does not put any constraint on b_μ [8, 24]. A fine-tuning is required to get a vanishing k_μ and non-vanishing b_μ at the same time since the CS term can be generated from quantum correction with $k_\mu \propto b_\mu$ [24].

Finally, a conclusion from an investigation on the anomalous magnetic moment and Lamb shift in QED with the extended fermionic sector (3) should be emphasized. It was found [28] that both the anomalous magnetic moment and Lamb shift receive additional IR divergent radiative correction proportional b^2. Furthermore, it was explicitly shown that even the IR divergence in the Lamb shift cannot be cancelled by the bremsstrahlung process as in the conventional QED [28], let alone eliminating the IR divergence in the anomalous magnetic moment. Since the anomalous magnetic moment and Lamb shift are two successful symbols of QED in describing electromagnetic interaction, the IR divergence embracing them seems thus to reflect physical inconsistency of QED with above extended fermionic sector, and therefore, ruins the mechanism of generating CS term from radiative correction by introducing an explicit Lorentz and CPT violating term in the fermionic sector. Of course, there is a possibility that the calculation on the vertex correction [28] has some drawbacks, since we only expanded the b-exact propagators to the second order, perhaps a summation to any order of b_μ might erase such an IR divergence.

ACKNOWLEDGMENTS

I would like to thank organizers of MRST 2001, Profs. V. Elias, D.G.C. McKeon and V.A. Miransky for hospitality. I am grateful to Prof. R. Jackiw for suggesting me this project and useful discussions. I am indebted to Prof. M. Chaichian, G. Kunstatter and Dr. R. González Felipe for collaboration. I would like to thank Profs. G. Bonneau, L.H. Chan, V.A. Kostelecký, G. Leibbrandt and R.B. Mann for communications and discussions. I am especially obliged to Dr. M. Pérez-Victoria for his continuous discussions and comments. This work was supported by NSERC of Canada.

REFERENCES

1. Nodland, B., and Ralston, J. P., *Phys. Rev. Lett.* **78**, 3043 (1997).
2. Carroll, S. M., Field, G. B., and Jackiw, R., *Phys. Rev.* **D41**, 1231 (1990).
3. Carroll, S. M., and Field, G.B., *Phys. Rev. Lett.* **79**, 234 (1997).
4. Colladay, D., and Kostelecký, V. A., *Phys. Rev.* **D58**, 116002 (1998).

5. For a review, see talks in *CPT ad Lorentz Symmetry*, ed. Kostelecký, V. A., World Scientific, Singapore, 1999.
6. Kostelecký, V. A., and Samuel, S., *Phys. Rev. Lett.* **63**, 224 (1989); *Phys. Rev.* **D39**, 683 (1989); *Phys. Rev.* D40, 1886 (1989).
7. Coleman, S., and Glashow, S. L., *Phys. Rev.* **D59**, 116008 (1999).
8. Bonneau, G., *Nucl. Phys.* **B593**, 398 (2001).
9. Adam, C., and Klinkhamer, F. R., *Causality and Radiatively Induced CPT Violation*, hep-th/0105037.
10. Jackiw, R., and Kostelecký, V.A., *Phys. Rev. Lett.* **82**, 3572 (1999).
11. Chung, J. M., and Oh, P., *Phys. Rev.* **D60**, 067702 (1999).
12. Pérez-Victoria, M., *Phys. Rev. Lett.* **83**, 2518 (1999).
13. Chung, J. M., *Phys. Lett.* **B461**, 318 (1999).
14. Freedman, D. Z., Johnson, K., and Latorre, J. I., *Nucl. Phys.* **B371**, 353 (1992).
15. Chen, W. F., *Phys. Rev.* **D60**, 085007 (1999).
16. Chung, J. M., *Phys. Rev.* **D60**, 127901 (1999).
17. Fujikawa, K., *Phys. Rev. Lett.* **42**, 1195 (1979); *Phys. Rev.* **D21**, 2848 (1980).
18. Chan, L. H., *Induced Lorentz-violating Chern-Simons Term in QED and Anomalous Contributions to Effective Action Expansions*, hep-ph/9907349.
19. Chan, L. H., *Phys. Rev. Lett.* **57**, 1199 (1986); Gaillard, M. K., *Nucl. Phys.* **B268**, 669 (1986).
20. Schwinger, J., *Phys. Rev.* **82**, 664 (1951).
21. Chaichian, M., Chen, W. F., and González Felipe, R., *Phys. Lett.* **B503** 215 (2001).
22. Chung, J. M., and Chung, B. K., *Phys. Rev.* **D63**, 105015 (2001).
23. Sitenko, Yu.A., *One-loop Effective Action for the Extended Spinor Electrodynamics with Violation of Lorentz and CPT Symmetry*, hep-th/0103215, v1.
24. Pérez-Victoria, M. *JHEP* **0104**: 032 (2001).
25. Haagensen, P. E., and Latorre, J. I., *Ann. Phys.* **221**, 77 (1993); Chen, W. F., *Phys. Lett.* **B459**, 242 (1999).
26. Jackiw, R., *Comments Mod. Phys.* **A1**, 1 (1999).
27. *There is a comment from Dr. M. Pérez-Victoria on this argument: According to the prescription of renormalization theory, the bare Lagrangian should consist of all the possible renormalizable terms allowed by gauge symmetry, while $\mathcal{L}_{CS}$ is just such a term. Thus the Lagrangian (33) is actually incomplete and one must start from the classical Lagragian (34) to discuss the quantum correction. This is the reason why the quantum theory based on (33) is not well defined. Therefore, there should exist no problem for a quantum field theory with the breaking of Lorenz symmetry if one start from the complete Lagrangian (34). The model (33) can be understood as a special case of (34) with the renormalized k_μ happening to to zero. My comprehension is that (33) and (34) are just two different model construction and they describe distinct classical electrodynamics. The quantum phenomena of these two models thus have different intepretation. Of course, classical dynamics is an approximation of the quantum counterpart. In this sense, the model (34) is more appealing since it has a well defined qunatum theory. However, in general, it is the model (33) rather than (34) that can arise from the low-energy limit of a more fundamental theory.*
28. Chen, W. F., and Kunstatter, G., *Phys. Rev.* **D62**, 105029 (2000).

Thermal Conductivity of the 2+1-Dimensional NJL Model in an External Magnetic Field

Efrain J. Ferrer[1], Valery A. Gusynin[2] and Vivian de la Incera[1,*]

[1]*Physics Department, SUNY-Fredonia, Fredonia, NY 14063*
[2]*Bogolyubov Institute for Theoretical Physics, Kiev, 03143 Ukraine*

Abstract. The thermal conductivity κ of a 2+1-dimensional NJL theory, proposed to modeling quasiparticle interactions around the nodes of a d-wave high-temperature superconductor, is calculated as a function of an externally applied magnetic field. Taking into account the dynamical generation of a gap that takes place in this theory at a critical field $B_c(T) \sim T^2$, the magnetic-field-dependence of the thermal conductivity is found to exhibit a sharp break at B_c, followed by a region where κ becomes field independent (plateau region). These results qualitatively reproduce Krishana's et. al. experimental observations of high Tc cuprates.

In this work it is shown how the description of the quasiparticle (QP) interactions about the nodal points of a d-wave superconductor in terms of a rather simple relativistic 2+1-dimensional four-fermion interaction model (Nambu-Jona-Lasinio (NJL) type) can numerically reproduce the intriguing behavior [1] of the thermal conductivity of high-Tc cuprates' in the presence of an externally applied magnetic field. The unusual feature, first observed in experiments done by Krishana et. al. [1-2], and later reproduced by an independent team [3], consists on the appearance of a sharp break (kink) in the profile of the thermal conductivity $\kappa(B)$ below T_c when a magnetic field perpendicularly applied to the cuprate planes reaches a critical value that depends on the temperature. The observed kink is then followed by a plateau region where κ becomes insensitive to the magnetic field strength. Typically, the critical magnetic field in all the measured samples is of order T^2.

As shown below, the kink may be related to the dynamical generation of a gap induced by the magnetic field, which is just the manifestation of a phenomenon of universal character known as magnetic catalysis (MC) [4-6] and to the enhancement of the transitions between the zeroth and the first Landau levels (LL). Most of the results here presented are based on our recent paper [7].

Previous to our own calculations, several attempts to explain Krishana's et al. observations were made [8-13], but they failed in reproducing the observed kink. Our main goal here is therefore to discuss a mechanism that generates the kink effect within the framework of the MC phenomenon. We present for the first time a consistent calculation of the thermal conductivity in the presence of an external magnetic field in a model with the simplest four-fermion interaction for quasiparticles. We use the same constant magnetic field approximation that was already explored in [8,9] to calculate the thermal conductivity. However, our calculation deviates

---
* Conference Speaker

considerably from what was done in [8,9]. Not only we take into account the contribution of all Landau levels, but the definition of the heat current itself is different. In the limit of narrow width of quasiparticles ($\Gamma \ll T, \sqrt{B}$), after a gap is opened, the thermal conductivity exhibits a new term proportional to σ^2 (σ is the gap). This term originates from the compensation of the matrix elements of transitions between the zeroth and the first LLs (behaving as 1/B) and the LL density of states which in turn is proportional to B. This is one more manifestation of the MC phenomenon: not only a gap is induced by the magnetic field, but the transitions between the zeroth and the first LLs are enhanced. In mean-field approximation and near the phase transition point the gap behaves like $\sigma \cong 0.523\sqrt{eB - eB_c}$ what yields a positive contribution into the slope of the thermal conductivity leading to a jump (kink effect) of $\kappa(B)$ at $B = B_c$.

Our starting point is the 2+1-dimensional NJL model in an external magnetic field, which is believed to possess the symmetries of the d-wave hamiltonians defined on a lattice [9]

$$L = \overline{\psi}_a [i\gamma^0 \hbar \frac{\partial}{\partial t} + iv_D \gamma^i (\hbar \frac{\partial}{\partial x^i} - \frac{e}{c} A_i)]\psi_a + \frac{g}{2N}(\overline{\psi}_a \psi_a)^2 \quad (1)$$

In (1) the vector potential is taken in the symmetric gauge $A_\mu = (0, -\frac{B}{2}x_2, \frac{B}{2}x_1), v_D = \sqrt{v_F v_\Delta}$ with v_F, v_Δ being the velocities perpendicular and tangential to the Fermi surface respectively. They originate from the quasiparticle excitation spectrum in the vicinity of the gap nodes that takes the form of an anisotropic Dirac cone $E(k) = \sqrt{v_F^2 k_1^2 + v_\Delta^2 k_{21}^2}$. After rescaling coordinates this leads to Eq. (1). In what follows we take $\hbar = v_D = k_B = 1$ and absorb c into the charge e. We will restore these constants when needed. We assume also that the fermions carry an additional flavor index a=1,...,N (N=2 for realistic d-wave superconductors). The Dirac γ matrices are taken in a reducible four-component representation.

The Lagrangian density (1) is invariant under the discrete (chiral) symmetry $\psi \to \gamma_5 \psi, \overline{\psi} \to -\overline{\psi}\gamma_5$, which forbids the fermion mass generation in perturbation theory. The mass generation can be studied introducing an auxiliary field $\sigma = -\left(\frac{g}{N}\right)\overline{\psi}_a \psi_a$ by means of the Hubbard-Stratanovich trick that permits one to integrate over fermion fields in the path integral representation of the partition function. The field σ has no dynamics at the tree level but it acquires the kinetic term due to fermion loops. Likewise, a nontrivial minimum of the effective potential (the expectation value of σ) gives mass to fermions and spontaneously breaks the discrete symmetry leading to a neutral condensate of fermion-antifermion pairs. Studying the minimum of the effective potential we find that, at a fixed temperature T, there is a critical value of the magnetic field $\sqrt{eB_c/T} \cong 4.148$ such that for subcritical fields

$eB \leq eB_c$ the gap is zero, while for $eB \geq eB_c$ a nontrivial gap appears. The critical curve equation has the form $(v_D/c)^2 10^{10} B = 21.5 T^2$ for B measured in Tesla. Using the estimated value of the characteristic velocity $v_D = 1.16 \times 10^7 \, cm/s$ reported in [10] we obtain the critical curve $B = 0.014 T^2$ that fits the experimental curve of Ref. [1].

To derive an expression for the static thermal conductivity in an isotropic system, we follow the familiar linear response method and apply Kubo's formula [14]

$$\kappa = -\frac{1}{TV} \operatorname{Im} \int_0^\infty dt\, t \int d^2 x_1 d^2 x_2 <u_i(x_1,0) u_i(x_2,t)> \qquad (2)$$

where V is the volume of the system and $u_i(x,t)$ is the heat-current density operator. The brackets denote averaging in the canonical ensemble. Physically, the thermal conductivity κ appears as a coefficient in the equation relating the heat current to the temperature gradient $\vec{u} = -\kappa \vec{\nabla} T$ under the condition of absence of particle flow. If we neglect the chemical potential, the heat density coincides with the energy density ε, hence the quantity $\vec{u}$ that satisfies the continuity equation $\dot{\varepsilon}(x) + \vec{\nabla} \cdot \vec{u}(x) = 0$ can be interpreted as the heat current. From the Lagrangian density we find $u_i = \frac{i}{2}(\bar{\psi}\gamma_i \partial_0 \psi - \partial_0 \bar{\psi} \gamma_i \psi)$.

Neglecting vertex corrections [15] the calculation of the thermal conductivity reduces to the evaluation of the bubble diagram [16]. One can use the spectral representation

$$A(\omega,\vec{k}) = e^{-\frac{\vec{k}^2}{eB}} \frac{\Gamma}{2\pi} \sum_{n=0}^\infty \frac{(-1)^n}{M_n} [\frac{(\gamma^0 M_n + \sigma) f_1(\vec{k}) + f_2(\vec{k})}{(\omega - M_n)^2 + \Gamma^2} + \frac{(\gamma^0 M_n - \sigma) f_1(\vec{k}) - f_2(\vec{k})}{(\omega + M_n)^2 + \Gamma^2}] \qquad (3)$$

for the fermion Green's function, to write the thermal conductivity as

$$\kappa = \frac{1}{32\pi T^2} \int_{-\infty}^\infty \frac{d\omega\, \omega^2}{\cosh^2 \frac{\omega}{2T}} \int d^2 k\, tr[\gamma^i A(\omega,\vec{k}) \gamma^i A(\omega,\vec{k})] \qquad (4)$$

In Eq. (3) the notation

$$M_n = \sqrt{\sigma^2 + 2eBn},$$
$$f_1(\vec{k}) = 2[P_- L_n(s) - P_+ L_{n-1}(s)], \qquad (5)$$
$$f_2(\vec{k}) = 4\vec{k} \cdot \vec{\gamma} L^1_{n-1}(s)$$

was introduced. Here $P_{\pm} = (1 \pm i\gamma^1\gamma^2)/2$ are projector operators, L_n, L'_n are Laguerre's polynomials, $s = 2\vec{k}^2/eB$, σ is the dynamical fermion mass obtained from the finite temperature gap equation in the magnetic field, and Γ is the QP width, which is due to interaction processes, in particular, scattering on impurities.

After several cumbersome calculations [7] we arrive at the following final expression for κ

$$\kappa = \frac{N\Gamma^2}{2\pi^2 T^2} \int_0^\infty \frac{d\omega \omega^2}{\cosh^2 \frac{\omega}{2T}} \frac{1}{(eB)^2 + (2\omega\Gamma)^2} \{ 2\omega^2 + \frac{(\omega^2+\sigma^2+\Gamma^2)(eB)^2 - 2\omega^2(\omega^2-\sigma^2+\Gamma^2)(eB)}{(\omega^2-\sigma^2-\Gamma^2)^2 + 4\omega^2\Gamma^2} - \frac{\omega(\omega^2-\sigma^2+\Gamma^2)}{\Gamma} \operatorname{Im}\psi\left(\frac{\sigma^2+\Gamma^2-\omega^2-2i\omega\Gamma}{2eB}\right) \} \qquad (6)$$

expressed through the digamma function $\psi(v)$.

Let us consider the case $\Gamma \ll T$ with $B \neq 0$. In this approximation we obtain

$$\kappa \cong \frac{N\Gamma}{4\pi T^2} \{ \frac{\sigma^2}{\cosh^2 \frac{\sigma}{2T}} + 4\sum_{n=1}^\infty \frac{n(\sigma^2 + 2eBn)}{\cosh^2 \frac{\sqrt{\sigma^2 + 2eBn}}{2T}} \} \qquad (7)$$

Asymptotically, at $\sqrt{eB} \gg \sqrt{eB_c} \cong T$, the dynamical mass behaves as $\sigma \sim \sqrt{eB}$ leading to an exponential decrease as in the case of gapless fermions. However, the term $\cosh^{-2} \sigma/2T$ in Eq. (7) is of order one when $\sigma \leq 2T$, thus, there is no suppression of this term for a certain range of fields where it is almost field independent (plateau region).

Near the phase transition point $\sigma \cong a\sqrt{eB - eB_c}$ and the first term in Eq. (7) gives positive contribution to the slope of the thermal conductivity at $eB \geq eB_c$ leading to the jump in the slope of κ (kink effect) at $eB = eB_c$. The parameter a is model-dependent. For the NJL model (1) we find, in mean-field approximation, $a \approx 0.5$.

The explicit appearance of the kink has been corroborated by numerical calculations as shown in Fig. 1. Notice the break in the slope of κ (kink effect) at the critical value B_c in the presence of σ. For $B > B_c$ the kink is followed by a region where κ is only weakly dependent on the field. While decreasing the temperature, the position of the kink moves to the left in accordance with the critical line $B_c = 0.014T^2$. Fig. 1 has one more similarity with the experimental behavior reported in Ref. [1]: with decreasing T the crossing of the curves occurs in such a way that the lower T-curve reaches the higher value at large fields. Another noticeable characteristic of the numerical results is that the presence of the kink-plateau feature is quite sensitive to the relation between Γ and T. The kink is obtained only for values of Γ smaller, but not much smaller than T. This may provide some insight on the apparent difficulty in reproducing Krishana's et al. data (for instance, Ando et al. [17] observed the kink in only two samples out of more than 30 samples measured and called attention to the need to understand what determines the occurrence of the plateau).

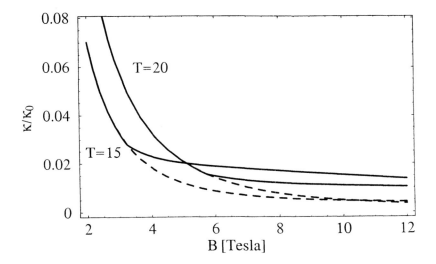

FIGURE 1. The magnetic field dependence of κ at $T = 20K$ and $T = 15K$ in the narrow width case ($\Gamma = 5K$). The solid lines represent κ/κ_0 (κ_0 being the zero-field thermal conductivity) when a QP gap σ is MC-induced at $B \geq B_c(T)$ ($B_c(20) = 5.75$ T, $B_c(15) = 3.23T$). The dashed lines represent the behavior of κ/κ_0 when σ remains zero at $B \geq B_c(T)$.

The first term in Eq. (7) is essential to get the kink effect. One can trace its origin to the contribution of transitions between the zeroth and the first LLs, which at large fields decrease as the first power of the field $\sim 1/eB$, in contrast to transitions between Landau levels with $n \geq 1$, which behaves as $1/(eB)^2$, hence giving a smaller contribution. Since the density of LLs that is proportional to eB multiplies these contributions, this implies that the transitions between the zeroth and the first LLs are not suppressed, even though the gap between the levels grows with the field.

In conclusion, we have shown that the modeling of the QP dynamics of a cuprate superconductor using a 2+1-dimensional NJL Lagrangian is enough to numerically reproduce the unusual behavior of the thermal conductivity in the presence of a magnetic field perpendicularly applied to the cuprate's plane. In the emergence of a kink followed by a plateau in the profile of κ versus B a basic role is played by the MC phenomenon, which gives rise to a second order phase transition with the generation of a dynamical fermion mass at a critical field $B_c(T)$. The MC is also responsible for the enhancement of the contribution of zeroth to first LL's transitions into the thermal conductivity.

ACKNOWLEDGMENTS

It is a pleasure to acknowledge V.A Miransky for useful insights and stimulating discussions. This research has been supported in part by NSF under Grant PHY-0070986.

REFERENCES

1. K. Krishana et al., Science, 277, 83 (1997).
2. N.P. Ong et al., cond-mat/9904160.
3. H. Aubin et al., Phys. Rev. Let., 82, 624 (1999).
4. K.G. Klimenko, Z. Phys. C, 54, 323 (1992).
5. V. P. Gusynin, V. A. Miransky, and I. A. Shovkovy, Phys. Rev. Lett., 73, 3499 (1994); Phys. Rev. D, 52, 4718 (1995). Phys. Let. B 349, 477 (1995); Phys. Rev. D 52, 4747 (1995).
6. C.N. Leung, Y.J. Ng, and A.W. Ackley, Phys. Rev. D, 54, 4181 (1996); D.K. Hong, Y. Kim, and S.-J. Sin, Phys. Rev. D, 54, 7879 (1996); D.S. Lee, C.N. Leung, and Y.J. Ng, Phys. Rev. D, 55, 6504 (1997); V.P. Gusynin and I. A.Shovkovy, Phys. Rev. D 56, 5251 (1997); E.J. Ferrer and V. de la Incera, Int. J. Mod. Phys., A 14, 3963 (1999); Phys. Lett. B 481, 287 (2000); E.J. Ferrer, V.P. Gusynin, and V. de la Incera, Phys. Lett. B 455, 217 (1999); G.W. Semenoff, I.A. Shovkovy, and L.C.R. Wijewardhana, Phys. Rev. D 60, 105024 (1999); V.Ch. Zhukovsky et al., hep-th/0012256.
7. J. Ferrer, V. P. Gusynin and V. de la Incera, hep-ph/0101308.
8. K. Farakos and N.E. Mavromatos, Int. J. Mod. Phys., B 12, 809 (1998).
9. G.W. Semenoff, I.A. Shovkovy, and L.C.R. Wijewardhana, Mod. Phys. Lett. A 13, 1143 (1998).
10. W. V. Liu, Nucl. Phys. B 556, 563 (1999).
11. A.V. Balatsky, Phys. Rev. Lett. 80, 1972 (1998).
12. R.B. Laughlin, Phys. Rev. Lett., 80, 5188 (1998); T.V. Ramakrishnan, J. Phys. Chem. Solids, 59, 1750 (1998).
13. M. Franz, Phys. Rev. Lett. 82, 1760 (1999).
14. V. Ambegaokar and A. Griffin, Phys. Rev. 137, 1151 (1965).
15. It has been argued that for small impurity densities the thermal conductivity is unaffected by vertex corrections, see A.C. Durst and P.A. Lee, Phys. Rev. B 62, 1270 (2000).
16. J.S. Langer, Phys. Rev. 127, 5 (1962); V. Ambegaokar and L. Tewordt, Phys. Rev. 134, A805 (1964).
17. Y. Ando, J. Takeya, Y. Abe, K. Nakamura, and A. Kapitulnik, cond-mat/9812265.

Example of an Asymptotically Free Matrix Model

A. Agarwal*
S.G. Rajeev†

University of Rochester. Dept of Physics and Astronomy.
Rochester. NY - 14627

Abstract

We construct an example of a two dimensional matrix field theory theory that exhibits asymptotic freedom. This model example is the matrix analogue of the two dimensional BCS super conductor. We analyze the Planar large N limit of this field theory and show that certain questions related to its ground state; namely the existence or the lack of a mass gap in the spectrum; can be answered by studying a 'One Matrix Model' with a non-polynomial potential. This one matrix model associated with the matrix field theory is shown to describe the effective potential of the full theory.

1 Introduction

The study of Matrix models or Matrix field theories is prompted by their ubiquitous presence in the world of gauge theories. QCD, for instance, where the gluonic fields (A_μ) belong in the adjoint representation of the gauge group, is perhaps the most natural example of a field theory where the dynamical degrees of freedom are matrix valued. While perturbative QCD is

---

*abhishek@pas.rochester.edu
†rajeev@pas.rochester.edu

well understood, answering questions related to the non-perturbative aspects of the theory, e.g. getting a handle on the nature of its ground state, comprehending the mechanism for the formation of a mass gap in its spectrum, or understanding of the confining phase of the theory etc remain interesting theoretical challenges to date. It is believed that the planar large N limit is the correct simplifying limit to consider while looking at such issues. The absence of generic methods for studying non-perturbative phenomena related to the large N limit of matrix field theories is part of the reason why these questions are largely unanswered. In this article, we present a toy matrix field theory that, like QCD, exhibits asymptotic freedom and we also discuss some simple methods for studying questions related to its spectrum.

Although methods for studying matrix field theories, or theories involving a continuously infinite number of matrices are rudimentary at present, the world of the large N limits of 'one matrix models' (theories involving a single matrix or zero dimensional field theories) [1, 2]is rather thoroughly understood. In the present article, we show that certain questions related to the ground state of the continuum theory can be answered by studying a one matrix model, and we investigate this one matrix model, which describes the effective potential for the full theory, using techniques used in the study of ensembles of random complex matrices.[3, 4, 5]

The matrix field theory which is investigated here is non-relativistic in nature, and its dynamical degrees of freedom are N×N matrices ($\Psi^\dagger$, Ψ), whose matrix elements are Grassman numbers. It is characterized by the following action functional:

$$S[\Psi^\dagger \Psi] = Tr[\int_0^\beta d\tau \int d^2x (\Psi^\dagger(\partial_\tau - \nabla^2 - \mu)\Psi - \alpha(\Lambda)\Psi^\dagger\Psi^\dagger\Psi\Psi)] \quad (1)$$

β and μ represent the inverse temperature and the chemical potential respectively.

When $N = 1$, the theory reduces to that of the two dimensional BCS super conductor, which is known to exhibit both asymptotic freedom and the appearance of a mass gap in its spectrum[6, 7]. The coupling constant $\alpha(\Lambda)$, which appears to be scale invariant by naive dimensional arguments, does undergo dimensional transmutation, $\alpha(\Lambda) \sim \frac{1}{\ln\frac{\Lambda}{\nu}}$, where, $\Delta(= 2\nu)$ has the physical interpretation of the mass gap. The question of principal interest to us is whether or not these phenomena persist in the large N limit. This can be answered by computing the one-loop effective potential for the model which we shall now focus on.

2 The effective potential

For the purpose of computing the effective potential, it is useful to introduce auxiliary (matrix) fields $\sigma, \sigma^\dagger$, and rewrite the interaction part of the Lagrangian in the following form;

$$L_{int} = -\alpha(\Lambda) Tr(\Psi^\dagger \Psi^\dagger \Psi \Psi) = -Tr[\Psi^\dagger \sigma \Psi^\dagger + \Psi \sigma^\dagger \Psi] + \frac{Tr}{\alpha(\Lambda)} \sigma^\dagger \sigma \quad (2)$$

We can also define the matrix equivalents of the Nambu spinors ($\bar{\Psi}^\dagger \bar{\Psi}$) used in the theory of super conductivity as;

$$\bar{\Psi} = \begin{pmatrix} \Psi \\ \Psi^\dagger \end{pmatrix}, \bar{\Psi}^\dagger = (\Psi^\dagger \Psi). \quad (3)$$

With the aid of these new variables, the action may be re-expressed as,

$$S = Tr \left(\int d^2x \int d\tau \bar{\Psi}^\dagger \begin{bmatrix} \partial_\tau - \nabla^2 - \mu & -\sigma \\ -\sigma^\dagger & \partial_\tau + \nabla^2 + \mu \end{bmatrix} \bar{\Psi} + \frac{1}{\alpha(\Lambda)} \sigma^\dagger \sigma \right) \quad (4)$$

The effective potential can now be obtained by integrating $\bar{\Psi}^\dagger$ and $\bar{\Psi}$ after making the ansatz that the auxiliary variables $\sigma^\dagger$ and σ are constant (but not diagonal) matrices. The integration over the Fermionic variables is a Gaussian one and can be carried out easily. It yields the following formula for the effective potential $V(\sigma^\dagger, \sigma)$.

$$V(\sigma^\dagger, \sigma) = -NTr \left[\int [d^2k][d\omega] \ln \left[1 + \frac{\sigma^\dagger \sigma}{\omega^2 + \epsilon^2(k)} \right] - \frac{1}{g} \sigma^\dagger \sigma \right] \quad (5)$$

In the formula above $g = N\alpha$ is held fixed as $N \to \infty$. In other words, our large N limit is the same as the planar one. $\epsilon(k) = k^2 + |\mu|$ is the non-relativistic energy - momentum dispersion relation, while ω is the Matsubara frequency (the variable conjugate to τ), which takes on continuous values in the zero temperature limit. The analysis in the limit of the temperature approaching zero is sufficient to address the issue of the formation of a mass gap, and so we shall stick to this limit for the rest of the calculation.

It is worth noting that unlike the large N limit of the vector models, the effective potential does not dominate the effective action, but it does contain important non-perturbative information about the spectrum.

Having obtained the effective potential let us now try to understand how one might salvage information about the spectrum from it. To understand that we shall appeal to the simplifications that take place when $N = 1$. In this limit, if one calculates the number density, for instance, one obtains, $\langle \Psi^\dagger \Psi \rangle \sim \int f(E(k))d^2k$, where $f(E(k)) = \frac{1}{e^{\beta E(k)}+1}$ is the Fermi distribution function with $E(k) = \sqrt{(k^2 + |\mu|)^2 + \sigma^\dagger \sigma}$. Here $\sigma^\dagger \sigma = \Delta^2$ is a 1×1 matrix and Δ is the gap in the spectrum. In the language of the world of matrices the number density can also be expressed as $\langle \Psi^\dagger \Psi \rangle \sim \int f[\sqrt{(k^2 + |\mu|)^2 + \lambda^2}]\rho(\lambda)d\lambda d^2k$ where $\rho(\lambda) = \delta(\lambda - \Delta)$ is the spectral density function for the distribution of eigenvalues of the random Hermitian matrix $\sqrt{\sigma^\dagger \sigma}$. The question of the appearance of a mass gap now translates into a property of the spectral density function. The theory will posses a gap in the spectrum if the spectral density function is identically zero for some finite interval along the positive real axis starting at the origin. This is indeed the case in the $N = 1$ case, the interesting question is that whether or not this persists as $N \to \infty$. To answer this question we shall have to find the spectral density function for the one matrix model given by the effective action of our theory, and note if it is identically zero in some finite interval on the positive real axis.

3 The Random Matrix model

The potential for the random matrix model of interest to us (5) is of the following generic form $V(\sigma^\dagger \sigma) = \sum_{i=o}^{\infty} g_i Tr(\sigma^\dagger \sigma)^i$. Here we regard the logarithm in (5) as a formal power series. Let us note that though σ is a random complex matrix, only the Hermitian combination $\sigma^\dagger \sigma$ appears in the potential. Let us briefly discuss the methods for analyzing matrix models of this type. The generating function of all the Feynman graphs is,

$$Z = \int d\sigma^\dagger d\sigma e^{-\sum_{i=0}^{\infty} g_i Tr(\sigma^\dagger \sigma)^i} = \int d\sigma^\dagger d\sigma e^{-V(\sigma^\dagger \sigma)} \quad (6)$$

$d\sigma$ above is the invariant measure on the space of random complex matrices [5]. Going over to space of eigenvalues (λ_i) of the positive Hermitian matrix $\sqrt{\sigma^\dagger \sigma}$, we have,

$$Z = \int \Pi_i d\lambda_i e^{-\sum_i g_i [\sum_k \lambda_k^{2i}] + \log \sum_{i \neq j} |\lambda_i^2 - \lambda_j^2|} \quad (7)$$

The logarithmic term in the exponential arises reflects the Jacobian of the change of variables from the matrix elements to the eigenvalues. In the large N limit, the random eigenvalues of the positive hermitian matrix can be replaced by a continuous positive random variable with an associated distribution function ρ. In this limit, Z may be expressed as the following functional integral;

$$Z = \int D[x] e^{-\sum_i \int g_i x^{2i} \rho(x) dx + \int_0^b \int_0^b \log|x^2 - y^2| \rho(x) \rho(y) dx dy} \tag{8}$$

ρ, which has support in $[0, b]$ is determined by the saddle point of the argument of the exponential, which yields the following integral equation for ρ:

$$\frac{\partial}{\partial x} V(x^2) = 2 \wp \int_0^b \frac{x \rho(y) dy}{x^2 - y^2} \tag{9}$$

ρ and b are to be found out as the solution to the above equation.

To solve this equation, we shall introduce the following generating function:

$$F(z) = \sum_{k=0}^{\infty} \frac{\langle Tr(\sigma^\dagger \sigma)^k \rangle}{z^{2k+1}} = \int_0^b \frac{z \rho(y) dy}{z^2 - y^2}, z > b; \tag{10}$$

And using the principal value prescription suggested by (9) this function has the following behavior in the support of ρ;

$$F(x + i\epsilon) = V'(x^2) - i\pi Sgn(\epsilon) \rho(x); [0 < x < b] \tag{11}$$

It is obvious from the equation above, that determining F is equivalent to finding ρ, which is the discontinuity of this generating function across the real line. To find $F(z)$ we make the following ansatz:

$$F(z) = V'(z) - \sqrt{z^2 - b^2} f(z) \tag{12}$$

Noting from (10) that $F(z) \sim \frac{1}{z}$, for large values of z, it follows from (12) that,

$$\frac{V'(z)}{\sqrt{z^2 - b^2}} - f(z) = \frac{1}{z^2} + O(\frac{1}{z^4}) \tag{13}$$

This equation above is nothing but the solution for $f(z)$; or to be precise, it is the statement that $f(z)$ is the part involving non-negative powers of z in the expansion of $\frac{V'(z)}{\sqrt{z^2-b^2}}$ around infinity. Putting this solution back in (13) and requiring that the coefficient of $\frac{1}{z^2}$ on the right hand side equal unity determines b, while it follows from (11) and (12) that,

$$\rho(x) = \frac{1}{\pi} f(x) \sqrt{b^2 - x^2} \tag{14}$$

This solves the problem of finding the spectral density function of eigenvalues (in the large N limit) for the kind of one matrix models of interest to us. Before we plunge into the specifics of our model, let us look at a few instructive examples that can help illustrate the qualitative relations between the shape of the potential V and the nature of ρ.

Example 1: The Gaussian:
$V(\sigma^\dagger \sigma) = \frac{1}{2} Tr(\sigma^\dagger \sigma) \Rightarrow V(x^2) = \frac{1}{2}x^2, x \geq 0$. For this simplest of models, it follows immediately from the discussion above that, $\rho(x^2) = \frac{1}{\pi}\sqrt{b^2 - x^2}$, $b^2 = 2$.

This is the celebrated semicircular distribution for the Gaussian ensemble, and it is worth noting that ρ is peaked at the origin which is also the minimum of the potential.

Example 2: The Quartic model:
$V(x^2) = -\frac{1}{2}x^2 + \frac{g}{4}x^4, x \geq 0$. The potential is the positive half of the 'Mexican hat' potential, and its minimum is shifted away from the origin. Solving for ρ in this model, we get $\rho(x) = \frac{1}{\pi}[gx^2 + \frac{g}{2}b^2 - 1]\sqrt{b^2 - x^2}$. It is evident from this solution that the ρ is no longer peaked at the origin, in fact it is easy to check (after working out the numerical value of b) that it is peaked exactly at the minimum of the potential, V.

It is easy to convince oneself that this is a general feature for matrix models of this sort. So, for the purposes of understanding qualitative features of ρ; which is what we need to understand whether or not there is a mass gap in the spectrum of the continuum theory; it is sufficient to understand the qualitative nature of the potential. In the present case the potential for the one matrix model is nothing but the effective potential for the continuum theory. Let us now turn our attention to this particular case.

4 The Verdict

The potential for the model of interest to us (5) is,

$$V(x^2) = -N\left[\int \frac{d^2k}{4\pi^2}\frac{d\omega}{2\pi}\log\left(1 + \frac{x^2}{\omega^2 + \epsilon^2(k)}\right) - \frac{x^2}{g}\right] \tag{15}$$

This expression for the effective potential is ill defined as it stands, as it is U.V divergent, which can be seen by expanding out the logarithm; i.e. $V(x^2) \sim \left[-N\int_\Lambda \frac{d^2k d\omega}{\omega^2 + \epsilon^2(k)} + \frac{N}{g}\right]x^2 +$ convergent terms. the subscript under the integral implies that the integral is cut off at Λ. This situation calls for a re normalization of the coupling constant. The appropriate re normalization of g is given below;

$$\frac{1}{g} = \frac{1}{16\pi} \log(\frac{\Lambda}{\nu}) \qquad (16)$$

ν, is a dimensional parameter that sets the scale for the theory. This equation suggests that the phenomenon of asymptotic freedom persists in the planar large N limit. Now let us address the more interesting issue of the possibility of formation of a mass gap.

after incorporating the re normalization in (15), the integral equation (9) determining ρ for this matrix model is;

$$\frac{\partial}{\partial x} V(x^2) = \frac{1}{8\pi} x \left[\log(\frac{|\mu|}{2\nu}) + \log\left(1 + \sqrt{1 + \frac{x^2}{\mu^2}}\right) \right] = \wp \int_0^b \frac{x\rho(y) dy}{x^2 - y^2} \qquad (17)$$

Let us now look at the qualitative nature of this potential.

1: $\mu > 2\nu$.

As the parameter ν measures the strength of the interaction, this inequality can be thought of as a weak coupling regime. In this regime, $V'(x) \sim x$ for small values of x, and $V'(x) \sim x \log x$ as x becomes large. So the potential grows faster than the Gaussian but slower than that of the Quartic theory. And it has its minimum at $x = 0$. Form the previous discussion, then, it is clear that ρ is peaked at the origin, and it is certainly not vanishing in some finite interval starting at the origin. This implies that the theory is gap less in this regime.

2: $\mu < 2\nu$.

This is the strong coupling regime. While this inequality holds, $V'(x) \sim -x$ for small x, and $V'(x) \sim x \log x$ for large x. This suggests that the qualitative nature of the potential is the same as that of the Mexican hat type potential discussed above. As the minimum of the potential is now shifted away from the origin, so is the peak of the ρ, but it still remains non-vanishing at the origin. Hence the spectrum is gap less in this regime too.

So the conclusion is that the spectrum for the matrix analogue of the two dimensional BCS super conductor is gap less in the planar large N limit. Or in other words, the gap is of order $\frac{1}{N}$, which explains why this result is not in contradiction with the well known fact that the theory does posses as gap when $N = 1$.

Acknowledgment: This work was supported in part by the US Department of Energy, Grant No. DE-FG02-91ER40685

References

[1] E.Brezin, C.Itzykson, G.Parisi, and S.B.Zuber, Commun. math Phys **59**, 35 (1978).

[2] M.L.Mehta, Commun. math Phys **79**, 327 (1981).

[3] J.Ambjorn, J.Jurkiewicz, and Y.M.Makeenko, Phys Lett B **251**, 517 (1990).

[4] E.M.Ilgenfritz, Y.M.Makeenko, and T.V.Shahabazyan, Phys Lett B **172**, 81 (1986).

[5] M.L.Mehta, *Random Matrices* (Academic Press; New York, 1967).

[6] J.Polchinski, Effective Field Theory and the Fermi Surface, in *Proceedings of the 1992 Theoretical Advanced Studies Institute in Elementary Particle Physics*, World Scientific, Singapore, 1992.

[7] Abrikosov, Gorkov, and Dzyaloshinshi, *Methods of Quantum Field Theory in Statistical Mechanics* (Dover, New York, 1963).

Variational Principle for Large N Matrix Models

L. Akant, G. S. Krishnaswami and S. G. Rajeev [1]

*Department of Physics and Astronomy,
University of Rochester, Rochester, New York 14627*

Abstract. We derive a variational principle for large N matrix models. The partition function and vacuum green functions are determined by the principle of minimization of a free energy. The Schwinger-Dyson equations are the conditions for the free energy to be an extremum. We obtain a parametric representation for the greens functions and ground state energy. The parameter is a change of variable that transforms a reference theory to the one of interest.

INTRODUCTION

Large-N matrix models have turned out to be a convenient testing ground for ideas on non-abelian gauge theories, quantum gravity, the theory of random surfaces and the statistical mechanics of spins on random lattices. The reviews by M. Douglas [4], Y. Makeenko [6] and W. Taylor [7] serve as a window to the vast literature on the subject. For connections to non-commutative probability, see the work of Voiculescu et. al. [3]. Here we report on a variational principle for large N matrix models.

---

[1] presented by 2nd author govind@pas.rochester.edu at MRST 2001

SCHWINGER-DYSON EQUATIONS

We consider multi-matrix models with a finite number of $N \times N$ hermitian matrices, $M^i, 1 \leq i \leq \Lambda$ in the large-N limit. The partition function is

$$Z = \int d\tilde{M} e^{-NS(\tilde{M})}$$
$$S(M) = S_J \operatorname{tr} M^J. \tag{1}$$

We are interested in calculating the greens functions and ground state energy (relative to that of a reference theory with action S^0 and partition function Z_0):

$$G^K \equiv \lim_{N \to \infty} Z^{-1} \int d\tilde{M} e^{-NS(\tilde{M})} \frac{\operatorname{tr}}{N} \tilde{M}^K \equiv \langle \frac{\operatorname{tr}}{N} \tilde{M}^K \rangle$$
$$E \equiv -\lim_{N \to \infty} \frac{1}{N^2} \log(\frac{Z}{Z_0}) \tag{2}$$

The greens functions G^K are cyclically symmetric tensors. One approach to determine them is to solve the Schwinger-Dyson (SD) equations. Note that the partition function is invariant under a change in the variables of integration. Consider the following infinitesimal variations, which preserve the hermiticity of the matrices:

$$\tilde{M}^j \mapsto M^j = \tilde{M}^j + \epsilon \delta_i^j \tilde{M}^I$$
$$\Rightarrow \tilde{M}^j = M^j - \epsilon \delta_i^j M^I \tag{3}$$

ϵ is an infinitesimal real parameter. If I is an empty string, then M^i is translated by a multiple of the identity. We can represent these variations as vector fields:

$$L_i^I M^j = \delta_i^j M^I, \tag{4}$$

which satisfy a generalization of the Virasoro algebra [5]:

$$[L_i^I, L_j^J] = \delta_{J_1 i J_2}^J L_j^{J_1 I J_2} - \delta_{I_1 j I_2}^I L_i^{I_1 J I_2}. \tag{5}$$

For a multi-index J,

$$L_i^I M^J = \delta_{J_1 i J_2}^J M^{J_1 I J_2} \tag{6}$$

so that

$$L_i^I S = S_{J_1 i J_2} \text{ tr } M^{J_1 I J_2}. \tag{7}$$

The partition function becomes

$$Z = \int dM |det(\frac{\delta \tilde{M}}{\delta M})| e^{-NS(M)}[1 + \epsilon N L_i^I S] \tag{8}$$

The measure is *not* invariant (except when I is the empty string) [1]. Its variation leads to an anomalous term in the SD equations. The Jacobian is:

$$\det(\frac{\partial \tilde{M}}{\partial M}) \approx 1 - \epsilon \delta^I_{I_1 i I_2} \text{ tr } M^{I_1} \text{ tr } M^{I_2}. \tag{9}$$

Using the factorization of greens functions in the large-N limit we get the SD equations:

$$S_{J_1 i J_2} G^{J_1 I J_2} - \delta^I_{I_1 i I_2} G^{I_1} G^{I_2} = 0 \tag{10}$$

The anomalous quadratic term vanishes for an empty string I.

CLASSICAL ACTION FOR MULTI MATRIX MODELS

Except in special cases, it has not been possible to solve these SD equations exactly. On the other hand, it is natural to ask whether there is a variational principle from which the SD equations follow. We seek a classical action, the conditions for whose extremum are the large-N SD equations. The first term in the SD equations (10) is obviously a gradient of the 'internal energy', $S_J G^J$

$$S_{J_1 i J_2} G^{J_1 I J_2} = L_i^I S_J G^J. \tag{11}$$

Here L_i^I act as vector fields on the space of greens functions $L_i^I G^J = \delta^J_{J_1 i J_2} G^{J_1 I J_2}$. In addition, we need an entropic term χ such that

$$L_i^I \chi = \eta_i^I = \delta^I_{I_1 i I_2} G^{I_1} G^{I_2}. \tag{12}$$

so that the free energy $F = S_J G^J - \chi$ can be taken as the classical action. However, the existence of such a χ is non-trivial, due to the anomaly [2]. For this, the integrability conditions

$$L_i^I \eta_j^J - L_j^J \eta_i^I = \delta_{J_1 i J_2}^J \eta_j^{J_1 I J_2} - \delta_{I_1 j I_2}^I \eta_i^{I_1 J I_2}. \tag{13}$$

must be satisfied. In other words η_i^I must be a closed 1-form. We find that the integrability conditions are indeed satisfied. Though χ exists, we find that it is not a single valued function of the G^K; i.e. though η_i^I is closed it is not exact. Moreover, we do not have any simple formula for χ in terms of the moments. Rather, we have a parametric solution to the PDEs $L_i^I \chi = \eta_i^I$. Both the greens functions and χ are expressed in terms of a change of variable.

G^K FROM A FINITE CHANGE OF VARIABLES

The path integrals for G^K and Z are unaltered by arbitrary changes of integration variable. The SD equations are the conditions for the invariance of Z under an infinitesimal transformation. Thus, to 'solve' the SD equations, we must make (an unknown) *finite* change of variable. Let

$$\begin{aligned} M^i &= \psi^i(M_0) \\ M_0^i &= \phi^i(M), \ \psi = \phi^{-1}. \end{aligned} \tag{14}$$

where

$$\psi^i(M_0) = \psi_j^i M_0^j + \sum_{|J| \geq 2} \psi_J^i M_0^J, \ \det \psi_j^i > 0. \tag{15}$$

Suppose ψ were chosen so as to transform the action $S(M)$ into the reference action $S^0(M_0)$, whose greens functions are known:

$$\begin{aligned} S^0(M_0) &= S(\psi(M_0)) \\ Z_0 &= \int dM_0 e^{-N S^0(M_0)} \\ G_0^I &= \langle \frac{\text{tr}}{N} M_0^I \rangle_0 \equiv \lim_{N \to \infty} Z_0^{-1} \int dM_0 e^{-N S^0(M_0)} \frac{\text{tr}}{N} M_0^I. \end{aligned} \tag{16}$$

Then with $J = \frac{\delta \psi(M_0)}{\delta M_0}$,

$$Z = \int dM_0 \det(J) e^{-N S^0(M_0)}$$

$$\frac{Z}{Z_0} = \langle \det J \rangle_0$$
$$\log \frac{Z}{Z_0} = \log \langle e^{\text{tr} \log J} \rangle_0. \tag{17}$$

Therefore, using factorization of greens functions in the large-N limit,

$$E \equiv \lim_{N\to\infty} -\frac{1}{N^2} \log(\frac{Z}{Z_0}) = \lim_{N\to\infty} -\frac{1}{N^2} \langle \text{tr} \log J \rangle_0. \tag{18}$$

In a longer paper, we will show directly that $\chi = -E$ satisfies the equations $L_i^I \chi = -\eta_i^I$, so that it coincides with the entropic term that was necessary for the classical action principle.

As for the greens functions,

$$G^K = \frac{\langle \det[J] \frac{\text{tr}}{N} \psi^K(M_0) \rangle_0}{\langle \det[J] \rangle_0} = \langle \frac{\text{tr}}{N} \psi^K(M_0) \rangle_0. \tag{19}$$

Thus, once ψ is known, it is straightforward to calculate the greens functions G^K in terms of those of the standard theory, G_0^L.

χ FROM A FINITE TRANSFORMATION

We can also express χ in terms of the transformation ψ and the known greens functions G_0^P. We must calculate the jacobian, $\det[J]$, for the finite change of variable ψ. We will display the $N \times N$ matrix indices a, b.

$$(\psi^i(M_0))_b^a = \psi_k^i M_b^{ka} + \psi_K^i M_{a_1}^{k_1 a} M_{a_2}^{k_2 a_1} \cdots M_b^{k_n a_{n-1}} \tag{20}$$

$$J^{ia\ d}_{\ bjc} = \frac{\partial (\psi^i(M_0))_b^a}{\partial M^{Jc}_d}$$
$$= \psi_k^i \delta_j^k \delta_c^a \delta_b^d + \sum_{|K|\geq 2} \psi_K^i \sum_{l=1}^n (M^{k_1\cdots k_{l-1}})_c^a \delta_j^{k_l} (M^{k_{l+1}\cdots k_n})_b^d. \tag{21}$$

Now we factor out the ψ_k^i so that we can use the power series for log. Since $\det \psi_k^i > 0$, it is invertible: $(\psi^{-1})_j^i = \phi_j^i$. Writing $\psi_K^i = \psi_m^i \phi_l^m \psi_K^l$, we get

$$J^{ia\ d}_{\ bjc} = \psi_m^i \Big[\delta_j^m \delta_c^a \delta_b^d$$

$$+\phi_l^m \sum_{|K|\geq 2} \psi_K^l \sum_{l=1}^{n} (M^{k_1\cdots k_{l-1}})_c^a \delta_j^{k_l} (M^{k_{l+1}\cdots k_n})_b^d \Big] \qquad (22)$$

Then,

$$\text{tr } \log J = N^2 \text{ tr } \log(\psi_j^i) + \text{ tr } \log[1+R] \qquad (23)$$

where

$$R_{jbc}^{mad} = \phi_l^m \sum_{\substack{|K|,|L|\geq 0 \\ |K|+|L|\geq 1}} \psi_{KjL}^l (M^K)_c^a (M^L)_b^d \qquad (24)$$

We evaluate tr log J as a power series, using

$$\text{tr } R^p = \phi_{l_1}^m \psi_{K_1 m_1 L_1}^{l_1} \phi_{l_2}^{m_1} \psi_{K_2 m_2 L_2}^{l_2} \cdots \phi_{l_p}^{m_{p-1}} \psi_{K_p m L_p}^{l_p}$$
$$\text{tr } M^{K_1 \cdots K_p} \text{ tr } M^{L_p \cdots L_1}. \qquad (25)$$

We get

$$\chi = \text{tr } \log \psi_j^i + \sum_{p=1}^{\infty} \frac{(-1)^{p+1}}{p} \sum_{\substack{|K_i|,|L_i|\geq 0 \\ |K_i|+|L_i|\geq 1}} \phi_{m_1}^{m_0} \psi_{K_1 m_2 L_1}^{m_1} \phi_{m_3}^{m_2} \psi_{K_2 m_4 L_2}^{m_3} \cdots$$
$$\cdots \phi_{m_{2p-1}}^{m_{2p-2}} \psi_{K_p m_0 L_p}^{m_{2p-1}} G_0^{K_1 \cdots K_p} G_0^{L_p \cdots L_1}. \qquad (26)$$

A direct verification that this formula for χ satisfies the PDEs $L_I^I \chi = \eta_i^I$ will be given in a longer paper.

CONCLUSION

To summarize, the greens functions G^K of a matrix model with action $S(M) = S_J \text{ tr } M^J$, are determined, in the large N limit, by the principle that the free energy $F = S_J G^J - \chi$ be a minimum. This can also be regarded as the condition for the maximization of the entropic term χ, holding fixed the greens functions G^J conjugate to the couplings S_J that appear in the action. The coupling constants S_J are then thought of as Lagrange multipliers enforcing this constraint.

We have a parametric representation for both the greens functions G^K and χ in terms of a change of variable ψ that brings the action $S(M)$ to a standard action $S^0(M_0)$, whose moments G_0^K are assumed known:

$$S(\psi(M_0)) = S^0(M_0)$$
$$G^K = \langle \frac{\text{tr}}{N} \psi^K(M_0) \rangle_0 \qquad (27)$$

The formula for χ in terms of ψ and G_0^K was given above in eqn (26). Moreover, the maximum value of χ is the negative of the ground state energy of the system.

Thus, we have found a classical action for large N matrix models. In a forthcoming paper, we will use it to obtain variational approximations.

ACKNOWLEDGEMENTS

The authors wish to thank Abhishek Agarwal for useful discussions and the organizers of MRST 2001 for the opportunity to present this work. This work was supported in part by U.S.Department of Energy grant No. DE–FG02-91ER40685

REFERENCES

1. K. Fujikawa, Phys. Rev. Lett. **42**, 1195 (1979).

2. E. Witten, Nucl. Phys. B **223** (1983) 422.

3. D. Voiculescu, K. J. Dykema and A. Nica, *Free Random Variables*, AMS (1992).

4. M. R. Douglas, Fields Institute Communications Vol 12, (1995)

5. S. G. Rajeev and O. T. Turgut, J. Math. Phys. **37**, 637 (1996) [hep-th/9508103].

6. Y. Makeenko, hep-th/0001047.

7. W. Taylor, hep-th/0101126.

BRANES, STRINGS, AND THINGS

Analytic semi-classical quantization of a QCD string with light quarks

T. J. Allen[1]*, C. Goebel†, M. G. Olsson† and S. Veseli**

*Dept. of Physics, Hobart & William Smith Colleges, Geneva, New York 14456
†Dept. of Physics, University of Wisconsin, 1150 University Avenue, Madison, Wisconsin 53706
**Fermi National Accelerator Laboratory, P.O. Box 500, Batavia, Illinois 60510

Abstract. We perform an analytic semi-classical quantization of the straight QCD string with one end fixed and a massless quark on the other, in the limits of orbital and radial dominant motion. Our results well approximate those of the exact numerical semi-classical quantization as well as our exact numerical canonical quantization.

INTRODUCTION

Linearly rising Regge trajectories are a prediction of both string and scalar confinement with one (or two) light quark(s). In a QCD string theory with one (or two) light spinless quark(s), the relation of the energy of the light degrees of freedom, E, to the the angular momentum and radial quantum numbers, J and n, is well approximated by

$$\frac{E^2}{(2)\pi a} \simeq J + 2n + \frac{3}{2}, \qquad (1)$$

where a is the tension (linear energy density) of the string. The same relation holds exactly in scalar confinement, though with a denominator of $2a$ instead of πa. We derive Eq. (1) analytically in semi-classical quantization of a straight string fixed at one end and with a massless and spinless quark at the other, in the limit of large radial quantum numbers n. The straight string approximation is an excellent approximation to the motion of a Nambu-Goto string; string curvature affects the energy and the angular momentum of the system very little [1]. The semi-classical result will follow from the evaluation of a single integral when one or both quarks are massless.

DYNAMICS & QUANTIZATION

The energy and angular momentum of the system are its only conserved quantities. We would like to relate the energy to the angular momentum but the relationship of the angular momentum to the quark's transverse velocity (or to its angular velocity) is

---

[1] talk presented by T. J. Allen

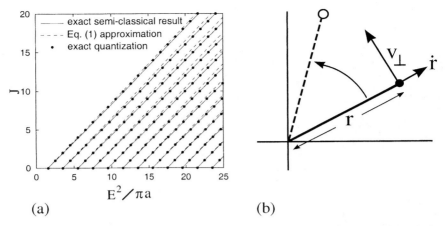

FIGURE 1. (a) Regge plot of angular momentum versus squared energy over πa. The numerical semi-classical quantized results agree nearly exactly with the exact numerical canonical results. Eq. (1) also agrees for $n \gg J$. (b) The configuration of a straight string with a quark on the end is described by the quark's position **r**, its radial velocity $\dot{r}$ and its transverse velocity $v_\perp$.

transcendental and the transverse velocity cannot be eliminated analytically. Despite this difficulty, we can carry out semi-classical quantization analytically in the limit of large quantum numbers. This is not an essential limitation as semi-classical quantization is only valid in this limit.

The energy and angular momentum of the quark–string system, shown in Fig. 1(b), is well-known and given by [2]

$$E = W_r \gamma_\perp + ar \frac{\arcsin(v_\perp)}{v_\perp}, \tag{2}$$

$$J = W_r \gamma_\perp v_\perp r + ar^2 \left(\frac{\arcsin(v_\perp)}{2v_\perp^2} - \frac{\sqrt{1-v_\perp^2}}{2v_\perp} \right), \tag{3}$$

where

$$W_r \equiv \sqrt{p_r^2 + m^2} = \sqrt{1-v_\perp^2} \cdot \sqrt{\vec{p}^2 + m^2}, \tag{4}$$

$$\gamma_\perp \equiv (1-v_\perp^2)^{-1/2}. \tag{5}$$

Dimensionless Variables

To simplify the system, it pays to work in dimensionless units. We choose our units to be those of a system in uniform circular motion. The orbital energy and radius are $E_0 = \sqrt{J\pi a}$ and $r_0 = 2\sqrt{J/(\pi a)}$ respectively. We take the dimensionless energy, radius,

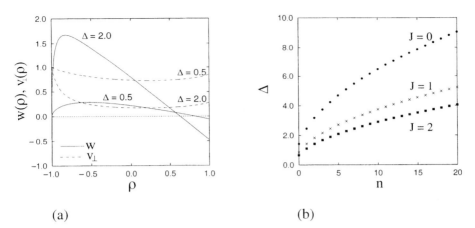

FIGURE 2. (a) The behavior of $w(\rho)$ and $v_\perp(\rho)$ for $\Delta = 2$ and $\Delta = 0.5$. (b) Δ versus n for $J = 0, 1, 2$.

and quark radial energy to be given in terms of dimensionless numbers Δ, ρ, and w by

$$E/E_0 \equiv \sqrt{1+\Delta^2}, \tag{6}$$
$$r/r_0 \equiv \sqrt{1+\Delta^2} + \rho\Delta, \tag{7}$$
$$W_r/E_0 \equiv w. \tag{8}$$

When $\Delta = 0$ the system is in uniform circular motion and lies along the leading Regge trajectory. Radial excitation corresponds to $\Delta > 0$.

We can find $v_\perp$ and w as functions of ρ and Δ from Eqs. (2) and (3), made suitably dimensionless,

$$w = \sqrt{1-v_\perp^2}\left[\sqrt{1+\Delta^2} - \frac{2}{\pi}S(v_\perp)\left(\sqrt{1+\Delta^2}+\rho\Delta\right)\right], \tag{9}$$

$$1 + \frac{4}{\pi}(v_\perp S(v_\perp) - f(v_\perp))\left[\sqrt{1+\Delta^2}+\rho\Delta\right]^2$$
$$= 2v_\perp\left[\sqrt{1+\Delta^2}+\rho\Delta\right]\sqrt{1+\Delta^2}, \tag{10}$$

where $S(v_\perp) = \frac{\arcsin(v_\perp)}{v_\perp}$ and $f(v_\perp) = \frac{1}{2v_\perp}\left(S(v_\perp) - \sqrt{1-v_\perp^2}\right)$. Figure 2(a) shows the dependence of w and $v_\perp$ on ρ for $\Delta = 0.5$ and $\Delta = 2.0$.

Turning Points

In order to carry out semi-classical quantization, one must know the locations of the turning points of the classical motion. The turning points are the radii at which the radial momentum vanishes, $p_r = 0$. For a massless quark, this implies that $w = 0$, but because the quark is massless, the radial velocity need not vanish. The inner turning point occurs at $\rho_- = -1$ and the outer turning point occurs in the interval

$$\frac{\pi}{2} - 1 \leq \rho_+ \leq 1 \,. \tag{11}$$

The lower limit is reached for large Δ and the upper limit for $\Delta = 0$. The turning points for $\Delta = 0.5$ and $\Delta = 2.0$ can be seen in Fig. 2(a).

Semi-Classical Quantization

The Bohr-Sommerfeld quantization condition with the Langer correction [3] is

$$\int_{r_-}^{r_+} p_r \, dr = \pi \left(n + \frac{1}{2} \right) \,, \qquad n = 0, 1, 2, \ldots \,, \tag{12}$$

where the radial momentum is a function of radius and shifted angular momentum, $p_r = p_r(r, J + \frac{1}{2})$. In the case of a massless quark, we define the integral

$$I(\Delta) = \frac{4}{\pi} \int_{-1}^{\rho_+(\Delta)} d\rho \, w(\rho, \Delta) \,, \tag{13}$$

which is proportional to the integral in Eq. (12). The spectrum is computed from the relation

$$\Delta I(\Delta) = \left(\frac{2n+1}{J + \frac{1}{2}} \right) \,, \tag{14}$$

once Δ is expressed in terms of the energy through Eq. (6).

Radial Dominant Motion

Semi-classical quantization should become exact in the limit of $n \gg J$. For small J, large n implies large Δ, as can be seen from Fig. 2(b), which shows the dependence of Δ on n for small values of J. This is the radial dominant regime.

Large Δ implies either $\rho = -1$, or $v_\perp \to 0$ and

$$w(\rho, \Delta) \longrightarrow \Delta \left[1 - \frac{2}{\pi} (1 + \rho) \right] \quad \Rightarrow \quad \rho_+ = \frac{\pi}{2} - 1 \,. \tag{15}$$

In the limit of large Δ, the integral $I(\Delta)$ in Eq. (13) becomes simply Δ.

$$I(\Delta) \simeq \int_{-1}^{\frac{\pi}{2}-1} d\rho \, w(\rho, \Delta) = \Delta \,. \tag{16}$$

The quantization condition (14) becomes

$$\Delta I(\Delta) = \Delta^2 = \left(\frac{E^2}{(J+\frac{1}{2})\pi a} - 1\right) = \left(\frac{2n+1}{J+\frac{1}{2}}\right), \qquad (17)$$

which is the same as Eq. (1).

Angular Dominant Motion

In the regime of radial dominant motion ($J \gg n$), Δ is small. Small Δ implies that the upper turning point is $\rho_+ \simeq 1$. In this approximation we can also find an analytical form for $I(\Delta)$;

$$\begin{aligned} I(\Delta) &= \frac{4}{\pi}\int_{-1}^{+1} d\rho\, w(\rho,\Delta) = \frac{8}{21}\left(\frac{3}{\pi}\right)^{5/6}\frac{\Gamma(\frac{2}{3})}{\Gamma(\frac{7}{6})}\Delta^{4/3} \\ &\simeq 1.3367 \Delta^{4/3}. \end{aligned} \qquad (18)$$

Numerical evaluation of $I(\Delta)$ shows that Eq. (18) is a good approximation for $\Delta \leq 0.4$. For larger Δ the radial dominant result of Δ is a better approximation.

CONCLUSIONS

We have performed a semi-classical quantization of the straight QCD string with massless and spinless quarks on the end and found an approximate spectrum that is similar in form to that of scalar confinement, though with a different Regge slope. In performing the quantization it is useful to use dimensionless variables related in an algebraic way to the energy and angular momentum.

In the case of massless quarks, a single integral function $I(\Delta)$ determines the spectrum. In the regime of radial dominant motion, this integral can be easily evaluated, and it is simply Δ. This leads to the simple relation (1) that agrees well in the radial dominant regime with a full numerical canonical quantization of the system as well as to the exact numerical semi-classical quantization, as can be seen in Fig. 1(a).

REFERENCES

1. T.J. Allen, M.G. Olsson, and S. Veseli, *Phys. Rev. D* **59**, 094011 (1999).
2. D. LaCourse and M. G. Olsson, *Phys. Rev. D* **39**, 2751-2757 (1989); C. Olson, M. G. Olsson, and K. Williams, *Phys. Rev. D* **45**, 4307-4311 (1992); M. G. Olsson and K. Williams, *Phys. Rev. D* **48**, 417-421 (1993); M. G. Olsson and S. Veseli, *Phys. Rev. D* **51**, 3578-3586 (1995); M. G. Olsson, S. Veseli, and K. Williams, *Phys. Rev. D* **53**, 4006-4012 (1996).
3. R.E. Langer, *Phys. Rev.* **51**, 669-676 (1937); M.S. Child, *Molecular Collision Theory*, Academic, New York, 1974.

String Webs from Field Theory

P. Argyres and K. Narayan

Newman Laboratory, Cornell University, Ithaca, NY 14853, USA

Abstract. The spectrum of stable electrically and magnetically charged supersymmetric particles can change discontinuously due to the decay of these particles as the vacuum on the Coulomb branch is varied. We show that this decay process is well described by semi-classical field configurations purely in terms of the low energy effective action on the Coulomb branch even when it occurs at strong coupling. The resulting picture of the stable supersymmetric spectrum is a generalization of the "string web" picture of these states found in string constructions of certain theories.

INTRODUCTION

Let us begin by considering a gauge theory, with gauge group G that is Higgsed to $U(1)^r$, r being the rank of the group. Then massive charged 1-particle states in this theory are labelled by their electric and magnetic charges with respect to the $U(1)^r$, i.e. $(\vec{Q}_e, \vec{Q}_m)$. An example is the familiar 't Hooft-Polyakov monopole in the $SU(2)$ gauge theory Higgsed to $U(1)$ at weak coupling. However, in general, this is a hard question, since the equations to be solved for the field configurations are complicated coupled nonlinear differential equations.

Supersymmetry yields a more powerful handle on these theories – due to a tighter analytical structure, quantum corrections are much more controlled. Four dimensional gauge theories with at least eight supercharges have a whole moduli space (or Coulomb branch) of inequivalent vacua in which the low energy theory has unbroken $U(1)$ gauge factors. The spectrum of charged particle states in this case lie in supersymmetry multiplets. The masses of those in short multiplets are related to their charges by the supersymmetry algebra [1]. These states, known as BPS states, leave some fraction of the supersymmetry unbroken. They saturate the Bogomolny-Prasad-Sommerfeld (BPS) bound :

$$M \geq M_{BPS} \qquad (1)$$

where M_{BPS} is a function of the couplings and vacuum expectation values (vevs) of the fields. However it is not clear if the states exist. Furthermore, even if such a state exists in some region of the Coulomb branch, it may be unstable to decay at curves of marginal stability (CMS) on the Coulomb branch [3]. We propose [4] a solution to the question of the multiplicity of BPS states for $N = 2$ supersymmetric theories in four dimensions just in terms of the low energy effective $U(1)^n$ action on the Coulomb branch.

The form of the answer we get coincides with the "string web" picture of BPS states developed in the context of the D3-brane construction of $N = 4$ $SU(n)$ superYang-Mills theory and some F-theory constructions of $N = 2$ $SU(2)$ gauge theory [6]. Moreover, our solution generalizes these constructions to arbitrary field theory data (gauge groups, matter representations, couplings and masses).

Our solution describes the stability of the BPS spectrum wholly in terms of the $U(1)^n$ low energy effective action. This is possible because the distance ΔX from a given vacuum on the Coulomb branch to a CMS acts as a new low energy scale which can be made arbitrarily small compared to the strong coupling scale Λ of the non-Abelian gauge theory as we approach the CMS. In particular, as we approach the CMS, the classical low energy field configuration describing a state which decays across the CMS will develop two or more widely separated charge centers (which appear as singularities in the low energy solution); the distance between these centers varies inversely with ΔX. In this limit the details of the microscopic physics become irrelevant for the decay of a BPS state across a CMS, which is described by a low energy field configuration with charge centers becoming infinitely separated.

The classical BPS field configuration of the scalar fields in the low energy $U(1)^n$ theory will have one or more singularities or sources where scalar field gradients and $U(1)$ field strengths diverge. Near these points the low energy description breaks down and should be cut off by boundary conditions reflecting the matching onto the microscopic physics of the non-Abelian gauge theory. These boundary conditions are determined by the BPS condition.

Previous discussions [7, 8] of the semi-classical description of BPS states in supersymmetric theories were the starting point for this work.

BPS STATES NEAR CMS

The Coulomb branch is parametrized by the vevs of the scalar fields in the theory. There are isolated singularities on the Coulomb branch where charged particles become massless.

The mass of a BPS state is $M = |Z|$ where the central charge Z is the sum of terms proportional to the charges. This mass is the minimum mass of any state (BPS or not, single particle or not) in this charge sector. A single BPS particle M in this charge sector is stable or at worst marginally stable against decay into two (or more) constituent particles, since by charge conservation $Z = Z_1 + Z_2$, so by the triangle inequality $M \leq M_1 + M_2$. The CMS are submanifolds of the Coulomb branch where this inequality is saturated. As one adiabatically moves on the Coulomb branch from a vacuum on one side of a CMS to a vacuum on the other, the 1-particle state M becomes more and more nearly degenerate with the two particle state $M_1 + M_2$. The 1-particle state thought of as a bound state of the constituent states becomes more and more loosely bound as one approaches the CMS from one side. The binding energy vanishes exactly at the CMS.

It follows from the triangle inequality that the two particle state $M_1 + M_2$ is not BPS, even for zero relative momentum, except precisely at the CMS. Thus there will generically be no BPS force cancellation between particles M_1 and M_2, so the zero relative momentum two particle state is classically approximated by two spatially infinitely separated one-particle states (to have a static configuration). Thus the transition across the

CMS of a one particle state to a widely separated, zero momentum, two particle state goes by way of field configurations with growing spatial extent, that eventually diverges as the transition is approached. Thus *if a BPS state does decay across a CMS, that decay will be visible semiclassically in the low energy effective action even if it takes place at strong coupling from a microscopic point of view.* The resulting picture is quite intuitive: a BPS state decaying across a CMS does so by becoming an ever-larger, more loosely bound state of its eventual decay products. Once the CMS is crossed, the bound state ceases to exist, and so, in particular, there will be no static BPS solutions to the low energy equations of motion and boundary conditions in this region of the Coulomb branch.

A $U(1)$ TOY EXAMPLE

The essential physics of the curve of marginal stability can be illustrated in a simple toy model. Consider a $U(1)$ effective action with two real scalars X, Y:

$$S = -\int d^4x \left(\frac{1}{4}F_{\mu\nu}F^{\mu\nu} + \frac{1}{2}\partial_\mu X \partial^\mu X + \frac{1}{2}\partial_\mu Y \partial^\mu Y \right) \tag{2}$$

Add to this theory charged sources, that become massless at singularities in the vacuum manifold. In this case the vacuum manifold (Coulomb branch) is the X-Y plane. Specifically, let us consider two singularities at points on the Coulomb branch with coordinates $(X,Y) = (L,0)$ where a particle with electric charge $(Q_E, Q_B) = (1,0)$ becomes massless, and $(X,Y) = (0,L)$ where a particle with magnetic charge $(Q_E, Q_B) = (0,1)$ becomes massless.

The low energy effective action (2) should be thought of as comprising the leading terms in a derivative expansion. The higher derivative terms are suppressed by powers of a cutoff energy scale Λ, or distance scale $r_\Lambda \equiv \frac{1}{\Lambda}$. We are justified in dropping these higher derivative terms since we will use this low energy action only away from the charge cores, imposing boundary conditions in the vicinity of the cores ($r \sim r_\Lambda$). The basic boundary condition on the electric and magnetic fields follows from Gauss' law, $\oint \mathbf{E} \cdot d\mathbf{a} = Q_E$, and $\oint \mathbf{B} \cdot d\mathbf{a} = Q_B$, where the integrals are over spheres enclosing the charge cores.

The X, Y scalar fields tend to approach their values at Coulomb branch singularities near the charge cores. Such field configurations are energetically favored because they approach those of the charged BPS states which become massless at the singularities $(X,Y) = (0,L)$ or $(L,0)$. Thus the scalar fields satisfy the approximate Dirichlet boundary conditions

$$(X,Y) \simeq (0,L) \text{ or } (L,0) \text{ within } B^3_\Lambda \tag{3}$$

where B^3_Λ is a ball of approximate radius r_Λ around each charge core and $(X,Y) \simeq (L,0)$ means that (X,Y) pass within a distance Λ of $(L,0)$ on the Coulomb branch. This "fuzzy ball" boundary condition is all we can physically demand of the low energy solution since it is not accurate on spatial resolutions less than r_Λ nor for field value resolutions less than Λ. In the limit as the vacuum approaches a CMS, as we have described qualitatively in the last section and will see explicitly below, the size of the field configuration grows, and so the relative size of the cutoff region r_Λ to the field configuration shrinks. In this limit the fuzzy ball Dirichlet boundary conditions become precise.

For simplicity, we choose the vacuum to be symmetrically placed at the point (X_0, X_0) on the Coulomb branch and choose the state to have electric and magnetic charges $(Q_E, Q_B) = (1, 1)$. There will then be separate electric and magnetic charge cores. The usual BPS equations for a static field configuration following from (2) are then simply $\nabla^2 X = \nabla^2 Y = 0$ while the electric and magnetic fields are determined in terms of X and Y by $\mathbf{E} = \nabla X$, and $\mathbf{B} = \nabla Y$. The Gauss constraint implies a solution of the approximate form

$$X - X_0 = \frac{1}{4\pi \,|\mathbf{x} - \mathbf{x}_E|}, \qquad Y - X_0 = \frac{1}{4\pi \,|\mathbf{x} - \mathbf{x}_B|} \tag{4}$$

so that the electric and magnetic charge cores are located at $\mathbf{x}_E$ and $\mathbf{x}_B$ respectively. To check if such a solution exists, we need to check whether there are values of $\mathbf{x}_E$ and $\mathbf{x}_B$ for which this solution satisfies the fuzzy ball Dirichlet boundary condition (3), which demands that the (X, Y) values taken by a solution approximately approach a small ball around $(L, 0)$ as $\mathbf{x} \to \mathbf{x}_E$ (and a small ball around $(0, L)$ as $\mathbf{x} \to \mathbf{x}_B$).

Now for $|\mathbf{x} - \mathbf{x}_E| \simeq (L - X_0)^{-1}$ it follows that $X \simeq L$ as desired near the electric singularity. But this then implies that $Y \simeq X_0 + r_{EB}^{-1}$ where $r_{EB} \equiv |\mathbf{x}_E - \mathbf{x}_B|$ is the spatial distance between the charge centers. The fuzzy Dirichlet boundary condition $Y \simeq 0$ then implies that the separation of the charge centers must be fixed to be

$$r_{EB} = -\frac{1}{X_0}. \tag{5}$$

This implies the following : (i) A solution exists only for $X_0 < 0$. This inequality is equivalent to saying that there is only a solution on one side of the CMS for our dyonic $(Q_E, Q_B) = (1, 1)$ state (as can be checked from the explicit BPS mass formula). (ii) As $X_0 \to 0^-$ (i.e. as we approach the CMS) the spatial separation of the electric and magnetic charge centers goes to infinity. This confirms the qualitative description of the decay of a supersymmetric state near the CMS given above.

As we describe in detail in [4], in the limit as the vacuum approaches the CMS, the projection on the Coulomb branch of the above low energy solution more and more accurately approximates a string web. Further, the charge conservation and tension balance properties of the string web stretched between D-branes in the string construction of the above theory [5] are reproduced by our low energy field theory solution above.

The low energy solution however has a new feature – the extended spatial structure, as is clear in the above toy model. While the string construction exhibits a pointlike junction where different legs of the string web meet, we do not see a pointlike junction in the above field theory solution. The difference between the two perspectives is the result of different orders of limits. Consider the Dirac-Born-Infeld action for the D3-brane corresponding to the decaying dyon, treating it as a probe in a background of other D3-branes. The brane prong corresponding to the decaying $(1, 1)$ dyon then ends on other D3-branes that are treated as a fixed background. Looking at the quadratic terms and comparing with the low energy effective action we have used above, we can see that the scalars X in the field theory with mass dimension unity and the coordinates in the brane transverse space, x, are related by $X = x/\alpha'$, where α' is the string length squared. Thus the separation (5) between the charge centers in the D-brane worldvolume is $r_{EB} = \alpha'/(-x_0)$. The low energy field theory is a good approximation in the $\alpha' \to 0$ limit holding the scalar vevs, including X_0, fixed. On the other hand, in the string junction/geodesic picture in string theory, the coordinates x_0 are what are held fixed : taking $\alpha' \to 0$ to suppress higher stringy corrections gives a vanishing separation r_{EB} between the charge centers, *i.e.* a pointlike junction. For $x_0 = 0$, the separation is indeterminate, which corresponds to the two strings ending anywhere on the D-brane. This recovers the string theory result and it further illustrates that a *spatially* point-like junction is not visible within the field theory approximation.

The essential features of the toy model described above carry over to and are generalized in field theories with higher rank gauge groups, possibly with less supersymmetry [4]. In these cases, the projection on the moduli space of the low energy solutions degenerates to a collection of segments of geodesics—"strings"—on the moduli space which can meet in string junctions or end at the vacuum or singularities on the moduli space. For example, in the $U(n)$ $N = 4$ SYM theory these "string" webs are thus stretched in the $6n$-dimensional moduli space and end on $6(n-1)$-dimensional singular submanifolds.

These moduli space webs are different from the string webs found in string constructions of the gauge theory, where the strings are stretched in the 6-dimensional space transverse to n D3-branes. It can be shown [4] that the 6-dimensional string theory webs are obtained from our $6n$-dimensional webs by a simple mapping. The basic reason this works is that the $6n$-dimensional moduli space $\mathcal{M}$ of the $U(n)$ theory is $\mathcal{M} = (R^6)^n/S_n$ where the permutation group S_n interchanging the R^6 factors is the Weyl group of $U(n)$.

ACKNOWLEDGMENTS

It is a pleasure to thank A. Buchel, J. Hein, R. Maimon, J. Maldacena, M. Moriconi, S. Pelland, M. Rangamani, V. Sahakian, and A. Shapere for helpful comments and discussions. This work was supported in part by NSF grant PHY95-13717.

REFERENCES

1. E. Witten and D. Olive, *Phys. Lett.* **78B** (1978) 97.
2. N. Seiberg and E. Witten, *Nucl. Phys.* **B426** (1994) 19; *Nucl. Phys.* **B431** (1994) 484.
3. S. Cecotti and C. Vafa, *Commun. Math. Phys.* **158** (1993) 569.
4. P. Argyres and K. Narayan, *JHEP* **0103** 047 (2001).
5. O. Bergman, *Three-pronged strings and 1/4 BPS states in N=4 super-Yang-Mills theory*, *Nucl. Phys.* **B525** (1998) 104.
6. A. Sen, *Nucl. Phys.* **B475** (1996) 562; T. Banks, M. Douglas and N. Seiberg, *Phys. Lett.* **B387** (1996) 278.
7. J. Gauntlett, C. Koehl, D. Mateos, P. Townsend and M. Zamaklar, *Phys. Rev.* **D60** (1999) 045004.
8. K. Lee and P. Yi, *Phys. Rev.* **D58** (1998) 066005; D. Bak, C. Lee, K. Lee and P. Yi, *Phys. Rev.* **D61** (2000) 025001; D. Bak, K. Lee and P. Yi, *Phys. Rev.* **D61** (2000) 045003; J. Gauntlett, N. Kim, J. Park and P. Yi, *Phys. Rev.* **D61** (2000) 125012.

Geometry of large extra dimensions versus graviton emission

Frédéric Leblond

Department of Physics, McGill University, Montréal, Québec, Canada H3A 2T8

Abstract. We study how the geometry of large extra dimensions may affect field theory results on a three-brane. More specifically, we compare cross sections for graviton emission from a brane when the internal space is an N-torus and an N-sphere for $N = 2-6$. We find that the ability of high energy colliders to determine the geometry of the extra dimensions is limited but there is an enhancement when both the quantum gravity scale and N are large.

INTRODUCTION

In recent years we have learned there could be more than meets the eye concerning gravity. While this is expected what is rather surprising is that we can appreciate this statement without entering the realm of M theory. For example, effective theories with a factorizable metric and a Planck scale of energy lower than the one associated with the usual four-dimensional gravitational coupling are not excluded [1]. In fact, there is a vast literature on the idea of using extra dimensions and a lower quantum gravity scale to devise effective gravitational models. Inspired by the D-brane concept of string theory, the standard model (SM) fields are assumed to be localized on a three-brane while gravity propagates in the entire spacetime. An interesting feature of this scenario is the potentially large size of the extra dimensions. For example, a brane world model with $N = 2$ transverse dimensions and a quantum gravity scale M_D on the order of 1 TeV leads to dimensions that can be as large as 1 mm. Although this particular set of parameters seems to be ruled out by astrophysical bounds [2], it is still worth investigating the large dimensions scenario for other values of N and M_D. When the effects of large extra dimensions on SM processes are studied these are usually compactified on an N-dimensional torus [3, 4, 5, 6]. In this work, we present some consequences of having the extra dimensions compactified on an N-sphere.

GRAVITON MODES ON THE INTERNAL SPACE

The degrees of freedom we consider are those of general relativity which we use as our effective theory. The quantized $(4+N)$-dimensional version of gravity we use follows the covariant approach (for example see Ref. [7]) in which the graviton field, h_{AB}, is a small perturbation,

$$g_{AB} = \bar{g}_{AB} + h_{AB} \qquad |h_{AB}| \ll 1, \tag{1}$$

where $A, B = 0, 1, ..., (D-1)$. The result is a perturbative theory on a background with metric $\bar{g}_{AB}$. We are considering quantization around $M^4 \times T^N$ and $M^4 \times S^N$, where M^4 is four-dimensional Minkowski space.

A detailed analysis of the Kaluza–Klein reduction and gauge fixing procedure for compactification on a N-torus can be found in Ref. [10]. Using a similar procedure one expects to find the following $(4+N)$–dimensional equations of motion for the graviton:

$$\left[\partial_\mu \partial^\mu + \nabla^2\right] h_{\mu\nu}(x, \mathbf{y}) = 0, \tag{2}$$

where ∇^2 is the Laplacian on B^N. Note that we have demonstrated this to be true explicitly only for $B^N = S^N$ and T^N in Ref. [9]. Because of the compact nature of B^N the gravitational field can be recast as an infinite tower of Kaluza–Klein (KK) modes,

$$h_{\mu\nu}(x, \mathbf{y}) = \sum_{\{n\}} h_{\mu\nu}^{\{n\}}(x) \psi^{\{n\}}(\mathbf{y}), \tag{3}$$

where $\{n\}$ is a set of quantum numbers related to the isometry group of the compact space and $\psi^{\{n\}}(\mathbf{y})$ is a normalized wave function.

The simplest compact geometry for the extra dimensions is a N–dimensional torus with a unique radius a. Cases with toric extra dimensions characterized by different length scales are studied in Ref. [11]. The wave function in transverse space is simply obtained by solving

$$\left[\partial_i \partial^i + m_\mathbf{n}^2\right] \psi^\mathbf{n}(\mathbf{y}) = 0, \tag{4}$$

which leads to

$$\psi^\mathbf{n}(\mathbf{y}) = \frac{1}{(2\pi a)^{\frac{N}{2}}} e^{i\frac{\mathbf{n} \cdot \mathbf{y}}{a}}, \tag{5}$$

where $\mathbf{n} = \{n_1, n_2, ..., n_N\}$ with the n_i integers running from $-\infty$ to $+\infty$ and $0 < y_i \leq 2\pi a$ are the components of the vector $\mathbf{y}$.

The derivation of the wave function for the graviton propagating on a N-sphere is more challenging. In that case we need to solve the following equation:

$$\left(\nabla_{S^N}^2 + m_{\{n\}}^2\right) \psi^{\{n\}}(\mathbf{y}) = 0, \tag{6}$$

where $\nabla_{S^N}^2$ is the Laplacian on a N-sphere of fixed radius a. An important quantity in our analysis is the wave function evaluated at a point on the sphere. For example, when $N = 6$ we get,

$$\psi^n(\mathbf{y} = \mathbf{0}) = \left[\frac{(5+2n)(n+4)(n+3)(n+2)(n+1)}{120 V_{S^6}}\right]^{1/2}, \tag{7}$$

where $V_{S^N} = \frac{2\pi^{(N+1)/2} a^N}{\Gamma((N+1)/2)}$ is the volume of a N-sphere.

PROBING THE EXTRA DIMENSIONS

We now consider how the geometry of the internal space affects the couplings of SM fields to the gravitational sector. The linear coupling, which is universally determined by general covariance, is of the form (see Ref. [10])

$$\frac{1}{M_D^{1+(N/2)}} \int d^D x \, h^{AB} T_{AB}, \tag{8}$$

where T_{AB} is the stress–energy tensor associated with SM fields on the three–brane. We are studying gravity on the product space $M^4 \times B^N$ with SM fields localized on the M^4 submanifold. It is a reasonable approximation for our purposes to use the following form for the stress–energy tensor:

$$T_{AB}(x, \mathbf{y}) = \delta_A^\mu \delta_B^\nu T_{\mu\nu}(x) \delta^{(N)}(\mathbf{y}). \tag{9}$$

This expression is written in the so–called static gauge which consists of ascribing four bulk coordinates to the three–brane ($A = 0, 1, 2, 3 \to \mu = 0, 1, 2, 3$) and the remaining N coordinates to the internal space ($A = 4, ..., D-1 \to i = 1, ..., N$).

Using Eqs. (8) and (9) ensures that everything coupling to the gravitational sector is located on the three–brane. Each KK mode is characterized by the coupling

$$\frac{1}{M_D^{1+(N/2)}} \psi^{\{n\}}(\mathbf{y}=\mathbf{0}) \int d^4x \, h_{\mu\nu}^{\{n\}}(x) T^{\mu\nu}(x). \tag{10}$$

When the compact geometry is a torus this expression becomes

$$\frac{1}{M_P} \int d^4x \, h_{\mu\nu}^{\mathbf{n}}(x) T^{\mu\nu}(x), \tag{11}$$

where use has been made of the fact that $M_P = V_N^{1/2} M_D^{1+(N/2)}$. When considering a spherical compact geometry, we obtain the same kind of expression:

$$\frac{f_N(n)}{M_P} \int d^4x \, h_{\mu\nu}^n(x) T^{\mu\nu}(x), \tag{12}$$

where $f_N(n)$ represents a family of polynomials in n related to the mutliplicity of the states propagating on the sphere at each Kaluza–Klein level.

Using Eq. (10) we can find the Feynman rules for processes involving gravitons coupled to SM fields on the three–brane [10, 3]. We restrict ourselves to studying the potential relevance of the process $e^+e^- \to \gamma h$ for probing the geometry of the transverse space. From the four–dimensional point of view on submanifold M^4 this corresponds to the emission of a kinematically cutoff tower of massive graviton modes during a high–energy collision. The differential cross section for the emission of a quantum electrodynamics photon and a massive graviton (denoted h_m) following an e^+e^- collision with c.m. energy $\sqrt{s}$ is

$$\frac{d\sigma_m}{dt}(e^+e^- \to \gamma h_m) = \frac{\alpha}{16} \frac{1}{M_P^2} F(x,y), \tag{13}$$

where α is the electromagnetic fine-structure constant, $x = t/s$, $y = m^2/s$ and $F(x,y)$ can be found in Refs. [5, 3, 9]. From our limited four–dimensional point of view we do not distinguish between gravitons of different transverse momenta (mass squared). Thus, the actual cross section for graviton emission from the brane is obtained by summing Eq. (13) over all kinematically allowed values of m^2, i.e., up to $m^2 = s$. Note that when we use the variable $y = m^2/s$ the sum conveniently runs from $y = 0$ to 1.

From an experimental perspective, the $e^+e^- \to \gamma h$ process is competing with the Standard Model background $e^+e^- \to \gamma \nu \bar{\nu}$. Of course, when there are either no or extremely small extra dimensions the gravitational process, being suppressed by a M_P^{-2} factor, is completely undetectable. With large extra dimensions the relatively important number of KK modes enhances the graviton signal, which leads to a potentially detectable departure from the SM signature. This corresponds to $\sigma(e^+e^- \to \gamma h)$ no longer being suppressed by a M_P^{-2} factor but by a $M_D^{-(2+N)}$ factor. Consequently, picking M_D as small as possible leads to larger gravitational signals. There is a fundamental limitation in our freedom to do that though. In fact, using the Gauss law one finds the following low energy constraint[1]:

$$M_P^2 = M_D^{2+N} V_N, \tag{14}$$

where V_N is the volume of the compact space. Requiring the effective low–energy four–dimensional gravitational coupling to be the observed Newton constant G_N is equivalent to imposing Eq. (14), which is a relationship between the size of the extra dimensions, their number N, and the true quantum gravity scale M_D. It is interesting to note that experiments have been performed probing gravity down to approximately 1 mm [12] without finding any discrepancies with the usual $1/r$ potential. Based on Eq. (14), this implies that for $N = 2$ the quantum gravity scale could be as low a 1 TeV. Although this particular set of parameters seems to be excluded by astrophysical constraints [2], it does not mean that other values of N and M_D leading to detectable signatures have to be rejected. It should be clear that the graviton signal is increasingly suppressed relative to the background process $e^+e^- \to \gamma \nu \bar{\nu}$ as M_D and N are increased.

Phase space integrals on T^N

For a T^N geometry the wave function for the transverse graviton modes at $\mathbf{y} = 0$ is independent of $\mathbf{n}$. Based on Eq. (11) this means that there is no restriction on the quantum numbers of the modes that are emitted at a given point on the torus (from the three–brane). Then the operator we use to sum over transverse momenta is [5]

$$O_{T^N} = \sum_{\mathbf{k}} \to \frac{1}{V_{|\mathbf{k}|}} \int d^N k = \frac{\Omega_N R^N}{2} \int m^{N-2} dm^2 = \frac{\Omega_N R^N s^{N/2}}{2} \int_0^1 dy \, y^{(N-2)/2}, \tag{15}$$

where $V_{|\mathbf{k}|}$ is the volume occupied by one state in $\mathbf{k}$-space, $y = m^2/s$, and

$$\Omega_N = \frac{2\pi^{N/2}}{\Gamma(N/2)} \tag{16}$$

is the volume of a unit sphere embedded in a N-dimensional space. Because the extra dimensions are assumed to be large (in TeV^{-1} units) with respect to the inverse center of mass energy of the process, it is reasonable to take the continuum limit when performing the sum over momenta (see Ref. [9] for more details).

Phase space integrals on S^N

In Ref. [9] we computed the operator for summing over modes propagating on the transverse sphere. For example, when $N = 6$ we obtain

$$O_{S^6} \to V_{S^6} \frac{s^3}{128\pi^3} \int_0^1 dy \left(y + \frac{4}{a^2 s} \right) \left(y + \frac{6}{a^2 s} \right). \tag{17}$$

The parameter playing a role in distinguishing a spherical from a toric geometry depends on the size of the internal space. We label it

$$d_a = \frac{1}{n_{max}^2}, \tag{18}$$

where $n_{max} = a\sqrt{s}$ is the maximum quantum number over which we integrate when performing the phase space sum on a sphere. If we integrate over an overwhelmingly large number of states ($d_a \to 0$) the $e^+e^- \to \gamma h$ cross section evaluated on T^N is expected to be very close to the one evaluated on S^N. This corresponds to a sector of the theory for which the typical size of the extra dimensions (in TeV^{-1} units) is large with respect to the inverse center of mass energy of the process ($a \gg 1/\sqrt{s}$). In this case the spacing between KK levels is small compared with the c.m. energy. For example, it can be seen that $a \sim 2 \times 10^5$ TeV^{-1} if $N = 6$, $M_D = 1$ TeV and $a \sim 200$ TeV^{-1} with $M_D = 30$ TeV. As we increase both M_D and N we expect the difference between cross sections on T^N and S^N to increase since this corresponds to taking larger values of d_a (a smaller number of KK modes are summed over). The numerical factors multiplying d_a in O_{S^N} are more important for large N which also contributes in enhancing the difference in cross sections between the two geometries.

When the internal space has a typical length scale which is extremely small (with respect to the inverse c.m. energy of the process), one expects processes taking place on a torus to be indistinguishable from processes on a sphere (or on any other smooth manifold for that matter). This is not reflected in our phase space integral procedure. In the limit when the extra dimensions are extremely small the procedure we are using is not valid anymore since then it is highly probable that, for the range of c.m. energies considered, only the zero mode of the graviton will be excited. The phase space integral procedure is useful only when numerous modes are excited, i.e., when the extra dimensions are large. The $N = 2$ case is special since it is then impossible, for large extra dimensions, to distinghish between the T^2 and the S^2 geometries using a process such as $e^+e^- \to \gamma h$ ($O_{T^2} = O_{S^2}$).

If d_a is not too small we expect differences in cross sections evaluated on T^N and S^N for $N > 2$. Models with a spherical transverse space lead to larger cross sections. As s

is augmented the difference between the cross sections increases slightly. This suggests that overall the number of KK modes excited on a $N > 2$ sphere is larger than on a torus. Based on Eq. (18) we see that s is related to the maximum quantum number (n_{max}) over which the integration is performed. Consequently, the larger the c.m. energy is, the larger we expect the cross section differences to be (a large s corresponds to integrating over more modes). Since highly energetic modes are not expected to differentiate between smooth geometries (their wavelength is assumed to be much smaller than the inverse curvature squared of the internal space), there exists a c.m. energy beyond which the multiplicity at each level is the same both for the sphere and the torus. Past this critical s value, the difference between cross sections becomes constant.

A parameter we can use to quantify the effect of the compact geometry on graviton emission from the three–brane is the ratio

$$D_N(s, M_D) = \frac{\sigma_{S^N}(e^+e^- \to \gamma h) - \sigma_{T^N}(e^+e^- \to \gamma h)}{\sigma_{S^N}(e^+e^- \to \gamma h)}, \quad (19)$$

which is a function of the size of the extra dimensions through N and M_D. This quantity characterizes relative rather than absolute differences. As shown earlier, the ratio D_N is zero for $N = 2$. For $N = 3$ and $N = 4$, all considered values of M_D and $\sqrt{s}$ lead to a function D_N which is a negligible fraction of a percent. D_5 is also negligible for $M_D = 1$ TeV but reaches 0.002% for $M_D = 10$ TeV (for small values of s). When $M_D = 30$ TeV the ratio D_5 goes as high as 0.04% for $\sqrt{s} = 0.1$ TeV but goes down to 0.002% for $\sqrt{s} = 0.5$ TeV. The most noticeable effects occur for $N = 6$. Then, with $M_D = 30$ TeV, D_6 varies from 5%–0.1% as $\sqrt{s}$ spans the 0.1 to 0.5 TeV range. Still for $N = 6$, when $M_D = 10$ TeV, D_N varies from 0.3% to 0.01% but is negligible for $M_D = 1$ TeV.

In conclusion, we find that the relative difference between cross sections (for a given s) in models with spherical and toric geometries takes larger values when the quantum gravity scale is large and the dimensions are numerous. While the absolute difference increases with s (for a given M_D and N) the relative difference D_N does just the opposite. This is expected as cross sections are rising functions of the c.m. energy.

CONCLUDING REMARKS

Having graviton modes propagating on an N–torus is exactly the same as having them existing in an N–dimensional box. Such a geometry has no intrinsic curvature; so whether the modes have low or high discretized momenta does not matter. By that we mean that all modes perceive the space as being T^N. If the compact geometry is S^N the situation is different. The high energy modes, having a small wavelength in transverse space, do not behave differently than when they are propagating on a N–torus (the wavelength squared is assumed to be much smaller that the inverse local curvature). It then makes sense to assume that the physics resulting from these high energy modes cannot be used to distinguish between processes taking place on different compact geometries (unless the associated curvature is large). The graviton modes associated with small quantum numbers (low–energy modes) are the ones that can be used to study the shape of the extra dimensions. In fact, their wavelength is presumably large enough to allow them to recognize a sphere from a torus for example. The multiplicity of states at each quantum level on the compact geometry lattice grows as the norm of the momentum is increased. We have shown that overall this mulitiplicity is larger on the sphere. This explains why the cross section for a process like $e^+e^- \to \gamma h$ is larger when the compact geometry has a spherical symmetry. We have seen that past a certain transerve momenta the multiplicity on a sphere and a torus become equal. This means that beyond some critical value for the c.m. energy, the difference between the cross sections evaluated for different geometries stabilizes to a constant value. This is what we have found for the $N=6, M_D=30$ TeV case but this is true in general. As M_D is augmented the geometry of the compact space plays an increasingly important role. While this is true, it is also worth noting that when M_D is increased, deviations from the $e^+e^- \to \gamma\nu\bar\nu$ process progressively become negligible. In fact, the size of the extra dimensions then becomes small, which allows only a limited number of modes to propagate in the extra dimensions.

In summary, high energy colliders are limited in their ability to determine the geometry of large extra dimensions. This is due to two factors:

- as $\sqrt{s}$ is increased the relative difference between cross sections on different geometries decreases because the contribution of the low energy modes becomes increasingly small;
- the impact of the geometry is more important for a large quantum gravity scale and numerous extra dimensions but this also corresponds to a smaller gravitational contribution to SM background processes.

In fact, the geometrical effect we find for the $e^+e^- \to \gamma h$ process is rather small. Nevertheless, it is still conceivable that it could be detected at the upcoming LHC, using the process $q\bar q \to \gamma h$, for certain values of N and M_D.

Acknowledgements: The author is indebted to R. C. Myers for many insightful comments during the realization of this work which was supported by the Natural Sciences and Engineering Research Council of Canada (NSERC).

REFERENCES

1. N. Arkani-Hamed, S. Dimopoulos and G. Dvali, Phys. Lett. B **429**, 263 (1998) [hep-ph/9803315]; I. Antoniadis, N. Arkani-Hamed, S. Dimopoulos and G. Dvali, Phys. Lett. B **436**, 257 (1998) [hep-ph/9804398].
2. N. Arkani-Hamed, S. Dimopoulos and G. Dvali, Phys. Rev. D **59**, 086004 (1999) [hep-ph/9807344]; S. Nussinov and R. Shrock, Phys. Rev. D **59**, 105002 (1998) [hep-ph/9811323]; S. Cullen and M. Perelstein, Phys. Rev. Lett. **83**, 268 (1999) [hep-ph/9903422]; V. Barger, T. Han, C. Kao and R. J. Zhang, Phys. Lett. B **461**, 34 (1999) [hep-ph/9905474]; D. Atwood, C. P. Burgess, E. Filotas, F. Leblond, D. London and I. Maksymyk, Phys. Rev. D **63**, 025007 (2001) [hep-ph/0007178]; C. Hanhart, D. R. Phillips, S. Reddy and M. J. Savage, Nucl. Phys. **B595**, 335 (2001) [nucl-th/0007016]; M. Fairbairn, "Cosmological constraints on large extra dimensions," hep-ph/0101131.
3. G. F. Giudice, R. Rattazzi and J. D. Wells, Nucl. Phys. **B544**, 3 (1999) [hep-ph/9811291].
4. J. L. Hewett, Phys. Rev. Lett. **82**, 4765 (1999) [hep-ph/9811356].
5. E. A. Mirabelli, M. Perelstein and M. E. Peskin, Phys. Rev. Lett. **82**, 2236 (1999) [hep-ph/9811337].
6. T. G. Rizzo, Int. J. Mod. Phys. A **15**, 2405 (2000).

7. C. W. Misner, K. S. Thorne and J. A. Wheeler, *Gravitation* (Freeman, San Francisco, 1973).
8. F. Leblond, R. C. Myers and D. J. Winters, JHEP **0107**, 031 (2001) [hep-th/0106140].
9. F. Leblond, Phys. Rev. Lett. D **64**, 045016 (2001) [hep-ph/0104273].
10. T. Han, J.D. Lykken and R.-J. Zhang, Phys. Rev. D **59**, 105006 (1999) [hep-ph/9811350].
11. J. Lykken and S. Nandi, Phys. Lett. B **485**, 224 (2000) [hep-ph/9908505].
12. C. D. Hoyle, U. Schmidt, B. R. Heckel, E. G. Adelberger, J. H. Gundlach, D. J. Kapner and H. E. Swanson, Phys. Rev. Lett. **86**, 1418 (2001) [hep-ph/0011014].

Quantum Myers effect and its Supergravity dual for D0/D4 systems

Pedro J. Silva

Physics Department, Syracuse University, Syracuse, New York 13244

Abstract. The presence of a distant D4-brane is used to further investigate the duality between M-theory and D0-brane quantum mechanics. A polarization of the quantum mechanical ground state is found. A similar deformation arises for the bubble of normal space found near D0-branes in classical supergravity solutions. These deformations are compared and are shown to have the same structure in each case.

In recent years the outlook within string theory has changed immensely. The new cornerstones of the theory are non-perturbative duality conjectures. Some of the most impressive such conjectures are those of matrix theory [1] and Maldacena's AdS/CFT conjecture [2].

These conjectures relate the physics of certain gravitating systems to that of specific non-gravitating gauge theories. The dynamics of the dual field theories are deduced from the low energy effective actions of the various non-abelian D-brane systems. The correspondences appear to rely on the particular form of the non-abelian interactions.

Recently, several investigations [3, 4] have uncovered the form of certain non-abelian couplings of D-branes to supergravity background fields. Our goal here is to invest the 'polarization' of the Dp-brane bound state in the background of a D(p+4)-brane. For definiteness, we shall concentrate on the D0/D4 context.

In certain cases, the application of a Ramond-Ramond background field to a D0-brane system induces a classical dielectric effect and causes the D0-branes to deform into a non-commutative D2-brane [4]. While the Ramond-Ramond fields of our D4-brane background will not be strong enough to induce such a classical effect, they do modify the potential that shapes the non-abelian character of the quantum D0 bound state. As a result, this bound state is deformed, or polarized. Fundamental to this studies will be the connection described by Polchinski [5] relating the size of the matrix theory bound state to the size of the bubble of space that is well-described by classical supergravity in the near D0-brane spacetime.

The near D0-brane spacetime is obtained by taking a limit in which open strings decouple from closed strings and the result is a ten-dimensional spacetime which has small curvature and small string coupling when one is reasonably close (though not too close) to the D0-branes. However, if one moves beyond some critical r_c, the curvature reaches the string scale. As a result, the system beyond r_c is not adequately described by the massless fields of classical supergravity.

Our goal is therefore to compare the deformations of the non-abelian D0-brane bound state with the deformations of this bubble of 'normal' space around a large stack of D0-

branes. As has become common in string theory, we find that the quantum mechanical effects of the non-abelian D0-brane couplings correctly reproduce the effects of classical supergravity in the large N limit.

Throughout this document a series of approximation and large N calculations together with hand-waving arguments will be given, with no justification, the actual justifications can be found on [6].

We begin with the world-volume effective field theory describing N D0-Branes in the standard D4-brane background. This action is a suitable generalization of the action for a single D0-brane, consisting of the Born-Infeld term together with appropriate Chern-Simon terms. However, the full action encodes the Chan-Paton factors or non-abelian degrees of freedom that arise from strings stretching between the D0-branes.

The first part of the non-abelian D0 effective action is the Born-Infeld term

$$S_{BI} = -T_0 \int dt\, STr \left(e^{-\phi} \sqrt{-\left(P\left[E_{ab} + E_{ai}(Q^{-1} - \delta)^{ij} E_{jb}\right]\right) det(Q^i{}_j)} \right) \quad (1)$$

with $E_{AB} = G_{AB} + B_{AB}$ and $Q^i{}_j \equiv \delta^i{}_j + i\lambda [\Phi^i, \Phi^k] E_{kj}$. In writing (1) we have used a number of conventions taken from Myers [4]:

The rest of the action is given by the non-abelian Chern-Simon terms

$$S_{CS} = \mu_0 \int dt\, STr \left(P\left[e^{i\lambda i_\Phi i_\Phi} (\sum C^{(n)} e^B) \right] \right). \quad (2)$$

The symbol i_Φ is a non-abelian generalization of the interior product with the coordinates Φ^i, $i_\Phi \left(\frac{1}{2} C_{AB} dX^A dX^A \right) = \Phi^i C_{iB} dX^B$.

The D4-brane background is defined by the metric G_{AB}, the dilaton ϕ, and the Ramond-Ramond 6-form field strength $F_{A_1 A_2 A_3 A_4 A_5 A_6}$:

$$ds_4^2 = \mathcal{H}_4^{-1/2} \eta_{\mu\nu} dX^\mu dX^\nu + \mathcal{H}_4^{1/2} \delta_{mn} dX^m dX^n$$
$$e^{-2\phi} = \mathcal{H}_4^{1/2}, \quad F_{01234m} = \partial_m \mathcal{H}_4^{-1}, \quad (3)$$

with all other independent components of the field strength vanishing. Here the space-time coordinates described by the index A have been partitioned into directions parallel to the D4-brane (which we will label with a Greek index μ) and directions perpendicular to the D4-brane (which we label with a Latin index m). The function $\mathcal{H}_4 = 1 + (r_4/|X|)^3$ is the usual harmonic function of the D4-brane solution with $|X|^2 = \delta_{mn} X^m X^n$ and with $r_4 = (gN_4)^{1/3} l_s$ being the constant that sets the length scale of the supergravity solution.

Expanding the Born-Infeld and the Chern-Simon action in this background we get,

$$S_{eff.} = -T_0 \lambda^2 \int dt\, STr \left\{ \frac{1}{2g_{tt}} \partial_t \Phi \partial_t \Phi + \frac{1}{4} [\Phi, \Phi]^2 + \right.$$
$$+ \lambda \left(\frac{1}{2} \partial_t \Phi^i \partial_t \Phi^j \Phi^k \partial_k (g_{tt})^{-1} + \frac{1}{2g_{tt}} \partial_t \Phi^i \partial_t \Phi^j \Phi^k \partial_k g_{ij} + \right.$$
$$\left. \left. + \frac{1}{2} [\Phi, \Phi^i][\Phi^j, \Phi] \Phi^k \partial_k g_{ij} + \frac{1}{10} \Phi^i \Phi^j \Phi^k \Phi^l \Phi^m F_{\tau ijklm} \right) \right\}. \quad (4)$$

The are also Fermion terms that are rather long, and little insight is gained by writing them explicitly here.

One might begin with a discussion of classical solutions corresponding to the above effective action. However, aside from the trivial commutative solution, one does not expect to find any static solutions.However, aside from the trivial commutative solution, one does not expect to find any static solutions[1]. Nevertheless, we may expect that the non-abelian couplings to the background affect the quantum bound state by altering the shape of the potential and thus the ground state wavefunction. Let us calculate the size of the ground state by considering the expectation value of the squared radius operator $R^2 \equiv \lambda^2 Tr(\Phi^2) = \lambda^2 Tr(\Phi^i \Phi^j g_{ij})$. Note that by passing to an orthonormal frame one finds a full SO(9) spherical symmetry in the $O(\lambda^2)$ terms in our action. Thus, to $O(\lambda^2)$ the expectation value of $Tr(\Phi^i \Phi^i g_{ii})$ is independent of i and R^2 is the radius of the corresponding sphere measured in terms of string metric proper distance.

Here we give a simple argument for the behavior of $\langle R^2 \rangle$ based on the usual 't Hooft scaling behavior. Our strategy is to treat the couplings to the D4-brane fields as perturbations to the D0-brane action in flat empty spacetime. Thus, we divide (4) into an 'unperturbed action' S_0 and a perturbation S_1. Note that as we place the N D0-branes far from the D4-branes, the Ramond-Ramond coupling term can be written in the form

$$\Phi^{\mu_1}\Phi^{\mu_2}\Phi^{\mu_3}\Phi^{\mu_4}\Phi^m F_{0\mu_1\mu_2\mu_3\mu_4 m} = \frac{f}{\lambda^{1/2}} \Phi^{\mu_1}\Phi^{\mu_2}\Phi^{\mu_3}\Phi^{\mu_4} \varepsilon_{\mu_1\mu_2\mu_3\mu_4} \Phi^m \frac{X^m}{|X|} \quad (5)$$

where $f = 3(r_4^3 \lambda^{1/2})/(z_\perp^4) \approx 3\mathcal{H}_4^{-2}(r_4^3 \lambda^{1/2}/(z_\perp^4))$ is a scalar dimensionless measure of the field strength. Here $z_\perp$ is the distance between the N D0-branes and the D4-brane, and $\varepsilon_{\mu_1\mu_2\mu_3\mu_4}$ is the antisymmetric symbol on four indices. In what follows we treat all effects of the D4-brane only to lowest order in $(\mathcal{H}_4 - 1)$ and $f = 3(r_4^3 \lambda^{1/2})/(z_\perp^4)$, so that $f \approx \mathcal{H}_4^{-2} f$. With this understanding, the other $O(\lambda^3)$ terms are also proportional to f.

It will be useful to express the dynamics in terms of rescaled fields and a rescaled time coordinate:

$$\tilde{\Phi}^i = \lambda^{1/2} \mathcal{H}_4^{1/12} (gN)^{-1/3} \Phi^i \; , \; \tilde{t} = \lambda^{-1/2} \mathcal{H}_4^{-1/3} (gN)^{1/3} t. \quad (6)$$

This yields the action [10]

$$S_0 = -N \int d\tilde{t} \, STr \left(-\frac{1}{2} \partial_{\tilde{t}} \tilde{\Phi} \partial_{\tilde{t}} \tilde{\Phi} + \frac{1}{4} [\tilde{\Phi}, \tilde{\Phi}]^2 \right) \quad (7)$$

and the perturbation

$$S_1 = -\left[(gN)^{1/3} \mathcal{H}_4^{-1/12} f \right] N \int d\tilde{t} \, STr \left(\frac{1}{10} \tilde{\Phi}^i \tilde{\Phi}^j \tilde{\Phi}^k \tilde{\Phi}^l \tilde{\Phi}^m \varepsilon_{ijklm} + \right.$$
$$\left. \frac{1}{2} \partial_{\tilde{t}} \tilde{\Phi}^i \partial_{\tilde{t}} \tilde{\Phi}^j \tilde{\Phi}^k \frac{\partial_k (g_{tt})^{-1}}{f} + \frac{1}{2} \partial_{\tilde{t}} \tilde{\Phi}^i \partial_{\tilde{t}} \tilde{\Phi}^j \tilde{\Phi}^k \frac{\partial_k g_{ij}}{f} + \frac{1}{2} [\tilde{\Phi}, \tilde{\Phi}^i][\tilde{\Phi}^j, \tilde{\Phi}] \tilde{\Phi}^k \frac{\partial_k g_{ij}}{f} \right),$$

---

[1] The literature [11, 12, 13] contains some examples of classical non-commutative solutions in similar (but non-supersymmetric) systems.

$$\equiv -[(gN)^{1/3}\mathcal{H}_4^{-1/12}f]\tilde{S}_1. \tag{8}$$

Note that in writing $\tilde{S}_1$ we have extracted a factor of f from S_1. The advantage of this form is that both S_0 and $\tilde{S}_1$ are manifestly independent of g, λ, and f while they depend on N only through the overall factor and the trace. The dependence of S_0 and $\tilde{S}_1$ on $\mathcal{H}_4$ is only though contractions with g_{ij}. These could be further eliminated by passing to an orthonormal frame, so the dynamics of scalar contractions such as $\Phi^i \Phi^j g_{ij}$ will be independent of $\mathcal{H}_4$.

Let us now consider the case $f = 0$ and the corresponding ground state $\langle R^2 \rangle_0$. We will think of this as the limit of small $\frac{\ell_s}{z_\perp}$, so that we preserve $\mathcal{H}_4 \neq 1$. Note that we have

$$\langle R^2 \rangle_0 = (gN)^{2/3}\mathcal{H}_4^{-1/6}\lambda \langle Tr\tilde{\Phi}^2 \rangle_0. \tag{9}$$

The factor $\langle Tr\tilde{\Phi}^2 \rangle_0$ is manifestly independent of g, $\mathcal{H}_4$, and λ, and the form of S_0 is the usual one associated with 't Hooft scaling for which $\langle Tr\tilde{\Phi}^2 \rangle_0$ is also independent of N in the limit of large N with gN fixed. This reproduces the results of [5, 1]:

$$\sqrt{\langle R^2 \rangle_0} \sim (gN)^{1/3}\mathcal{H}_4^{-1/12}\lambda^{1/2}, \tag{10}$$

where the product of $(g\mathcal{H}_4^{-1/4})^{1/3}$ can be viewed as the natural dependence on the local string coupling $g_{local} \equiv ge^\phi = g\mathcal{H}_4^{-1/4}$ of the D4-brane background.

Let us now turn to the perturbed system. Considering the ground state expectation value as the low temperature limit of a thermal expectation value gives a Euclidean path integral for $\langle R^2 \rangle$. We wish to expand the factor $e^{-S_1} = e^{-\left((gN)^{1/3}\mathcal{H}_4^{-1/12}f\right)\tilde{S}_1}$ as $1 - (gN)^{1/3}\mathcal{H}_4^{-1/12}f\tilde{S}_1 + (gN)^{2/3}\mathcal{H}_4^{-1/6}f^2\tilde{S}_1^2 - \ldots$. Note that the order zero term gives just $\langle R^2 \rangle_0$, the expectation value in the unperturbed ground state. The contribution from the first order term then vanishes because $\tilde{S}_1$ is anti-symmetric under a total inversion of space while R^2, S_0, and the integration measure are invariant. Thus, the leading contribution is of second order in $\tilde{S}_1$, in the 't Hooft limit we express our final result as

$$\frac{\langle R^2 \rangle - \langle R^2 \rangle_0}{\langle R^2 \rangle_0} \sim (gN)^{2/3}\mathcal{H}_4^{-1/6}f^2. \tag{11}$$

where we have isolated the dependance on g, N, f.

Having considered the quantum mechanical description of the non-abelian D0-brane bound state, we now wish to compute a corresponding effect in classical supergravity. We seek a BPS solution containing both D0's and D4's with the D0's being both fully localized and separated from the D4-branes. It is conceptually simplest to discuss the full D0/D4 solution and then take a suitable decoupling limit. Such full solutions are known exactly, but only as an infinite sum over Fourier modes [15]. As a result, we find it more profitable here to follow a perturbative method as suggested by the quantum mechanical calculation above. We therefore expand the supergravity solution in f, the magnitude of the Ramond-Ramond 4-form field strength at the location of the zero-branes.

Let us consider a BPS system of D4-branes and N D0-branes with asymptotically flat boundary conditions. Using the usual isotropic ansatz in the appropriate gauge reduces the problem to solving the equations [16, 17, 18, 19]

$$\left(\partial_\perp^2 + \mathcal{H}_4 \partial_\parallel^2\right) \mathcal{H}_0 = 0 \, , \quad \mathcal{H}_4 = 1 + \left(\frac{r_4}{|X|}\right)^3, \tag{12}$$

where as before the D4-brane lies at $X^m = 0$, $|X|^2 = \delta_{mn} X^m X^n$, and $\mathcal{H}_4$ and $\mathcal{H}_0$ are the 'harmonic' functions for the D4-brane and D0-brane respectively. The two relevant derivative operators are a flat-space Laplacian ($\partial_\parallel^2 \equiv \sum_{\mu=1}^{\mu=4} \partial_\mu \partial_\mu$) associated with the directions parallel to the D4-brane and another ($\partial_\perp^2 \equiv \sum_{m=5}^{m=9} \partial_m \partial_m$) associated with the perpendicular directions.

In order to treat the D4-branes as a perturbation, we place them far away from the D0-branes. It is convenient to change to new coordinates x^m (lowercase) whose origin is located at the D0 singularity. One of these coordinates is distinguished by running along the line connecting the D0- and D4-branes. Let us call this coordinate $x_\perp$. The other four x^m coordinates will play a much lesser role. Introducing the distance $z_\perp$ between the D0- and D4-branes and expanding $\mathcal{H}_4$ to first order about the new origin yields

$$\mathcal{H}_4 \approx \mathcal{H}_4(x=0) - 3 \left(\frac{r_4}{z_\perp}\right)^3 \left(\frac{x_\perp}{z_\perp}\right) \equiv \mathcal{H}_4(0) + \delta \mathcal{H}_4. \tag{13}$$

Here we have used $z_\perp \gg (r_4, x_\perp)$, since the D4-branes are located far away. Note that fixing $z_\perp$ sets the location of the D0 singularity relative to the D4-brane. However, as we will see, it is not clear in general that the position of the singularity corresponds precisely to the center of mass. Equation 12 can be solved by expanding $\mathcal{H}_0$ in terms of $\delta \mathcal{H}_4$ i.e. $\mathcal{H}_0 = \mathcal{H}_{00} + \mathcal{H}_{01} + ...$ where $\mathcal{H}_{0n} = O(\delta \mathcal{H}_4^n)$. We find

$$(\partial_\perp^2 + \mathcal{H}_4(0) \partial_\parallel^2) \mathcal{H}_{00} = 0, \quad \text{so that} \quad \mathcal{H}_{00} = 1 + \left(\frac{r_0}{r}\right)^7,$$
$$(\partial_\perp^2 + \mathcal{H}_4(0) \partial_\parallel^2) \mathcal{H}_{01} = \delta \mathcal{H}_4 \partial_\parallel^2 \mathcal{H}_{00} \quad \text{and} \quad (\partial_\perp^2 + \mathcal{H}_4(0) \partial_\parallel^2) \mathcal{H}_{02} = \delta \mathcal{H}_4 \partial_\parallel^2 \mathcal{H}_{01} \tag{14}$$

Here we have introduced $r^2 = |x|^2 \equiv \delta_{mn} x^m x^n + \mathcal{H}_4^{-1}(0) \delta_{\mu\nu} x^\mu x^\nu$, a sort of coordinate distance from the D0-brane. Note that this r does in fact label spheres of symmetry for the unperturbed solution $\mathcal{H}_{00}$.

Since the D4-brane background has altered the background metric for the D0-brane system, this will change certain familiar normalizations. We therefore note that total electric flux from the D0-branes must equal the number N of D0-brane charge quanta $(g\ell_s^7)/(60\pi^3)$. Since the D4-brane is far away, we may compute this flux in a region where $\mathcal{H}_0 \approx 1$ but where $\mathcal{H}_4 = \mathcal{H}_4(0)$. The result yields $(gN\ell_s^7)/(60\pi^3) = \mathcal{H}_4^2 r_0^7$.

The above equations (14) are easily solved in terms of Green's functions. We stress that under a rescaling of coordinates $y, x \to \beta y, \beta x$ the function $\mathcal{H}_{01}$ scales homogeneously as β^{-6}. As a result, we may write

$$\mathcal{H}_{01} = \frac{\omega_1 f}{x^6} r_0^7 \lambda^{-1/2}, \tag{15}$$

where ω_1 is an unknown dimensionless function of the angles associated with the direction cosines x^A/r and $f = (3\lambda^{1/2}r_4^3)/(z_\perp^4)$. Furthermore, $\mathcal{H}_{01}$ is even under any $x^\mu \to x^\mu$ and under any $x^m \to x^m$ for $x^m \neq x_\perp$. However, we see that $\mathcal{H}_{01}$ is odd under $x_\perp \to -x_\perp$. Thus, the $\mathcal{H}_{01}$ term provides an (angle dependent) *shift* of the bubble so that it is no longer centered on the D0-brane center of mass. In particular, this has no effect on the *size* of the bubble. For the second order term, we have

$$\mathcal{H}_{02} = \frac{\omega_2 f^2}{x^5}, \tag{16}$$

where ω_2 is again a dimensionless function of the angles. This time, however, $\mathcal{H}_{02}$ is even under $x^A \to x^A$ for any A. As a result, $\mathcal{H}_{02}$ directly encodes a change in the size of the bubble.

Let us now calculate the size of this solution. We follow Polchinski [5] and use the measure that successfully reproduces the size of the unperturbed D0-brane bound state. This means that we should locate the surface enclosing the D0-brane singularity where the string metric is so strongly curved that it has structure on the string scale. Inside this surface is a bubble of space that is well described by classical supergravity. However, when r is large the proper distance $2\pi\mathcal{H}_4^{1/4}\mathcal{H}_0^{1/4}r$ around the sphere enclosing the origin may still be on the order of ℓ_s so that the metric clearly has structure on the string scale. In particular, one might think of strings encircling the bubble itself. The region inside this surface is to be associated with the quantum D0-bound state, and one expects the bubble of 'normal' space and the bound state to have corresponding sizes and shapes.

Now, given the correspondence between the r of supergravity isotropic coordinates and the non-abelian Φ^2 in D0 quantum mechanics in the absence of the D4-brane, it is natural to expect a correspondence between the $R^2 = Tr(\Phi^i\Phi^j g_{ij})$ on the QM calculation and the supergravity quantity $R^2 = \mathcal{H}_4^{1/2}\delta_{mn}x^m x^n + \mathcal{H}_4^{-1/2}\delta_{\mu\nu}x^\mu x^\nu = \mathcal{H}_4^{1/2}r^2$. The edge of the bubble lies at the value of $R \equiv \mathcal{H}_4^{1/4}r$ for which $\ell_s \sim r\mathcal{H}_0^{1/4}\mathcal{H}_4^{1/4} = R\mathcal{H}_0^{1/4}(r)$. This yields the relation

$$R \sim \ell_s[\mathcal{H}_0(r)]^{-1/4} \sim \ell_s\mathcal{H}_{00}^{-1/4}\left(1 - \frac{\mathcal{H}_{01}}{4\mathcal{H}_{00}} - \frac{\mathcal{H}_{02}}{4\mathcal{H}_{00}} + \frac{5}{20}\left(\frac{\mathcal{H}_{01}}{4\mathcal{H}_{00}}\right)^2 + O(f^3)\right). \tag{17}$$

In order to make connections with the quantum mechanical calculations, we wish to consider this system in the decoupling limit $g \to 0$ with fixed gN and $r/\ell_s \sim g^{1/3} \to 0$. Note that this scaling may seem more familiar when expressed in terms of the eleven-dimensional plank mass $M_{11} = g^{-1/3}\ell_s^{-1}$ as it holds fixed the dimensionless quantity rM_{11}. We see that the corresponding asymptotic behavior of $\mathcal{H}_{00}$ is given by

$$\mathcal{H}_{00} \approx \left(\frac{r_0}{r}\right)^7 = \frac{gN\ell_s^7}{\mathcal{H}_4^{1/4}R^7}. \tag{18}$$

As a result, keeping only the order zero term yields an unperturbed value R_0 of R given by $R_0 \sim (gN)^{-1/4}\ell_s^{-3/4}\mathcal{H}_4^{7/16}R_0^{7/4}$, or $R_0 \sim (gN)^{1/3}\ell_s\mathcal{H}_4^{-1/12}$ in agreement with (9).

Recall that the effect of the $\mathcal{H}_{01}$ term is to shift the bubble by an angle-dependent amount but not to change the size of the bubble. Due to the angle-dependence, concepts like the radius R of the bubble also become angle-dependent. However, it is the average $\langle R^2 \rangle$ of R^2 over the bubble that we expect to compare with expectation values in the quantum mechanical ground state. Taking such an average, it is clear that the term in (17) that is linear in $\mathcal{H}_{01}$ has no effect on $\langle R^2 \rangle$. Of course, a shift of the bubble away from the origin will contribute to $\langle R^2 \rangle$ at second order, and this effect is governed by the term $\frac{5}{20}(\mathcal{H}_{01})/(4\mathcal{H}_{00})^2$. This is in agreement with our quantum mechanical calculation where the effect on $\langle R^2 \rangle$ appeared only at second order in f. The result after averaging, and taking the decoupling limit in the necessary order of accuracy gives

$$\frac{\langle R \rangle^2 - R_0^2}{R_0^2} \sim f^2 \mathcal{H}_4^{-1/6} (gN)^{2/3}, \tag{19}$$

in agreement with the quantum mechanical results of (11).

ACKNOWLEDGMENTS

The author would like to thank Mark Bowick, Amanda Peet, Joe Polchinski, and Joel Rozowsky for useful discussions. This work was supported in part by NSF grant PHY97-22362 to Syracuse University, the Alfred P. Sloan foundation, and by funds from Syracuse University.

REFERENCES

1. T. Banks, W. Fischler, S.H. Shenker and L. Susskind, Phys.Rev. D55 (1997) 5112-5128, hep-th/9610043.
2. J. Maldacena, Adv. Theor. Math. Phys. 2 (1998) 231, hep-th/9711200.
3. W. Taylor, M. Van Raamsdonk, Nucl.Phys. B573 (2000) 703-734, hep-th/9910052.
4. R.C. Myers, JHEP 9912 (1999) 022, hep-th/9910053.
5. J. Polchinski, Prog.Theor.Phys.Suppl. 134 (1999) 158-170, hep-th/9903165.
6. D. Marolf and P. Silva, hep-th/0105298.
7. A.A. Tseytlin, hep-th/9908105.
8. P. Bain, hep-th/9909154.
9. W. Taylor, M. Van Raamsdonk, Nucl.Phys. B558 (1999) 63-95, hep-th/9904095.
10. M Clauson and B. Halpern, Nucl. Phys. B250 (1985)689.
11. S. P. Trivedi, S. Vaidya, JHEP 0009 (2000) 041, hep-th/0007011.
12. C. Gomez, B. Janssen, P. J. Silva, hep-th/0011242
13. J. Castelino, S. Lee, W. Taylor, Nucl.Phys. B526 (1998) 334-350, hep-th/9712105.
14. S. Surya and D. Marolf, Phys. Rev. D **58**, 124013 (1998), hep-th/9805121.
15. D. Marolf, A. W. Peet, Phys.Rev. D60 (1999) 105007, hep-th/9903213.
16. G. T. Horowitz and D. Marolf, Phys. Rev. D **55**, 3654 (1997), hep-th/9610171.
17. A. A. Tseytlin, Class. Quant. Grav. **14**, 2085 (1997), hep-th/9702163.
18. E. Bergshoeff, M. de Roo, E. Eyras, B. Janssen and J. P. van der Schaar, Class. Quant. Grav. **14**, 2757 (1997), hep-th/9704120.
19. I. Y. Aref'eva, M. G. Ivanov, O. A. Rytchkov and I. V. Volovich, Class. Quant. Grav. **15**, 2923 (1998), hep-th/9802163

AFTERWORD

Roger Migneron
(1937 - 1999)

Roger Migneron (1937-1999): *Reminiscences on Sharing Life with a Physicist*

Ina Pakkert Migneron

When Victor asked me if I could say a few words, I assumed that he meant, "say a few words about Roger." Sorry, Victor, if you meant that I should speak about physics. In the nearly 26 years that I have known Roger, I have never been able to sum up Roger in just a few words, but this time I tried very hard to get a reasonable facsimile together. As a friend told me once, Roger had a truly unique view of things. Perhaps this was because of his classical education, or because he was a chess player, or because he studied philosophy prior to studying science. He was a humanitarian, but first and foremost he was a scientist.

How do I best describe Roger to people who have not known him? I have often thought about this since his death. Some of the words that come to mind are Passionate, Dedicated, Bright, Compassionate, Gregarious, A GOOD FRIEND and someone with a great zest for life. Having watched Roger for many years, I have always been amazed at the effect he had on people. Through the years we shared doctors, a dentist and dermatologist, as well as various other professional people. Roger always came home with stories these people told him. *I* was never told any of these stories when I had to go and see them. Perhaps it was because he was a professional and I was not. I don't believe it was only that. He had a way of drawing people out and let them talk about themselves.

Roger seemed to have a natural charisma that attracted young and old. One example: while he was in the hospital, he frequently was visited by the gastro-enterologist he had met only a few weeks prior to his operation. This doctor lent him some cd's in French while he was recuperating. Children were attracted to Roger, because he treated them with respect and did not talk down to them. I once watched him try to explain to a 5-year old how to put a jigsaw puzzle together..." you look for shape and color".... Needless to say, the 5-year old was not sure what Roger was talking about. Roger was a good listener, most of the time, and that helped him in his relationships. When asked, he would really try to help people find a solution to their problem.

I said Roger was a good listener *most* of the time. When we would be talking with a friend, he would be participating, but as soon as we started talking about mundane things, his mind would wander and his attention to the conversation would slip until he caught a word or two which tweaked his curiosity. Then he would say, "What was that?" and I would have to repeat the last few sentences to bring him up-to-date. In the beginning I was wont to let him get away with it and I would often repeat the conversation. When it started to happen too often, I told him that if the conversation was not important enough for him to listen to in the first place, then it was not important enough for me to repeat it. He usually was able to reconstruct the conversation, and he would pay attention from then on.

Roger had difficulty sitting completely still - some part of him was usually moving. The result is that when I first met him, the impression I had was of someone lively and full of energy. I also remember thinking at our first encounter, " This man is going places and I want to go with him." Well, he went places and I ended up going with him - mostly to conferences. Little did I know at the beginning of our relationship, that I would have to share Roger with a mistress for the rest of his life - that mistress was his research.

Research was very important to Roger, as he told me at the beginning of our relationship. It was so important in fact, that I was always competing with it. I remember when Roger was on sabbatical in California in the beginning of the 80's. I had always wanted to visit San Francisco, even way back when I was still in Europe and a visit to California seemed to be a far-flung dream. In 1980, Roger was on sabbatical in California for 6 months. The plan was that I should visit him, and we would make a trip which included the Grand Canyon and, of course, San Francisco. I made my travel arrangements and all was set for me to visit Roger for three weeks. One week before my departure from Canada, Roger phoned and asked me how I would feel if he took off the last week that I was in California. I hit the roof and said that I thought it was very callous of him to leave me alone for the last week of "our" vacation. When I inquired where he wanted to go and why, he told me that he had just heard about a conference that was going to take place in Italy during that time and which he really wanted to attend. Having been told many times already how important his research was to him, I knew that I had little chance to talk Roger out of it. And so it happened that after having made a quick trip to the Grand Canyon, I took Roger to the San Francisco airport and went exploring San Francisco on my own. That, however, is another story.

For at least the first 10 years of our relationship, Roger was gone most of the summers, either going to conferences or visiting collaborators. He always found someone with whom he could work, preferably not in London. Conferences I did not mind so much, because these often gave me an opportunity to visit interesting places. We often combined a conference with a vacation. Doing research in another place usually meant that he would be gone for almost three months in the summer - my preferred season in Canada. Eventually I got so fed up with always being by myself in summer that I gave him an ultimatum - either you stay here for a few summers, or else I will not be here when you come back. I think he got the message because, from then on, he would invite collaborators to come to London. It gave me an opportunity to meet some of his colleagues while allowing me to observe first-hand how intense he was when fully engaged in research.

Research was his first love and his great passion. He would be enthralled when he found (what he called) an elegant solution to a problem. He worked hard to build up the theoretical physics sector in the Department of Applied Mathematics at U.W.O. so that there would be a viable group with which to work. He was dedicated to his specialty, high-energy physics. I am truly, truly sorry that Roger was not able to learn about the Perimeter Institute. He would have loved to be involved with it - it would have been right up his alley.

What Roger was like as a teacher I cannot say. From what I can deduce however, I think he was demanding but good. Often he would say that the students did not know how good a course he was giving. I think that students, looking for a "bird-course," would have felt ill at ease in his classes. If, however, a student worked hard and encountered difficulties, Roger would give of his time generously to help the student.

His greatest pleasure, however, was to have a very bright student in his class. This pleasure was often bittersweet, because most of these guys would disappear into medicine. In the beginning I was curious to see what Roger was like as a teacher, and I threatened that I would come and sit in on his courses. He told me that if ever I did, he would walk out. I never dared sit in on his classes, because he would have done exactly that - walk out.

Roger was a passionate man. Aside from his research, he was also passionate about music, theatre, sport, chess and bridge. Music was very important to him. Through Roger I have really come to appreciate classical music - that has been one of his gifts to me. For many years he was a staunch supporter of Orchestra London Canada as well as The Grand Theatre. As a matter of fact he liked theatre so much that in the 50's, when The Stratford Festival had just started, he would hitchhike along Highway 2 (the 401 had not yet been built) in order to see plays in Stratford. That was when The Stratford Festival did not yet have the theatre building and the plays were performed under the tent. That was also when Sirs John Gielgud and Alec Guinness came to Stratford to perform.

When he was in England working toward his Ph.D., Roger managed to find a place in the house of an actress, Rosalie Crutchley. She often appeared on British TV, especially in the Miss Marple series. Whenever she performed in one of those pieces, Roger immediately recognized her voice and would get excited saying, "That is Rosalie!" Roger, somehow, would always seem to connect with artists. I often suspected he was 'un artiste manqué'.

Sport was also important; he loved skiing and regretted the lack of hills in southwestern Ontario. He loved going to the Banff Winter Institute, because it gave him an opportunity to get some real skiing done. Aside from skiing, he also loved to play tennis. He followed the international tennis matches religiously. In the 70's, when we were much younger and more active, aerobics became very popular. There was a class here at Western "for ladies only." I would go every lunch hour and when Roger saw the class in action, he wanted to join too. As there were no men in the group, I had to ask special permission for Roger to join the class. The leader (a woman) told me that she would ask the "ladies" if it was O.K. with them. If they went along
with it, she would allow Roger to participate in the class, provided Roger would not ogle the "ladies." So, Roger became the first male to liberate aerobics classes at Western. He continued with jogging and cycling through the years. As a matter of fact, the year we spent in Holland, Roger would cycle to work every day. This was a trip of about 1 hour, which was gradually reduced to 40 minutes as he got in better shape.

Chess was also one of his passions. He would never pass up an opportunity to play, whether it be with an experienced player or a beginner; it did not matter to him. He loved to play but he restrained himself because, as he said, he would otherwise spend too much time on it. He would go and buy the Saturday papers, the *Globe* and the *Star,* because they published chess games on Saturday. He would then spend most of Saturday morning playing the games as indicated. For years he harassed *The London Free Press* because they discontinued the chess column. Wouldn't you know, shortly after Roger's death they started publishing a chess column again. He collected chess books and would use them regularly to look up chess problems. He did not care too much for chess on the computer. I guess it was because he had a love/hate relationship with PCs.

Roger loved to play games, be they tennis or backgammon or any other kind of game. He was good at most of the games and would often win, except at *Scrabble*. That's when I usually had a good chance to beat
him. He was a generous loser however. To my regret I must say that I was far less gracious.

Speaking about generous, one of the things I admired most about Roger was his generosity of spirit. He was most willing to admit that someone was brighter or better than he was. This does not mean that he hid his
light under the bushel, he knew that he did good work and was justifiably proud of it. His generosity and encouragement has helped me to finish my Master's degree and in general reach higher and extend myself more than I otherwise might have done. That is another legacy he left me.

When I was working on my Master's degree in German language and literature, I had to write many essays. As I don't have a natural aptitude for writing, I would procrastinate. Roger saw my struggle, and, in order to help me, Roger suggested that he would look after the cooking, thus giving me more time to write. I told him that when I was ready, I would take him up on his offer. One day, after dinner, I told him that as of tomorrow he would have to do the cooking. He said O.K. and all was set ... or so I thought. The first day was not too bad, because I had bought something extra the day before. The second day was not too bad either because there were leftovers. The third day he took something out of the freezer and we ate that. After we went to bed on the third day Roger asked me what I thought we should eat the next day. I told him that it was his problem, because he said that he would take care of the cooking. If I had to plan what we should eat the next day, I might as well do the cooking, because three quarters of meal preparation is in the planning. After a while I noticed that Roger was restless, and when I asked him what was the matter, he said he could not sleep. When I asked why, he said it was because he could not think of what to eat the next day. Needless to say, he was very relieved when shortly after that I took on the task of cooking again. His intentions were very good, but he found out that he had bit off more than he was prepared to chew.

Roger also had a good sense of humor. Anyone who has heard Roger tell a joke knows that he enjoyed doing that immensely. Sometimes he enjoyed it so much that he could not finish the joke because he laughed so hard. When he was telling a story however, that was a totally different performance. Roger liked to cross all the *t*'s and dot all the *i*'s -twice. His stories tended to meander from one point to the next. He felt it was necessary that we - his audience - should be informed of all the details. Often we wondered if he would ever get back to the main point of the story. His stories often were like Russian dolls - a story within a story, within a story, within a story..... I had a habit of telling him to make a long story short, but I did not always achieve the desired result.

Writing letters was for Roger a totally different affair. He did not write me too many letters when he was away. Those that he did write were no romantic epistles. He usually summarized whatever he had written in point form at the end. It was in his letters that he made long stories short.

On the path of life, Roger was a good traveling companion. He was willing to try everything at least once and if he was wrong, he would willingly admit it. I had wanted to see Newfoundland, and when Roger told me that CAP planned to have their annual conference in St. John's, I said that I would like to come with him and see some of Newfoundland. I must have mentioned it at a time when he was preoccupied because when the time drew near for him to order his ticket, he asked me when he should leave. We had an argument about whether or not I had mentioned going with him. He asked me, "Who would want to go to Newfoundland anyway?" I said that *I* would want to go and had wanted to go for some time. The long and the short of it was that, he agreed to a holiday in Newfoundland, albeit reluctantly. When we got there, there was an iceberg floating across the entrance to the harbour. Roger had his conference and we made various trips in the neighbourhood of St. John's. We then spent a few days in the Trinity Bay area. The weather was rainy and cold but we both had a very positive impression of our visit to Newfoundland. Roger later admitted that he had been wrong about his initial image of Newfoundland. He was like that; not afraid to admit he had been wrong.

Roger was by no means perfect and our twenty-five years together certainly have not always gone smoothly, but we always managed to iron things out - mostly thanks to Roger, for he was more forgiving than I was. We would argue a lot, but I think we both enjoyed it. Neither one of us was willing to let the other get away with anything. I often compared it to being in a canoe together: Roger always steered the canoe towards white water - he loved the excitement of it - whereas I was more inclined towards calmer waters.

Many of you have lost a good friend and colleague, but I have lost my traveling companion and my Number One supporter. Life with Roger was not always easy, but life without Roger is not always easy either - besides, it is boring. Roger had such a zest for life that he made *my* life sparkle. With some people, you know the world is going to be all right, as long as they are in it - once they're gone, the world is not as bright and we are all diminished. I think that Roger was one of these people.

May I say in closing that we were all enriched by Roger's dedication, we were all challenged by his inquiring mind, and we were all warmed by his friendship. Thank you.

LIST OF PARTICIPANTS

A. Abdel-Rahim (Syracuse)
A. Agarwal (Rochester) — abhishek@pas.rochester.edu
M. R. Ahmady (Ochanomizu U., Tokyo/ Mt. Allison University, N.B.) — mahmady@mta.ca
T. J. Allen (Hobart and Smith) — tjallen@hws.edu
G. Alexanian (Syracuse) — garnik@physics.syr.edu
D. Black (Syracuse) — black@physics.syr.edu
A. Bourque (McGill) — bourquea@hep.physics.mcgill.ca
C. Burrell (Toronto) — burrell@physics.utoronto.ca
H. Burton (Perimeter Institute) — hburton@perimeterinstitute.ca
B. Cabrera Palmer (Syracuse)
W. Chen (Guelph) — wchen@uoguelph.ca
F. A. Chishtie (UWO) — fachisht@julian.uwo.ca
H. Collins (Toronto) — hael@physics.utoronto.ca
A. Datta (U. Montreal) — datta@lps.umontreal.ca
M. Duff (Michigan) — mduff@umich.edu
V. Elias (UWO) — velias@uwo.ca
R. Epp (Perimeter Institute) — epp@perimeterinstitute.ca
A. Fariborz (SUNIT Utica) — fariboa@sunyit.edu
E. J. Ferrer (SUNY Fredonia) — ferrer@fredonia.edu
H. Firouzjahi (McGill) — firouzh@physics.mcgill.ca
M.-P. Gagne-Portelance (UWO) — mgagnepo@uwo.ca
C. R. Hagen (Rochester) — hagen@pas.rochester.edu
B. Holdom (Toronto) — bob.holdom@utoronto.ca
S. Homayouni (UWO) — shomayou@uwo.ca
V. de la Incera (SUNY Fredonia) — incera@fredonia.edu
A. Kager (UWO) — akager@uwo.ca
G. Karl (Guelph) — gk@physics.uoguelph.ca
N. Kiriushcheva (UWO) — nkiriusc@uwo.ca
G. Krishnaswami (Rochester) — govind@pas.rochester.edu
S. Kuzmin (UWO) — skuzmin@uwo.ca
S. Kruglov (Int. Education Centre, Toronto) — skruglov23@hotmail.com
F. Leblond (McGill) — fleblond@hep.physics.mcgill.ca
R. Leigh (Illinois) — rgleigh@uiuc.edu
M. Lepage (Saskatchewan) — martin@stainless.usask.ca
M. Luke (Toronto) — luke@physics.utoronto.ca
A. Majumder (McGill) — majumder@hep.physics.mcgill.ca
R. B. Mann (Waterloo) — rbmann@sciborg.uwaterloo.ca
L. C. Martin (UWO) — lcmartin@uwo.ca
D. G. C. McKeon (UWO) — dgmckeo2@uwo.ca
V. A. Miransky (UWO) — vmiransk@uwo.ca

S. Moussa (Syracuse)
K. Narayan (Cornell) — narayan@mail.lns.cornell.edu
W. van Neerven (Leiden, Netherlands) — neerven@lorentz.leidenuniv.nl
K. Nguyen (UWO) — knguyen@uwo.ca
V. Novikov (ITEP, Moscow, Russia)
P.J. O'Donnell (Toronto) — pat@medb.physics.utoronto.ca
A. W. Peet (Toronto) — peet@physics.utoronto.ca
J. Rozowsky (Syracuse) — Rozowsky@physics.syr.edu
M. D. Scadron (Arizona) — scadron@physics.arizona.edu
J. Schechter (Syracuse) — schechte@gluon.phy.syr.edu
G. Semenoff (UBC) — semenoff@physics.ubc.ca
M. A. Shifman (Minnesota) — shifman@physics.spa.umn.edu
P. Silva (Syracuse)
D. Spector (Hobart and William Smith) — spector@hws.edu
T. G. Steele (Saskatchewan) — Tom.Steele@usask.ca
O. Teodorescu (McGill) — octavian@physics.mcgill.ca
M. Trott (Toronto) — mrtrott@physics.utoronto.ca
J. Trudeau (McGill) — trudeau@physics.mcgill.ca
S. R. Valluri (UWO) — valluri@uwo.ca
E. Waled (Syracuse)
A. Williamson (Toronto) — awilliamson@physics.utoronto.ca
B. Ydri (Syracuse) — badisydri@yahoo.com

Author Index

A

Agarwal, A., 259
Ahmady, M. R., 105
Akant, L., 267
Alexanian, G., 226
Allen, T. J., 277
Argyres, P., 282

B

Bauer, C. W., 97
Black, D., 182, 188

C

Chen, W. F., 242
Chishtie, F. A., 175
Cline, J. M., 88
Collins, H., 82

D

de la Incera, V., 235, 253
Duff, M. J., 3

E

Elias, V., 140

F

Fariborz, A. H., 182, 188
Ferrer, E. J., 235, 253
Firouzjahi, H., 88

G

Gale, C., 168
Goebel, C., 277
Gusynin, V. A., 253

H

Hagen, C. R., 197
Holdom, B., 75

K

Karl, G., 148
Krishnaswami, G. S., 267
Kruglov, S. I., 206

L

Leblond, F., 287
Leigh, R. G., 161
Ligeti, Z., 97
Luke, M., 97

M

Majumder, A., 168
McKeon, D. G. C., 60
Migneron, I. P., 305
Miransky, V. A., 129
Moussa, S., 182, 188

N

Nagashima, M., 105
Narayan, K., 282
Nasri, S., 182, 188
Novikov, V., 148

O

Olsson, M. G., 277

P

Pinzul, A., 226

R

Rajeev, S. G., 259, 267
Romeo, A., 235
Rozowsky, J. S., 212

S

Scadron, M. D., 66
Schechter, J., 182, 188
Semenoff, G. W., 19
Sherry, T. N., 60
Silva, P. J., 295
Steele, T. G., 151
Stern, A., 226
Sugamoto, A., 105

V

van Neerven, W. L., 40
Veseli, S., 277

W

Williamson, A. R., 115

Y

Ydri, B., 219